D0626154

Edexcel GCSE

Mathematics B
Modular
Foundation

Student Book
Unit 2

Series Director: Keith Pledger
Series Editor: Graham Cumming

Authors:
Chris Baston
Julie Bolter
Gareth Cole
Gill Dyer
Michael Flowers
Karen Hughes
Peter Jolly
Joan Knott
Jean Linsky
Graham Newman
Rob Pepper
Joe Petran
Keith Pledger
Rob Summerson
Kevin Tanner
Brian Western

A PEARSON COMPANY

Published by Pearson Education Limited, a company incorporated in England and Wales, having its registered office at Edinburgh Gate, Harlow, Essex, CM20 2JE. Registered company number: 872828

Edexcel is a registered trademark of Edexcel Limited

Text © Chris Baston, Julie Bolter, Gareth Cole, Gill Dyer, Michael Flowers, Karen Hughes, Peter Jolly, Joan Knott, Jean Linsky, Graham Newman, Rob Pepper, Joe Petran, Keith Pledger, Rob Summerson, Kevin Tanner, Brian Western and Pearson Education Limited 2010

The rights of Chris Baston, Julie Bolter, Gareth Cole, Gill Dyer, Michael Flowers, Karen Hughes, Peter Jolly, Joan Knott, Jean Linsky, Graham Newman, Rob Pepper, Joe Petran, Keith Pledger, Rob Summerson, Kevin Tanner and Brian Western to be identified as the authors of this Work have been asserted by them in accordance with the Copyright, Designs and Patent Act, 1988.

First published 2010

12 11 10
10 9 8 7 6 5 4 3 2

British Library Cataloguing in Publication Data
A catalogue record for this book is available from the British Library

ISBN 978 1 84690 805 7

Copyright notice
All rights reserved. No part of this publication may be reproduced in any form or by any means (including photocopying or storing it in any medium by electronic means and whether or not transiently or incidentally to some other use of this publication) without the written permission of the copyright owner, except in accordance with the provisions of the Copyright, Designs and Patents Act 1988 or under the terms of a licence issued by the Copyright Licensing Agency, Saffron House, 6–10 Kirby Street, London EC1N 8TS (www.cla.co.uk). Applications for the copyright owner's written permission should be addressed to the publisher.

Typeset by Tech-Set Ltd, Gateshead
Picture research by Rebecca Sodergren
Printed in Great Britain at Scotprint, Haddington

Acknowledgements
The publisher would like to thank the following for their kind permission to reproduce their photographs:
(Key: b-bottom; c-centre; l-left; r-right; t-top)

Alamy Images: Aflo Foto Agency 305tr, Bildagentur-online.com 49, CandyBox Photography 211, FogStock 168, H. Mark Weidman Photography 305tc, MBI 146, roboxford 306-308, Stephen Shepherd 305tl, Mark Titterton 66; **Corbis:** Deborah Betz Collection 93, Alan Schein 195, Paul Seheult 288; **Getty Images:** Ko.fujiwara 229, Mike Powell 30, Antonio M. Rosario 113; **iStockphoto:** Sergey Ivanov 305bl, Michael Krinke 305br, Stephen Strathdee 1, Peeter Viisimaa 22; **Photolibrary.com:** Ron Chapple Stock 182; **Press Association Images:** Anthony Devlin 268, Ezra Shaw 76; **Rex Features:** Michael Fresco 166; **Science Photo Library Ltd:** 126, Detlev Van Ravenswaay 238; **Shutterstock:** Foodpics 308t

All other images © Pearson Education

Disclaimer

This material has been published on behalf of Edexcel and offers high-quality support for the delivery of Edexcel qualifications. This does not mean that the material is essential to achieve any Edexcel qualification, nor does it mean that it is the only suitable material available to support any Edexcel qualification. Edexcel material will not be used verbatim in setting any Edexcel examination or assessment. Any resource lists produced by Edexcel shall include this and other appropriate resources.

Copies of official specifications for all Edexcel qualifications may be found on the Edexcel website: www.edexcel.com

Contents

About this book

All set to make the grade!

Edexcel GCSE Mathematics is specially written to help you get your best grade in the exams.
Remember this is a non-calculator unit.

> Section objectives show what you'll be learning.

> Recap with a skills check at the start of a section – make sure you're up to speed.

> Crystal-clear worked examples – step-by-step guides to answering questions correctly, with helpful hints and reminders.

> Loads of practice to help you feel secure before you move on.

> Graded questions – so you know what you're achieving.

> Full coverage of the new-style assessment objective questions – A02 and A03.

> 'Focus on A02/3' pages demystify the new assessment objectives.

> A fully worked example of an A02/3 question... ...makes other A02/3 questions on the same topic easy to tackle.

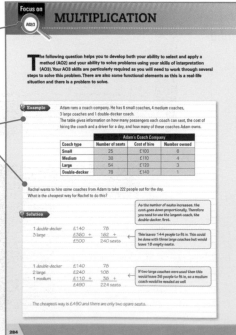

And:

- A pre-check at the start of each chapter helps you recall what you know.

- Functional elements highlighted – within ordinary exercises and on dedicated pages – so you can spend focused time polishing these skills.

- End-of-chapter graded review exercises consolidate your learning and include past exam paper questions indicated by the month and year.

About ActiveTeach

Use **ActiveTeach** to view and present the course on screen with exciting interactive content.

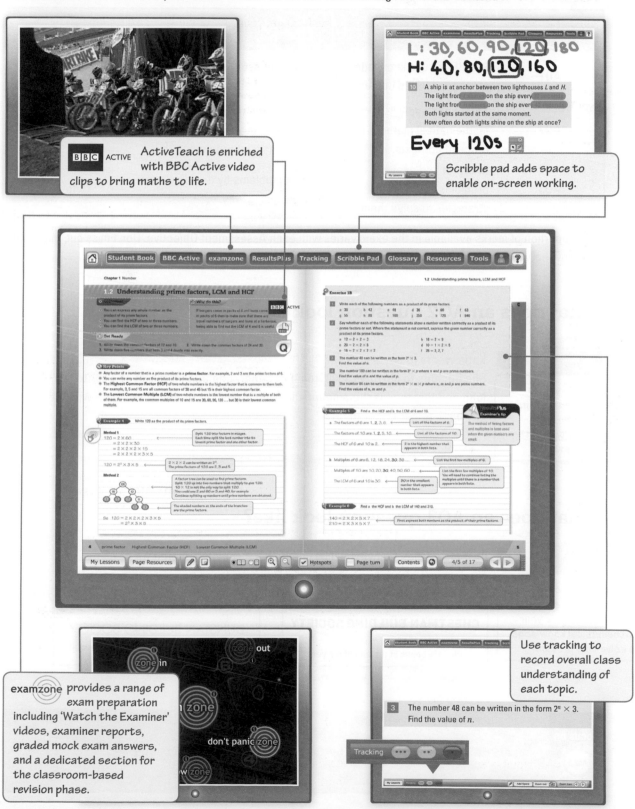

BBC ACTIVE — ActiveTeach is enriched with BBC Active video clips to bring maths to life.

Scribble pad adds space to enable on-screen working.

examzone provides a range of exam preparation including 'Watch the Examiner' videos, examiner reports, graded mock exam answers, and a dedicated section for the classroom-based revision phase.

Use tracking to record overall class understanding of each topic.

About Assessment Objectives

Assessment Objectives define the types of question that are set in the exam.

Assessment Objective	What it is	What this means	Range % of marks in the exam
AO1	**Recall** and use knowledge of the prescribed content.	Standard questions testing your knowledge of each topic.	45-55
AO2	**Select** and apply mathematical methods in a range of contexts.	Deciding what method you need to use to get to the correct solution to a contextualised problem.	25-35
AO3	**Interpret** and analyse problems and generate strategies to solve them.	Solving problems by deciding how and explaining why.	15-25

The proportion of marks available in the exam varies with each Assessment Objective. Don't miss out, make sure you know how to do AO2 and AO3 questions!

What does an AO2 question look like?

D **AO2**

> This just needs you to
> (a) read and understand the question and
> (b) decide how to get the correct answer.

16 Katie wants to buy a car.
She decides to borrow £3500 from her father. She adds interest of 3.5% to the loan and this total is the amount she must repay her father. How much will Katie pay back to her father in total?

What does an AO3 question look like?

D **AO3**

> Here you need to read and analyse the question. Then use your mathematical knowledge to solve this problem.

17 Rashida wishes to invest £2000 in a building society account for one year. The Internet offers two suggestions. Which of these two investments gives Rashida the greatest return?

CHESTMAN BUILDING SOCIETY	DUNSTAN BUILDING SOCIETY
£3.50 per month Plus **1% bonus** at the end of the year	**4%** per annum. Paid yearly by cheque

Focus on

AO2/3

We give you extra help with AO2 and AO3 on pages 300–303.

About functional elements

What does a question with functional maths look like?

Functional maths is about being able to apply maths in everyday, real-life situations.

GCSE Tier	Range % of marks in the exam
Foundation	30-40
Higher	20-30

The proportion of functional maths marks in the GCSE exam depends on which tier you are taking. Don't miss out, make sure you know how to do functional maths questions!

In the exercises…

20 The Wildlife Trust are doing a survey into the number of field mice on a farm of size 240 acres. They look at one field of size 6 acres. In this field they count 35 field mice.

a Estimate how many field mice there are on the whole farm.

b Why might this be an unreliable estimate?

> You need to read and understand the question. Follow your plan.
>
> Think what maths you need and plan the order in which you'll work.
>
> Check your calculations and make a comment if required.

…and on our special functional maths pages: 304–307!

Quality of written communication

There will be marks in the exam for showing your working 'properly' and explaining clearly. In the exam paper, such questions will be marked with a star (*). You need to:

- use the correct mathematical notation and vocabulary, to show that you can communicate effectively
- organise the relevant information logically.

ResultsPlus

ResultsPlus features combine exam performance data with examiner insight to give you more information on how to succeed. ResultsPlus tips in the **student books** show students how to avoid errors in solutions to questions.

Watch Out!

Some students use the term average – make sure you specify mean, mode or median.

This warns you about common mistakes and misconceptions that examiners frequently see students make.

Exam Question Report

91% of students scored poorly on this question because they did not use the midpoint of the range to find the mean of grouped data.

This gives a breakdown of how students did on real past exam questions.

Examiner's Tip

Make sure the angles add up to 360°.

This gives exam advice, useful checks, and methods to remember key facts.

ResultsPlus in the **ActiveTeach** provides interactive practice for AO2 and AO3 questions...

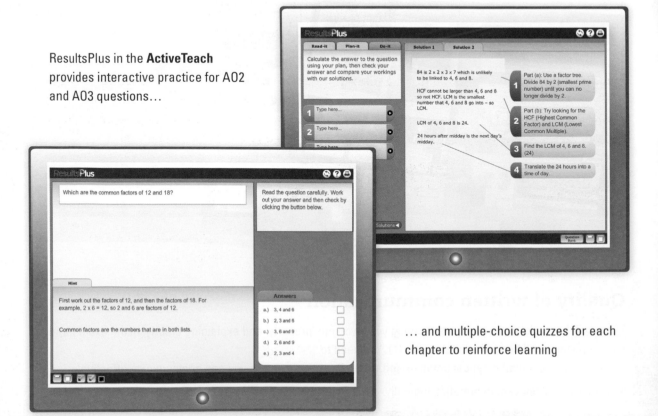

... and multiple-choice quizzes for each chapter to reinforce learning

1 NUMBER

During his in-flight announcement, a pilot normally gives the altitude at which the plane is flying. This is generally somewhere between 30 000 ft and 40 000 ft, but the pilot will give it correct to the nearest thousand. An air traffic controller needs to know the precise altitude at which the plane is flying, but for the passengers an approximation is enough.

◉ Objectives

In this chapter you will:
- see the importance of place value
- see the usefulness of a number line
- see why whole numbers can be negative as well as positive
- use the rules of addition, subtraction, multiplication and division for combining numbers
- learn some of the language of mathematics connected with numbers.

⬦ Before you start

You need to know:
- the addition number bonds to $9 + 9$
- the times tables to 10×10.

1.1 Understanding digits and place value

Objective

○ You understand the number system.

Why do this?

To carry out an everyday task such as buying something in a shop you need to understand how the number system works.

Get Ready

1. What numbers come before and after 509?
2. Write down the answer to 7 + 6.
3. Write down the answer to 7 × 8.

Key Points

● Although you can keep on counting to very high numbers, you only use ten **digits**, often called figures.

0 1 2 3 4 5 6 7 8 9

● Each digit has a **value** that depends on its position in the number. This is its **place value**.

Example 1

Draw a place value diagram for:

a a three-digit number with a 6 in the tens column

b a five-digit number with a 6 in the thousands column.

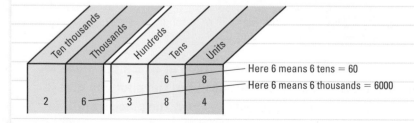

Here 6 means 6 tens = 60
Here 6 means 6 thousands = 6000

Ten thousands | Thousands | Hundreds | Tens | Units
| | 7 | 6 | 8
2 | 6 | 3 | 8 | 4

Exercise 1A

Questions in this chapter are targeted at the grades indicated.

G

1 Draw a place value diagram and write in:

a a four-digit number with a 4 in the thousands column

b a two-digit number with a 3 in the tens column

c a five-digit number with a 1 in the hundreds column

d a three-digit number with a 9 in the units column

e a four-digit number with a 0 in the tens column

f a five-digit number with a 4 in the hundreds column

g a four-digit number with a 7 in every column except the tens column

h a five-digit number with a 6 in the thousands column and the units column.

2 For each teacher, write down five different numbers that they could be thinking about.

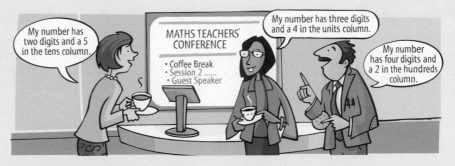

3 Write down the value of the 6 in each of these numbers.

a 63　　　　b 3642　　　　c 63 214　　　　d 2546　　　　e 56 345

1.2 Reading, writing and ordering whole numbers

Objectives

- You can read and write down whole numbers.
- You can put numbers in order of size.

Why do this?

You need to be able to compare the size of numbers when you want to decide which mobile phone handset is cheaper.

Get Ready

1. Write down the value of the 4 in each of these numbers.

a 46　　　　b 2034　　　　c 65 403

Key Points

- You can read and write numbers by thinking about the place value of each of the digits.
- You can use your knowledge of place value to put numbers in order of size.

Example 2

Write in figures the numbers shown in these newspaper cuttings.

Two thousand eight hundred and twenty-two walkers took part in Saturday's charity event.

Some of the thirty-one thousand two hundred and eight people at an outdoor rock concert

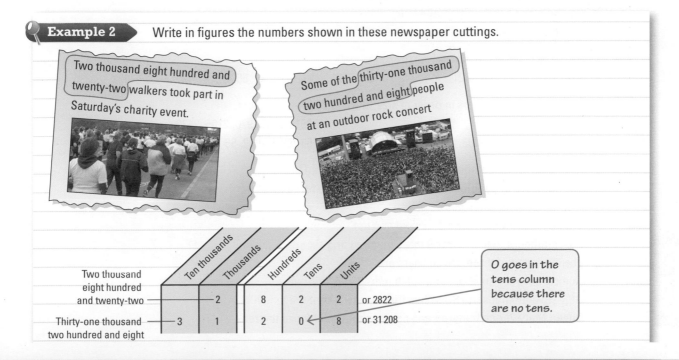

	Ten thousands	Thousands	Hundreds	Tens	Units	
Two thousand eight hundred and twenty-two		2	8	2	2	or 2822
Thirty-one thousand two hundred and eight	3	1	2	0	8	or 31 208

O goes in the tens column because there are no tens.

Example 3 ▶ Write in words the numbers shown in the place value diagram.

In 16 412 the thin space separates thousands from hundreds and makes it easier to read the number.

Ten thousands	Thousands	Hundreds	Tens	Units	
		9	8	7	— Nine hundred and eighty-seven
1	6	4	1	2	— Sixteen thousand four hundred and twelve

Example 4 ▶ Write the numbers 8400, 6991, 15, 2410, 84 000, 2406 in order of size.

Start with the smallest.

15 is the smallest. ← It has no digits above the tens column.

84 000 is the largest. ← It is the only one with a digit in the ten thousands column.

This leaves 8400, 6991, 2410 and 2406.
2000 is smaller than 6000, which is smaller than 8000.
Lastly, 2406 is smaller than 2410.
The order is 15, 2406, 2410, 6991, 8400, 84 000.

Exercise 1B

G

1 Write these numbers in figures.
   a Three hundred and twenty-five
   b One thousand seven hundred and eighteen
   c Six thousand two hundred and four
   d Nineteen thousand four hundred and twenty

2 Write these numbers in words.
   a 237   b 321   c 1792   d 6502   e 1053

3 Write each set of numbers in order of size, starting with the smallest.
   a 183, 235, 190, 73, 179   b 2510, 2015, 970, 2105, 2439
   c 30 300, 3033, 3000, 3003, 2998 d 56 762, 59 342, 56 745, 56 321

4 Write in figures the numbers that have been highlighted in blue.
   a The numbers of people employed by a local police force are:
      Traffic wardens: sixty-nine
      Civilian support staff: one thousand and ten
      Police officers: two thousand three hundred and six
   b The tonnages of three cruise liners are:
      Aurora: seventy-six thousand one hundred and fifty-two
      QE2: seventy thousand three hundred and sixty-three
      Queen Mary 2: one hundred and fifty-one thousand four hundred

5 This table gives the populations of five member states of the European Union in 2009.
Write the numbers in words.

	Country	Population
a	Czech Republic	10 467 542
b	Cyprus	793 963
c	Estonia	1 340 415
d	France	64 351 000
e	Portugal	10 627 250

6 The table gives the prices of some second-hand cars.

Car	Price
Peugeot 505	£7995
Focus	£11 495
Ka	£4835
Mini	£6549
Sharan	£13 205

a Write down the price of each car in words.
b Rewrite the list in price order, starting with the most expensive.

1.3 The number line

◎ Objectives

- You can see numbers in context.
- You can use a simple number line to increase or decrease numbers and to work out the difference between numbers.

⟳ Why do this?

You need to be able to read scales you meet, such as rulers, thermometers and speedometers.

⟳ Get Ready

Place these numbers in order, starting with the lowest.
1. £5490   £3645   £5250   £4190
2. 10 348 276   462 690   40 280 780   10 524 145   60 424 213
3. 76 152   70 363   150 400

Key Points

- You can use a number line to help you to increase or decrease a number.
- Number lines might be read:
  - from left to right
  - from bottom to top
  - clockwise.

Example 5 ▷ Use a number line to: a increase 6 by 4   b decrease 23 by 8.

a 10 — 10 ——— Answer 10

Increase by 4

6 ←

Start at 6

5 —

0 —

b 25 —

23 ← Start at 23

20 —

← Decrease by 8

15 — 15 ——— Answer 15

Exercise 1C

G

1 Draw a number line from 0 to 30.
   Mark these numbers on your number line.
   a 6   b 23   c 15   d 0   e 29

2 Use a number line from 0 to 25 to
   a increase 6 by 3   b decrease 15 by 7   c increase 11 by 7
   d decrease 19 by 13   e increase 17 by 8   f decrease 24 by 17.

3 For each of these moves, write down the difference between them, and whether it is an increase or decrease.
   a 6 to 10   b 15 to 21   c 10 to 3   d 29 to 24   e 5 to 27   f 19 to 2

1.4 Adding and subtracting

⊙ Objective

○ You can add and subtract without a calculator.

❓ Why do this?

When you are shopping you might want to know how much you have spent before you get to the checkout. You might not always have a calculator on hand to help you!

⬆ Get Ready

1. Use a number line to:
    a increase 29 by 33   b decrease 34 by 27   c find the difference between 20 and 3.

🕐 Key Points

◉ Words that show you have to add numbers are: add, plus, total and sum.
◉ Words that show you have to subtract are: take away, subtract, minus and find the difference.

Example 6   Work out 23 + 693 + 8.

```
    23
   693
 +   8
  ————
   724
   11
```

Step 1 Put the digits in their correct columns.
Step 2 Add the units column: 3 + 3 + 8 = 14.
        Write 4 in the units column and carry the 1 ten into the tens column.
Step 3 Repeat for the tens column: 2 + 9 = 11 plus the 1 that was carried across = 12.
        Write 2 in the tens column and carry the 1 into the hundreds column.
Step 4 Add the 6 and the 1 that was carried across: 7.

Exercise 1D

1 Find the total of 26 and 17.

2 Work out 58 plus 22.

3 Work out 236 + 95.

4 In four maths tests, Anna scored 61 marks, 46 marks, 87 marks and 76 marks.
   How many marks did she score altogether?

5 Find the sum of all the single-digit numbers.

6 In a fishing competition, five competitors caught 16 fish, 31 fish, 8 fish, 19 fish and 22 fish.
   Find the total number of fish caught.

7 The number of passengers on a bus was 36 downstairs and 48 upstairs.
   How many passengers were on the bus altogether?

8 On her MP3 player, Lena had 86 pop songs, 58 rock songs and 72 dance songs.
   How many songs did she have in total?

9 Work out 38 + 96 + 127 + 92 + 48.

10 On six days in June, 86, 43, 75, 104, 38 and 70 people went bungee jumping over a gorge.
   How many people jumped in total?

Example 7   Work out 423 − 274.

Method 1

```
   3 11 1
   4 2 3
 − 2 7 4
  ——————
   1 4 9
```

Step 1 Put the digits in their correct columns.
Step 2 In the units column try taking 4 from 3 (not possible). So exchange 1 ten for 10 units.
        You now have 3 + 10 = 13 units.
Step 3 13 − 4 = 9 in the units column.
Step 4 The tens column has 1 take away 7, which is not possible, so exchange 1 hundred for
        10 tens.
Step 5 11 − 7 = 4 in the tens column.
Step 6 Finally there is 3 − 2 = 1 in the hundreds column.

Method 2

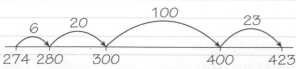

274 280   300   400   423

Step 1 Count on from 274 to 280 =   6
Step 2 Count on from 280 to 300 =   20
Step 3 Count on from 300 to 400 =   100
Step 4 Count on from 400 to 423 =   23
Step 5 Add   149

Exercise 1E

G

1   Work out 611 − 306.

2   How much is 7260 − 4094?

3   Take 1007 from 2010.

4   In a car boot sale Alistair sells 17 of his 29 CDs.
    How many does he have left?

A03

5   When playing darts, James scored 111 with his first three darts, Sunita scored 94 with her first three
    darts and Nadine scored 75 with her first three darts.
    a   How many more than Nadine did Sunita score?
    b   What is the difference between James's and Sunita's scores?
    c   How many less than James did Nadine score?

A03

6   The winner of the darts match is the first one to reach 501.
    James has now scored 413, Sunita 442 and Nadine 368.
    a   How many does James need to win the game?
    b   How many more does Sunita need to win the game?
    c   How many is Nadine short of winning the game?

1.5 Multiplying and dividing

◎ Objective

● You can multiply and divide without a calculator.

◈ Why do this?

If you buy several T-shirts or CDs that are the same price, it is useful if you can work out an estimate of the cost without a calculator.

◈ Get Ready

1. Work out
   a $123 + 23 + 23$
   b $57 − 19 − 19 − 19$

◈ Key Points

● Words that show you have to multiply numbers are: product, times and multiply.

● Divide and share are words that show you have to divide.

● You multiply by 10, 100 and 1000 by moving digits 1, 2 or 3 places to the left.

● You divide by 10, 100 and 1000 by moving digits 1, 2 or 3 places to the right.

Example 8   Work out 23×10.

H	T	U	
	2	3	
2	3	0	= 230

The 3 moves from the units into the tens column because $3 \times 10 = 30$.
The 2 moves from the tens into the hundreds column because $20 \times 10 = 200$.

Example 9   a Work out 43×6.   b Work out 256×37.

$$\begin{array}{r} 43 \\ \times\ \ 6 \\ \hline 2{\scriptstyle 1}58 \end{array}$$

$3 \times 6 = 18$ (1 ten and 8 units)
$40 \times 6 = 240$
Total 258

$$\begin{array}{r} 256 \\ \times\ \ 37 \\ \hline 17{\scriptstyle 3}9{\scriptstyle 4}2 \\ {\scriptstyle 1}7680 \\ \hline 9472 \\ {\scriptstyle 1} \end{array}$$

Step 1 256×7
Step 2 256×30
Step 3 add

Exercise 1F

1 Multiply each of these numbers by i 10   ii 100   iii 1000
   a 27   b 8   c 301   d 60   e 5020

2 Work out
   a 35×20 b 26×200 c 122×30
   d 43×500 e 115×40 f 214×3000

3 Work out
   a 65×7 b 53×3 c 314×6
   d 523×8 e 237×5 f 399×4

4 Work out
   a 34×12 b 65×15 c 53×33 d 314×16
   e 523×47 f 221×64 g 146×53 h 29×38

5 Find the product of 71 and 13.

6 How many is 321 multiplied by 14?

7 John travels 17 miles to work and 17 miles back each day.
   In July he went to work and back 23 times. How far did he travel altogether?

8 Bocton Football Club hires 23 coaches to take supporters to an away match.
   Each coach can take 63 passengers. How many supporters can be taken to the match?

9 Each box of matches contains 43 matches.
   How many matches are there in 36 boxes?

10 Tariq buys tins of soup which are packed in cases of 48.
   He buys 8 cases. How many tins of soup does he buy?

G

A03

Example 10 Work out 3200 ÷ 100.

```
Th   H   T   U
3    2   0   0
         3   2   = 32
```

The 3 moves from the thousands into the tens column because 3000 ÷ 100 = 30.

The 2 moves from the hundreds into the units column because 200 ÷ 100 = 2.

Example 11 Work out 256 ÷ 8.

Traditional short division

```
      3 2
8)2 5¹6
```

Step 3   8 into 16 goes 2 times.

Step 1   8 into 2 does not go.

Step 2   8 into 25 goes 3 times remainder 1. Carry remainder into units column.

Example 12 Work out 8704 ÷ 17.

Traditional long division

Step 1
17 divides into 87 five times remainder 2.

Step 2
17 divides into 20 one time remainder 3.

Step 3
17 divides into 34 two times exactly.

```
                    5              51            512
17)8704      17)8704       17)8704       17)8704
              − 85          − 85          − 85
                2            20            20
                          − 17          − 17
                             3            34
                                          34
                                           0
```

So 17 divides into 8704 512 times.

Short division method

```
      5 1 2
17)8 7²0³4
```

This is a shorter way of setting out the steps than in the long division method.

Exercise 1G

1 Work out
   a 3660 ÷ 10   **b** 4300 ÷ 100   **c** 9000 ÷ 10   **d** 87 000 ÷ 1000

2 Work out
   a 48 ÷ 2   **b** 69 ÷ 3   **c** 56 ÷ 4   **d** 96 ÷ 6
   e 640 ÷ 4   **f** 565 ÷ 5   **g** 72 ÷ 4   **h** 712 ÷ 8
   i 828 ÷ 9   **j** 637 ÷ 7   **k** 408 ÷ 2   **l** 1020 ÷ 3

3 Work out
  a 256 ÷ 16   b 660 ÷ 15   c 512 ÷ 32   d 861 ÷ 21
  e 756 ÷ 36   f 1020 ÷ 30   g 1440 ÷ 36   h 7500 ÷ 25

4 a Work out 315 ÷ 15.   b How many 50s make 750?   c Work out $\frac{680}{17}$.
   d Work out 600 divided by 30.   e Divide 8 into 112.

5 Five people shared a prize draw win of £2400 equally.
How much did each person receive?

6 In an online computer game tournament, players are put into groups of 24, with the group winners going through to the final.
How many finalists will there be if there are
  a 240 players   b 720 players   c 864 players?

7 An aeroplane can hold 18 parachute jumpers at a time.
How many trips does the plane have to make for
  a 126 jumps   b 234 jumps   c 648 jumps?

8 A packing case will hold 72 economy-size boxes, 24 large-size boxes or 12 family-size boxes.
How many packing cases would be needed to pack
  a 864 economy-size boxes   b 984 large-size boxes   c 960 family-size boxes?

1.6 Rounding

Objective
You can write numbers to a suitable accuracy.

Why do this?
When you want to estimate a shopping bill, it is useful to be able to round the numbers to make the calculation easier.

Get Ready

Work out the total cost of these.
1. Two T-shirts at £15 each and two pairs of shorts at £4 each
2. Four cookies at 75p each and two cups of coffee at £1 each

Key Points

* There are situations where an estimate of a number may be good enough.
    * If you are working out how long a journey is going to take, then an answer to the nearest hour may be good enough.
    * If you are recording the number of customers a supermarket has in a day, then the nearest 100 may be good enough.
* **Rounding** means reducing the accuracy of a number by replacing right-hand digits with zeros.
* To round a number, look at the digit before the place value you are rounding to (so for rounding to the nearest 10 you would look at the units digit). If it is less than 5, round down. If it is 5 or more, round up.

Example 13 Round 7650 to the nearest   a thousand   b hundred.

a 8000 ⟵ 7650 is nearer 8000 than 7000 on the number line. Rounded to the nearest thousand 7650 is 8000.

b 7700 ⟵ 7650 is exactly halfway between 7600 and 7700. When this happens you round up.

Example 14 Round 314 to the nearest 100.

314 to the nearest 100 is 300. ⟵ Look at the tens digit. It is less than 5 so round down.

Exercise 1H

1 Round these numbers to the nearest ten.
   a 57   b 63   c 185   d 194   e 991   f 2407

2 Round these numbers to the nearest hundred.
   a 312   b 691   c 2406   d 3094   e 8777   f 29 456

3 Round these numbers to the nearest thousand.
   a 2116   b 36 161   c 28 505   d 321 604   e 717 171   f 2 246 810

4

	Length (ft)	Cruising speed (mph)	Takeoff weight (lb)
Airbus A310	153	557	36 095
Boeing 737	94	577	130 000
Saab 2000	89	403	50 265
Dornier 228	54	266	12 566
Lockheed L1011	177	615	496 000

For each of these aircraft, round:
a the length to the nearest ten feet
b the takeoff weight to the nearest hundred pounds
c the cruising speed to the nearest ten mph.

1.7 Negative numbers

⊙ Objective

● You can use negative numbers to represent quantities that are less than zero.

⌗ Why do this?

Negative numbers are used to record temperatures below zero and distances below sea level.

⬆ Get Ready

1. Place the numbers 557, 577, 403, 266 and 615 in order. Start with the smallest.

Key Points

- The further a number is from zero on the left-hand side of a number line, the smaller it is, so -25 is smaller than -3.
- For **negative numbers** you count backwards from zero on the number line.
    - $-31°C$ is 31 degrees below zero.
    - $-396\,m$ is 396 m below sea level.

Example 15   Write down: **a** the highest   **b** the lowest number in this list.

$-19, 7, -10, 0, 4, -3$

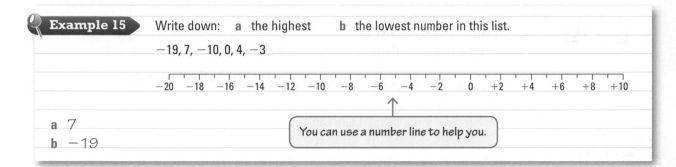

a 7

b -19

You can use a number line to help you.

Exercise 1I

1 For each list of numbers write down **i** the highest and the lowest number

ii the numbers in order, starting with the lowest.

a $5, -10, -3, 0, 4$

b $-7, -2, -9, -13, 0$

c $-3, 6, 13, -15, -6$

d $-13, -2, -20, -21, -5$

2 Write down the two missing numbers in each sequence.

a $4, 3, 2, 1, \_, \_, -2$

b $10, 7, 4, 1, \_, \_, -8$

c $-13, -9, -5, -1, \_, \_, 11$

d $13, 8, 3, -2, \_, \_, -17$

e $21, 12, 3, -6, \_, \_, -33$

f $-13, -10, -7, -4, \_, \_, 5$

3 Use the number line to find the number that is:

a 5 more than 2

b 4 more than -7

c 7 less than 6

d 2 less than -3

e 6 less than 0

f 10 more than -7

g 6 more than -6

h 4 less than -3

i 10 less than 5

j 1 less than -1.

4 What number is:

a 30 more than -70

b 50 less than -20

c 80 greater than -50

d 90 smaller than 60

e 130 smaller than -30

f 70 bigger than 200

g 170 bigger than -200

h 100 bigger than -100

i 140 more than -20

j 200 less than -200?

G

F

9
8
7
6
5
4
3
2
1
0
-1
-2
-3
-4
-5
-6
-7

F

5 The table gives the highest and lowest temperatures recorded in five cities during one year.

	New York	Brussels	Tripoli	Minsk	Canberra
Highest temperature	27°C	32°C	34°C	28°C	34°C
Lowest temperature	−9°C	−6°C	8°C	−21°C	7°C

a Which city recorded the lowest temperature?
b Which city recorded the biggest difference between its highest and lowest temperatures?
c Which city recorded the smallest difference between its highest and lowest temperatures?

6 The temperature of the fridge compartment of a fridge-freezer is set at 4°C.
The freezer compartment is set at −18°C.
What is the difference between these temperature settings?

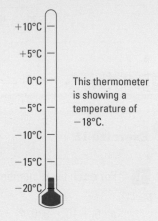

This thermometer is showing a temperature of −18°C.

7 The temperature of a shop freezer should be set at −18°C. It is set to −12°C by mistake.
What is the difference between these temperature settings?

1.8 Working with negative numbers

Objective

- You can work with negative numbers to find differences and make changes.

Why do this?

You may wish to find the difference between the temperatures at two ski resorts to see which is colder. To do this you need to work with negative numbers.

Get Ready

1. Which ski resort has the highest temperature?
2. Which resort has the lowest wind chill?

Resort	Kitzbühel	Val d'Isère	Civetta
Max (°C)	−5	−10	−3
Wind chill (°C)	−9	−20	−10

Key Point

- You can add and subtract negative numbers using a number line to help you.

Example 16   At 12 noon the temperature at the top of a mountain was 2°C.

By 6 pm it had fallen by 8°C.

What was the new temperature at 6 pm?

$2°C - 8°C = -6°C$

+4°
+3°
+2°
+1°
0°
−1°
−2° ▾ −8
−3°
−4°
−5°
−6°
−7°
−8°

ResultsPlus

Examiner's Tip

A number line can help you when working with negative numbers.

Example 17   a The temperature is 5°C. It falls by 8 degrees.

What is the new temperature?

b What is the difference in temperature between 4°C and −4°C?

a $-3°C$ ⟵ From 5°C count 8 degrees down to −3°C.

b $8°C$ ⟵ From 4°C count to −4°C. There is a difference of 8 degrees between the two temperatures.

°C
5
0
−3
−5
−8

Exercise 1J

Use this number line going from −10°C to +10°C to help you with these questions.

+10°
+9°
+8°
+7°
+6°
+5°
+4°
+3°
+2°
+1°
0°
−1°
−2°
−3°
−4°
−5°
−6°
−7°
−8°
−9°
−10°

F

1   Find the number of degrees between each pair of temperatures.

a −3°C, 2°C   b −4°C, −1°C

c 2°C, 8°C   d −6°C, 4°C

e 7°C, −3°C   f 1°C, 9°C

g −3°C, −8°C   h −7°C, 6°C

2   Find the new temperature after:

a a 2° rise from −4°C   b a 7° fall from 4°C

c 8°C falls by 15°   d −4°C rises by 7°

e −5°C rises by 8°   f 4°C falls by 10°

g −3°C falls by 6°.

1.9 Calculating with negative numbers

◉ **Objective**

○ You know the rules to use when working with positive and negative numbers.

Why do this?

Using rules is quicker than using a number line.

Get Ready

1. Write down the value of
   a an 8° rise from −10°C   b an 8° fall from −2°C.

Key Points

◉ Adding a negative number is the same as subtracting a **positive number**.

◉ Subtracting a negative number is the same as adding a positive number.

◉ The rules for multiplying and dividing are:
   ⊙ if the signs are the same the result is positive
   ⊙ if the signs are different the result is negative.

Example 18   Work out

a $2 - +3$ b $-3 - -2$ c $4 + -2$ d $-3 + +1$

a $2 - +3 = -1$

> $2 - +3$ is the same as $2 + -3$.
> Start at 2 and go down 3 to get to −1.

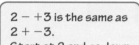

b $-3 - -2 = -1$

> $3 - -2$ is the same as $-3 + +2$.
> Start at −3 and go up 2 to get to −1.

c $4 + -2 = 2$

> $4 + -2$ is the same as $4 - 2$.
> Start at 4 and go down 2 to get to +2.

d $-3 + +1 = -2$

> Start at −3 and go up 1 to get to −2.

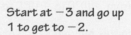

Exercise 1K

F

A03

1 Work out
   a $-4 + -3$ b $9 - +5$ c $8 - -2$ d $5 + +4$
   e $-7 - -6$ f $-2 + +4$ g $6 + -8$ h $-3 - +7$

2 A diver dives to a depth of −27 metres. A second diver dives to a depth of −16 metres.
   What is the difference in the depths of the dives?

3 The temperature at the Arctic Circle is recorded as −18°C one night. The following day it rises by 6°C.
   What is the temperature during the day?

Example 19 Work out

a 15×-3 b -7×-3 c -6×4 d $-12 \div -2$

a $+15 \times -3 = -45$ ← signs are different

b $-7 \times -3 = +21$ ← signs are the same

Where no sign is given the number is positive.

c $-6 \times +4 = -24$ ← signs are different

d $-12 \div -2 = +6$ ← signs are the same

Exercise 1L

Work out these multiplications and divisions.

1 a $+3 \times -1$ b $+24 \div -8$ c $+4 \div +1$
   d $+2 \times +6$ e $-12 \div +3$ f $-3 \times +4$

2 a $-9 \times +10$ b $-32 \div -8$ c $-20 \div -4$
   d $-2 \times +7$ e $+10 \div -5$ f -3×-4

3 a $-5 \times +4$ b $-16 \div -8$ c $-4 \times +5$
   d $-18 \div -3$ e $+18 \div +2$ f $-6 \times +7$

4 a -8×-3 b $-30 \div +2$ c $-16 \div +4$
   d -3×-9 e $+5 \times -8$ f $+24 \div +8$

5 a $-50 \div -5$ b $-7 \times +8$ c $+6 \times +6$
   d 3×7 e $9 \div 3$ f 7×6

F

Chapter review

- Although you can keep on counting, you only use ten **digits**, often called figures.
- Each digit has a **value** that depends on its position in the number. This is its **place value**.
- You can read and write numbers by thinking about the place value of each of the digits.
- You can use your knowledge of place value to put numbers in order of size.
- You can use a number line to help you to increase or decrease a number.
- Number lines might be read:
  - from left to right
  - from bottom to top
  - clockwise.
- Words that show you have to add numbers are: add, plus, total and sum.
- Words that show you have to subtract are: take away, subtract, minus and find the difference.
- Words that show you have to multiply are: product, times and multiply.
- Divide and share are words that show you have to divide.

- You multiply by 10, 100 or 1000 by moving digits 1, 2 or 3 places to the left.
- You divide by 10, 100 or 1000 by moving digits 1, 2 or 3 places to the right.
- There are situations where an estimate of a number could be good enough.
- **Rounding** means reducing the accuracy of a number by replacing right-hand digits with zeros.
- To round a number, look at the digit before the place value you are rounding to (so for rounding to the nearest 10 you would look at the units digit). If it is less than 5, round down. If it is 5 or more, round up.
- The further a number is from zero on the left-hand side of a number line, the smaller it is.
- For **negative numbers** you count backwards from zero on the number line.
- You can add and subtract negative numbers using a number line to help you.
- Adding a negative number is the same as subtracting a **positive number**.
- Subtracting a negative number is the same as adding a positive number.
- When multiplying or dividing numbers:
  - if the signs are the same the result is positive
  - if the signs are different the result is negative.

Review exercise

G

1. Draw a place value diagram and write in a number with five digits and a 2 in the thousands column and a 3 in the units column.

2. Write down the value of the 5 in the following numbers.
   a 651   b 5302   c 253 101   d 10 050   e 175

3. Write down these numbers in words.
   a 3723   b 107   c 2007   d 15 071

4. Write these numbers in figures.
   a twenty-one thousand two hundred and thirty-one
   b five hundred and seven
   c seventy thousand two hundred and three

5. This table shows the number of people who were seriously injured in road accidents in a part of Britain.

Year	2001	2002	2003	2004	2005
Number	37 346	33 645	31 456	29 788	26 466

   In which year were:
   a the smallest number of people seriously injured
   b more than 35 000 seriously injured
   c between 30 000 and 32 000 seriously injured
   d fewer than 28 000 seriously injured?

6. A shop's takings for March were £34 176.
   The takings for April were £58 358.
   Work out the total takings for the two months.

7. A school has 1321 students. 738 are boys. How many are girls?

8 Cans of cola are delivered in packs of 24.

Copy and complete the table.

Number of packs	Number of cans
1	24
2	
3	
4	
5	

If you need at least 75 cans of cola, how many packs should you order?

9 Round these numbers to the nearest multiple of 10 given in the brackets.

a 27 (10)   b 349 (100)   c 2047 (100)

d 78 939 (10)   e 7 813 076 (million) f 83.7 (10)

10 There are 376 passengers on a train.

At its first stop 27 passengers get off.

295 passengers get on.

How many passengers are now on the train?

11 Peter buys 273 stamps costing 38p each.

How much do they cost altogether?

12 There are 14 winners in a lottery.

They share the winnings of £10 332 equally.

How much does each get?

13 a The temperature during an Autumn morning went up from −3°C to 6°C.

By how many degrees did the temperature rise?

b During the afternoon the temperature fell by 11°C from 6°C.

What was the temperature at the end of the afternoon?

14 Work out

a $3 - 7$ b $-3 + 5$ c $-11 - 4$ d $4 - (-6)$ e $(-5) + (+3)$

15 Work out

a $+3 \times -7$ b -4×-5 c $16 \div -2$ d $-15 \div -3$ e $-28 \div +4$

16 Find three different numbers each below 10 which have a sum of 20.

17

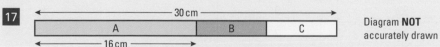

Diagram **NOT** accurately drawn

Here is a picture of a stick. The stick is in three parts, A, B and C.

The total length of the stick is 30 cm.

The length of part A is 16 cm.

The length of part B is the same as the length of part C.

Work out the length of part C.

May 2009 adapted

F

A03

18

	?	
	27	
28		
18	12	

In this set of squares, each number in a square is obtained by adding the two numbers immediately underneath.

What number should go in the top square?

A02

19 The table shows the cost of two different models of the Eiffel Tower.

Small	£2.40
Large	£4.50

Pierre buys 10 small models and 5 large models.

He pays with a £50 note.

Work out how much change he should get.

A02

20

City	Temperature
Cardiff	−2°C
Edinburgh	−4°C
Leeds	2°C
London	−1°C
Plymouth	5°C

The table gives information about the temperatures at midnight in 5 cities.

a Write down the lowest temperature.

b Work out the difference in temperature between Cardiff and Plymouth.

c Which city has a temperature halfway between London and Plymouth?

May 2009

E

A02
A03

21 Two whole numbers are each less than 10 and greater than 0. Work out the greatest possible difference between their product and their sum.

A02
A03

22 Here is some information about the coaches in a coach hire company.

Type of coach	Number of passengers	Number of coaches available
Small	14	5
Medium	32	3
Large	44	2
Touring	56	3

Jim hires two medium and two large coaches to take people to a show.

a Work out the total number of passengers that could go to the show.

Becky wants to take 300 people to a show.

b Work out the smallest number of coaches she would need to hire from the company.

23 Perfume bottles are sold in boxes which measure 4 cm by 5 cm by 20 cm.

They are supplied to shops in cartons containing 12 bottles.

Work out two arrangements of the bottles in these cartons. (There are lots of sensible answers.)

2 FACTORS, MULTIPLES AND PRIMES

A polyrhythm in drumming is when the drummer plays two different rhythms at the same time, one with each hand. A common polyrhythm is 3 : 2. The lowest common multiple of 2 and 3 is 6, so the drummer counts from 1 to 6. On the first, third and fifth beat he plays with his left hand; on the first and fourth he uses his right. The drum is silent on the sixth beat.

left: **1** 2 **3** 4 **5** 6
right: **1** 2 3 **4** 5 6

⊙ Objectives

In this chapter you will:
- ⊙ recognise types of numbers and know their properties
- ⊙ find the highest common factor (HCF) and lowest common multiple (LCM)
- ⊙ calculate the square and cube of a whole number.

◈ Before you start

You need to know:
- ⊙ how to add, subtract, multiply and divide whole numbers.

2.1 Factors, multiples and prime numbers

◎ Objectives

- ◉ You can recognise types of numbers and know their properties.
- ◉ You can separate the prime factors that make up a number.

◈ Why do this?

Credit card companies use prime numbers to protect your card details when you shop online.

◈ Get Ready

1. Write down the value of

    a 7×6 **b** 9×7 **c** 8×8 **d** 3×6

◈ Key Points

- ◉ Even numbers are whole numbers that divide exactly by 2.
- ◉ Any number that ends in 2, 4, 6, 8 or 0 is even.
- ◉ Odd numbers do not divide exactly by 2 and always end in 1, 3, 5, 7 or 9.
- ◉ The **factors** of a number are whole numbers that divide exactly into the number. They include 1 and the number itself.
- ◉ **Multiples** of a number are the results of multiplying the number by a positive whole number.
- ◉ A **common multiple** is a number that is a multiple of two or more numbers.
- ◉ A **prime number** is a whole number greater than 1 whose only factors are 1 and the number itself.
- ◉ A **prime factor** is a factor that is also a prime number.
  For example, the prime factors of 18 are 2 and 3.
- ◉ A **common factor** is a number that is a factor of two or more numbers.
- ◉ A number can be written as a product of its prime factors.

Example 1   Separate 3, 11, 14, 22, 23, 36, 39, 40, 52, 57, 60 into odd and even numbers.

14, 22, 36, 40, 52 and 60 end in an even number (2, 4, 6, 8, 0). These are the even numbers.
3, 11, 23, 39 and 57 end in an odd number (1, 3, 5, 7, 9). These are the odd numbers.

⚙ Exercise 2A

Questions in this chapter are targeted at the grades indicated.

1   Write down all the even numbers from this list.
    42, 18, 37, 955, 1110, 73 536, 500 000

2   Write down all the odd numbers from this list.
    105, 537, 9216, 811, 36 225, 300 000

3   Write down the next two even numbers after:
    a 12   **b** 28   **c** 196.

G

---

factor   multiple   common multiple   prime number   prime factor   common factor   **23**

G

4 Write down the odd number that comes before:

    a 5   **b** 31   **c** 200.

5 Write down all the even four-digit numbers that can be made using only the numbers on the cards below.

        4   7   6   5

A03

6 Using the digits 2, 7, 3 and 8 only once each,

    a write down the largest even number that can be made

    b write down the smallest odd number that can be made.

F

A03

7 A postman has letters to deliver to house numbers 16, 9, 4, 3, 22, 17, 14, 8, 16, 3, 12, 14, 17, 1, 42, 15, 16, 22, 9, 23, 31, 15 and 12.

He is going to walk down the side with even numbers first, starting at number 2, and come back on the opposite side with odd numbers. Arrange the house numbers in order for him.

Example 2   Find the factors of 12.

12 can be made from 1×12, 2×6, 3×4. The factors of 12 are 1, 2, 3, 4, 6 and 12.

Example 3   Write down the multiples of 3 that are between 20 and 29.

$7 \times 3 = 21$ is the first multiple of 3 after 20.

Then comes $21 + 3 = 24$ and $24 + 3 = 27$.

As the next multiple of 3 is 30 the answer is 21, 24 and 27.

ResultsPlus

Watch Out!

Students sometimes get confused between factors and multiples – remember multiples are from multiplying.

Exercise 2B

G

1 Write down all the factors of the following numbers.

    a 15   **b** 20   **c** 24   **d** 18   **e** 13   **f** 90

2 List the first five multiples of the following numbers.

    a 2   **b** 5   **c** 10   **d** 7   **e** 13

3 Write down three multiples of 10 that are larger than 50.

4 Write down the numbers in the cloud that are:

    a factors of 24

    b multiples of 5

    c factors of 16

    d multiples of 3.

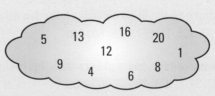

5 Find the two prime numbers that are between 30 and 40.

6 Find the next prime number after 91.

7 Find the largest number with a factor of 4 that can be made using the digits 3, 2, 4, 5.

8 A florist has 216 roses.
She makes these into bunches with an equal number of roses.
Each bunch has more than 10 roses.
Find a possible number for the roses in her bunches:
   a if all the roses are used
   b if 6 roses are left over.

Example 4 ▶ Find the common factors of 12 and 18.

The factors of 12 are 1, 2, 3, 4, 6, 12.
The factors of 18 are 1, 2, 3, 6, 9, 18.
1, 2, 3 and 6 are all factors of both 12 and 18.
They are the common factors of 12 and 18.

Example 5 ▶ Write 36 as a product of its prime factors.

Method 1
$36 = 2 \times 18$
$ = 2 \times 2 \times 9$
$ = 2 \times 2 \times 3 \times 3$

In this method, use 2 as often as possible, then move onto 3, then 5 and work through the prime numbers in order.

Method 2

This is called a factor tree.

$36 = 2 \times 3 \times 2 \times 3$

Exercise 2C

1 Find the common factors of:
   a 4 and 6
   b 10 and 15
   c 24 and 36
   d 10, 30 and 60
   e 16 and 24
   f 15 and 40
   g 12 and 28
   h 30 and 42
   i 18 and 25.

2 Find all the prime factors of the following numbers.
   a 30
   b 25
   c 42
   d 39
   e 105

3 Write these numbers as products of their prime factors.
   a 45
   b 36
   c 28
   d 80
   e 72

2.2 Finding lowest common multiple (LCM) and highest common factor (HCF)

◉ Objectives

- ○ You can find the highest common factor (HCF).
- ○ You can find the lowest common multiple (LCM).

⊘ Why do this?

If you know the number of seats per row at a cinema, you can use multiples to work out the total number of seats.

◈ Get Ready

1. Write these numbers as products of their prime factors.

    a 99   **b** 324   **c** 175

⊙ Key Points

- ◉ The **lowest common multiple (LCM)** is the lowest multiple that is common to two or more numbers.

- ◉ The **highest common factor (HCF)** is the highest factor that is common to two or more numbers.

Example 6   **a** Find the highest common factor (HCF) of 36 and 24.

                   b Find the lowest common multiple (LCM) of 6 and 8.

a $36 = 2 \times 2 \times 3 \times 3$

    $24 = 2 \times 2 \times 2 \times 3$ ←   Write each number in prime factor form. Pick out the factors common to both numbers.

    $2 \times 2 \times 3 = 12$

    12 is the HCF

b 6: 6, 12, 18, **24**, 30, 36, 42, **48**, 54

    8: 8, 16, **24**, 32, 40, **48**, 56   ←   Write a list of multiples for each number. The LCM is the lowest number that appears in both lists.

    There are two common multiples so far, but 24 is the lower.

    The LCM of 6 and 8 is 24.

⚙ Exercise 2D

D

1 Find the highest common factor of:

    a 4 and 8   **b** 9 and 12   **c** 18 and 24

    d 14 and 30   **e** 21 and 35.

2 Find the lowest common multiple of:

    a 3 and 4   **b** 4 and 6   **c** 12 and 15

    d 36 and 16   **e** 50 and 85.

3 Find the LCM and HCF of:

    a 12 and 18   **b** 120 and 180   **c** 24 and 84

    d 91 and 130   **e** 72 and 96   **f** 40 and 60.

4 Two lighthouses off the Cornish coast can be recognised by the different intervals between their flashes. One flashes every 24 seconds and the other every 40 seconds. A ship's captain sees them flash at the same time. How long will it be before this happens again?

A02

5 The light on a motorway service vehicle flashes every 20 seconds. The light on a tractor flashes every 10 seconds. As they pass each other on a road, the lights flash together. How long will it be before this happens again?

A02

2.3 Finding square numbers and cube numbers

◉ Objective

◉ You can calculate the square and cube of a whole number.

◈ Why do this?

Square and cube numbers are found in everyday life. There are 8 × 8 squares on a chessboard, and 3 × 3 × 3 cubes on a Rubik's Cube.

◈ Get Ready

1. Find these.

    a 11 × 11   **b** 13 × 13   **c** 3 × 3 × 3

Key Points

◉ A **square number** is the result of multiplying a whole number by itself.

    4 × 4 can be written as the square of 4, 4 squared or 4^2. (For more on indices, see Section 8.1.)

◉ You need to be able to recall the squares of all whole numbers up to 15 × 15.

◉ A **cube number** comes from multiplying a number by itself and then multiplying the result by the original number.

    4 × 4 × 4 can be written as the cube of 4, 4 cubed or 4^3.

◉ You need to be able to recall the cubes of 2, 3, 4, 5 and 10.

Example 7   Find   **a** the first four square numbers,   **b** the first four cube numbers.

a The first four square numbers are

    1 × 1 = 1, 2 × 2 = 4, 3 × 3 = 9, 4 × 4 = 16.

b The first four cube numbers are

    1 × 1 × 1 = 1, 2 × 2 × 2 = 8, 3 × 3 × 3 = 27, 4 × 4 × 4 = 64.

Results Plus
Watch Out!

Students often double rather than square or multiply by 3 rather than cube.

Exercise 2E

G

1 Work out the square of the following numbers.
   a 3   b 6   c 10   d 2   e 20

F

2 Work out the cube of the following numbers.
   a 2   b 5   c 7   d 20   e 12

3 Work out
   a 5 squared   b the cube of 6   c 9^2 d 3^3

Chapter review

- Even numbers are numbers that divide by 2. They always end in 2, 4, 6, 8 or 0.
- Odd numbers are numbers that do not divide by 2. They always end in 1, 3, 5, 7 or 9.
- The **factors** of a number are whole numbers that divide exactly into the number. They include 1 and the number itself.
- **Multiples** of a number are the results of multiplying the number by a positive whole number.
- A **common multiple** is a number that is a multiple of two or more numbers.
- A **prime number** is a whole number greater than 1 whose only factors are 1 and the number itself.
- A **prime factor** is a factor that is also a prime number.
- A **common factor** is a number that is a factor of two or more numbers.
- A number can be written as a product of its prime factors.
- The **lowest common multiple (LCM)** is the lowest multiple that is common to two or more numbers.
- The **highest common factor (HCF)** is the highest factor that is common to two or more numbers.
- A **square number** is the result of multiplying a whole number by itself.
- You need to be able to recall the squares of all whole numbers up to 15×15.
- A **cube number** comes from multiplying a whole number by itself and then multiplying the result by the original number.
- You need to be able to recall the cubes of 2, 3, 4, 5 and 10.

Review exercise

G

1 Using only the numbers in the cloud, write down:
   a all the multiples of 6
   b all the square numbers
   c all the factors of 12
   d all the cube numbers.

   8   12   5   6
   3   27   4   16

2 4, 5, 8, 9, 12, 14, 16, 20, 27, 35, 36, 37
   From the numbers in the list write down:
   a the odd numbers   b the multiples of 3   c the factors of 48
   d the prime numbers   e the square numbers   f the cube numbers.

3 Using the numbers 4, 1, 6, 7 no more than once each, make a number which is
  a a multiple of 3   **b** a multiple of 4.

4

| cube | multiple | factor | product |

Use a word from the box to complete this sentence correctly.

12 is a .... of 36

5 Dan writes down the numbers from 2 to 30.
He crosses out all the multiples of 2.
He crosses out all the multiples of 3 and then all the multiples of 5.
  a How many numbers have been crossed out more than once?
  b How many numbers have not been crossed out at all?

6 Write down all the multiples of 15 that are less than 100.

7 Jill says
'If you multiply any two prime numbers together, the answer will always be an odd number'.
Write down an example to show that Jill is wrong.   *June 2006*

8 Work out 10^3.

9 Charlie writes down the numbers from 1 to 50. Ben puts a red spot on all the even numbers.
Alex puts a blue spot on all the multiples of 3.
  a What is the largest number that has both a red and a blue spot?
  b How many numbers have neither a blue nor a red spot?
Sophie puts a green spot on all the multiples of 5.
  c How many numbers have exactly two coloured spots on them?

10 A buzzer buzzes every 4 seconds and a bell rings every 6 seconds. The buzzer and the bell start at the same time. How many times in the first minute will they make a sound at the same instant?

11 **a** Express 108 as the product of its prime factors.
  b Find the highest common factor of 108 and 24.

12 **a** Express the following numbers as products of their prime factors.
    i 60   **ii** 96
  b Find the highest common factor of 60 and 96.
  c Work out the lowest common multiple of 60 and 96.

13 Doughnuts are sold in packs of 8. Cakes are sold in packets of 6.
What is the smallest number of packs of doughnuts and the smallest number of packets of cakes that can be bought so that the number of doughnuts is equal to the number of cakes?

14 A car's service book states that the air filter must be replaced every 10 000 miles and the diesel fuel filter every 24 000 miles. After how many miles will both need replacing at the same time?

15 A chocolate company wishes to produce a presentation box of 36 chocolates for Valentine's Day.
It decides that a rectangular shaped box is the most efficient, but needs to decide how to arrange the chocolates.
How many different possible arrangements are there:
  a using one layer   **b** using two layers   **c** using three layers?
Which one do you think would look best?

3 DECIMALS AND ROUNDING

BBC ACTIVE

In slalom racing only one skier can be on the slope at any time, so each racer is timed, as the winner can't be decided by watching them cross the finish line. The timing has to be very accurate because only fractions of a second separate competitors, so results are given to 2 decimal places.

◎ Objectives

In this chapter you will:
- ◎ work with decimals
- ◎ give values to a suitable degree of accuracy
- ◎ work out an approximate answer to a calculation quickly in your head.

◈ Before you start

You need to:
- ◎ know about digits and place value
- ◎ understand decimal places
- ◎ be able to do simple arithmetic in your head.

3.1 Understanding place value

◎ **Objective**

◎ You can use decimals to achieve greater accuracy than whole numbers can give.

? Why do this?

You need to understand place value, including decimals, in order to carry out tasks such as shopping, measuring and timing.

◈ Get Ready

1. Write out these numbers in words:

  a 273   **b** 4076   **c** 3753

⚲ Key Point

◉ In a decimal number, the decimal point separates the whole number from the part that is smaller than 1.

Example 1   A Formula One Grand Prix driver has his lap time recorded as 123.398 seconds.

Put 123.398 in a place value diagram.

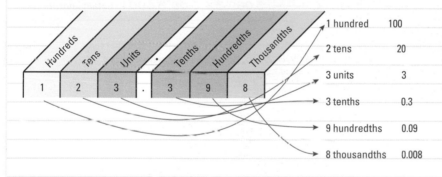

1 hundred	100
2 tens	20
3 units	3
3 tenths	0.3
9 hundredths	0.09
8 thousandths	0.008

You can better understand what 123.398 seconds means by drawing a decimal place value diagram.

Read the whole number and then read the decimal digits in order: one hundred and twenty-three point three nine eight.

Example 2   Write down the place value of the underlined digit in each number.

  a 3<u>2</u>.8   **b** 0.38<u>5</u>   **c** 10.<u>0</u>3   **d** 4.2<u>9</u>0

  a 2 units   **b** 5 thousandths   **c** 0 tenths   **d** 9 hundredths

⚙ Exercise 3A

Questions in this chapter are targeted at the grades indicated.

1 Draw a place value diagram like the one in Example 1 and write in these numbers.

  a 41.6   **b** 4.16   **c** 734.6   **d** 1.463

  e 0.643   **f** 1.005   **g** 5.01   **h** 0.086

2 What is the place value of the underlined digit in each number?

  a 2<u>5</u>.4   **b** 2.<u>5</u>4   **c** 25.4<u>6</u>   **d** 3.5<u>4</u>6

  e <u>1</u>8.07   **f** 9.66<u>9</u>   **g** 216.03<u>1</u>   **h** 2.135<u>7</u>

  i 9.1<u>0</u>2   **j** 3.<u>3</u>36   **k** 2.59<u>1</u>   **l** 0.0<u>2</u>7

G

3.2 Writing decimal numbers in order of size

Objective

- You understand the importance of place value in determining size.

Why do this?

You need to order decimal numbers to identify the winner of a race.

Get Ready

1. Write down the value of the 1 in each of these.
   a 0.102   b 0.418   c 1.002   d 10.07   e 42.001

Key Point

- To put decimal numbers in order of size, first compare the whole number parts, then the digits in the tenths place, then the digits in the hundredths place, and so on.

Example 3

Write these numbers in order of size, starting with the largest: 3.069, 5.2, 3.4, 3.08, 3.0901.

Step 1: Look at the whole-number parts.

5 is bigger than 3, so 5.2 is the biggest number. ← 3.069, 3.4, 3.08, 3.0901 remain unordered.

Step 2: Look at the tenths place.

4 is bigger than 0 so 3.4 comes next. ← 3.069, 3.08, 3.0901 remain unordered.

Step 3: Look at the hundredths place.

Here the digits are 6, 8 and 9.
The order is 5.2, 3.4, 3.0901, 3.08, 3.069.

Exercise 3B

G

1. The table gives the price of packets of dried apricots in different shops.

Shop	Stall	Corner	Market	Main	Store	Super
Price	£1.29	£1.18	£1.09	£1.31	£1.20	£1.13

   Write the list of prices in order. Start with the lowest price.

2. Rearrange these decimal numbers in order of size. Start with the largest.
   a 0.62, 0.71, 0.68, 0.76, 0.9   b 3.4, 3.12, 3.75, 2.13, 2.09
   c 0.42, 0.065, 0.407, 0.3, 0.09   d 3.0, 6.52, 6.08, 3.58, 3.7
   e 0.06, 0.13, 0.009, 0.105, 0.024   f 2.09, 1.08, 2.2, 1.3, 1.16, 1.1087

3. The fastest lap times, in seconds, of six drivers were:
   Ascarina   53.072   Bertollini   53.207
   Rascini   52.037   Alloway   54.320
   Silverman   53.702   Killim   53.027
   Write down the drivers' times in order. Start with the fastest.

3.3 Adding and subtracting decimals

⊙ Objective

⊙ You can add and subtract decimals in the same way as whole numbers.

⑦ Why do this?

You add and subtract decimals when paying for your shopping.

◈ Get Ready

1. Work out

a 24 + 37   **b** 109 − 64

🕐 Key Point

⊙ When adding and subtracting decimals you need the decimal points in line so that the place values match.

Example 4 ▸ Two children weigh 24.5 kg and 35.75 kg. What is their combined weight?

Combined weight is 24.5 kg + 35.75 kg.

Keep digits in their columns as in a place value diagram.

```
 24.5
 35.75
_____
```

← Put the decimal points under each other.

Then add:

```
  24.5
+ 35.75
 60.25
  11
```

← Decimal point in the answer should be in line.

⚙ Exercise 3C

Work these out, showing all your working.

1	1.5 + 4.6	**2**	3 + 0.25	**3**	26.7 + 42.2
4	25.7 + 0.32	**5**	0.1 + 0.9	**6**	16.1 + 2.625
7	9.9 + 9.9	**8**	10 + 1.001	**9**	0.005 + 1.909
10	117 + 1.17	**11**	6.3 + 17.2 + 8.47	**12**	13.08 + 9.3 + 6.33
13	0.612 + 3.81 + 14.7	**14**	8.6 + 3.66 + 6.066	**15**	7 + 3.842 + 0.222
16	23.43 + 5.36 + 2.216	**17**	3.07 + 12 + 0.0276	**18**	5.02 + 31.5 + 142.065

F

Example 5

Fiona buys a kettle costing £12.55. She pays with a £20 note.
How much change should she receive?

£20 − £12.55

$^{1}2^{9}\emptyset. \,^{9}\emptyset^{1}\emptyset$
− 1 2. 5 5
_____
    7. 4 5

> You need to write 20 as 20.00.

> Shopkeepers often give change by counting on:
> £12.55 + £0.05 = £12.60
> £12.60 + £0.40 = £13.00
> £13.00 + £7   = £20.00
> Change is
> £7 + 40p + 5p = £7.45

She receives £7.45 in change.

Example 6

Bill earns £124.65 per week but needs to pay £33.40 in tax and national insurance.
How much does he take home?

£124.65 − £33.40

$\cancel{1}^{1}24.65$
−   33.40
_____
   91.25

> Remember to put the decimal points under each other.

Bill takes home £91.25.

Exercise 3D

1 Work out these money calculations, showing all your working.

a £19.90 − £13.70   b £5.84 − £1.70   c £23.50 − £9.40
d £100.70 − £3.40   e £0.59 − £0.48   f £1 − £0.65
g £16.90 − £10.71   h £21.64 − £10.50   i £2.50 − £1.60
j £5.84 − £1.77   k £23.50 − £9.47   l £14 − £0.75

2 Work out these calculations, showing all your working.

a 6.125 − 4.9   b 14.01 − 2.361
c 3.29 − 1.036   d 204.06 − 35.48

3.4 Multiplying decimals

Objective

○ You can use the rule about the total number of decimal places in the answer.

Why do this?

In the supermarket meat is weighed in kilograms and priced in pounds and pence. To work out the price of 1.5 kg you would use decimal multiplication.

Get Ready

1. Work out
a 464 × 4   b 857 × 25   c 68 × 42

Key Point

⦿ When multiplying, the total number of decimal places in the answer is the same as the total number of decimal places in the question.

Example 7   Find the cost of 5 books at £4.64 each.

```
    464
×     5
  2320
   3 2
```

Multiply the numbers together, ignoring the decimals.

5 × 4.64

Count the total number of decimal places (d.p.) in the numbers you are multiplying.

0 d.p. + 2 d.p. = 2 d.p.

The answer must have 2 d.p.

The answer must have the same number of decimal places.

So the cost is £23.20.

Example 8   Work out 7.59 × 3.8.

```
    759          7.59 × 3.8
×    38
  6072
  4 7
 22 770
   2
 28 842
    1
```

2 d.p. + 1 d.p. = 3 d.p.

The answer must have 3 d.p. so it is 28.842.

Exercise 3E

Work these out, showing all your working.

1   Find the cost of:
- **a** 6 books at £2.25 each
- **b** 4 tins of biscuits at £1.37 each
- **c** 8 ice creams at £0.65 each
- **d** 1.5 kilos of pears at £0.80 per kilo.

2   Work out
- **a** 0.045 × 100
- **b** 0.45 × 100
- **c** 4.5 × 100
- **d** 0.0203 × 100
- **e** 0.203 × 100
- **f** 2.03 × 100

What do you notice about your answers to question 2?

3
- **a** 7.6 × 4
- **b** 0.76 × 4
- **c** 0.76 × 0.4
- **d** 2.25 × 5
- **e** 2.25 × 0.5
- **f** 0.225 × 0.5
- **g** 22.5 × 0.05
- **h** 2.25 × 0.005
- **i** 0.225 × 0.005

G

F

4   **a** 6.42×10 **b** 64.2×10 **c** 0.642×10

       d 56.23×10 **e** 5.623×10 **f** $0.056\,23 \times 10$

       Look carefully at your answers to question 4.

       What do you notice?

5   Work out

       a $24.6 \times 7\,\text{kg}$ **b** $3.15 \times 0.03\,\text{seconds}$ **c** $0.12 \times 0.12\,\text{m}$

       d $0.2 \times 0.2\,\text{miles}$ **e** $1.5 \times 0.6\,l$ **f** $0.03 \times 0.04\,\text{hours}$

6   A book costs £4.65. Work out the cost of buying:

       a   25 copies   **b**   36 copies   **c**   55 copies.

7   It costs £7.85 for one person to enter the Fun Beach. How much does it cost for:

       a   15 people   **b**   25 people   **c**   43 people?

8   A bucket holds 4.55 litres of water. How much water is contained in:

       a   15 buckets   **b**   25 buckets   **c**   65 buckets?

3.5   Squares and square roots, cubes and cube roots

◎ Objectives

- ○ You can extend your understanding of square and cube numbers.
- ○ You understand that finding the square root of a number is the opposite of squaring.
- ○ You understand that finding the cube root is the opposite of cubing.

❓ Why do this?

Surveyors, engineers and architects use square and cube numbers in their jobs.

⬆ Get Ready

1. Work out:

     a 3×3 **b** 6×6 **c** 11×11 **d** $4 \times 4 \times 4$

🌐 Key Points

- ◉ **Squares** are the result of multiplying any number by itself.
- ◉ $4 \times 4 = 16$, so we say that 4 is a **square root** of 16; it is a number which multiplied by itself gives 16. You can write the square root of 16 as $\sqrt{16}$.
- ◉ Notice that $-4 \times -4 = 16$, so -4 is also a square root of 16.
- ◉ Finding a square root of a number is the opposite (inverse) of squaring. A square root of 64 (written $\sqrt{64}$) is 8, since $8^2 = 64$.
- ◉ You need to know the squares of numbers up to 15×15 and their corresponding square roots.
- ◉ **Cubes** come from multiplying any number by itself and then multiplying the result by the original number again.
- ◉ You need to know the cubes of 2, 3, 4, 5 and 10.
- ◉ $2 \times 2 \times 2 = 8$, so we say that 2 is a **cube root** of 8; it is a number which multiplied by itself, then multiplied by itself again, gives 8. You can write the cube root of 8 as $\sqrt[3]{8}$.
- ◉ Finding a cube root is the opposite (inverse) of cubing.

Example 9 Find: **a** the square of 2.4 **b** 1.2^3

a The square of $2.4 = 2.4 \times 2.4 = 5.76$

b $1.2^3 = 1.2 \times 1.2 \times 1.2 = 1.728$

Exercise 3F

1 Work out the square of the following numbers.
   a 3.1   **b** 4.2   **c** 5.3   **d** 2.03   **e** 0.4

2 Work out the cube of the following numbers.
   a 1.5   **b** 2.5   **c** 3.2   **d** 0.2   **e** 0.5

E

3.6 Dividing decimals

Objective

● You can adjust decimal division so that you divide by a whole number.

Why do this?

You divide decimals whenever you work out how many calls at a certain price you can make for a £20 top up on your mobile phone.

Get Ready

1. Work out
   a $123 \div 3$ **b** $585 \div 9$ **c** $162 \div 27$

Key Point

● To divide decimals, multiply both numbers by 10, 100, 1000 etc. until you are dividing with a whole number.

Example 10 Five friends win £216.35 in a charity lottery. They share the money equally. How much do they each get?

$216.35 \div 5$ ← *Because 5 is a whole number, divide straight away.*

  4 3. 2 7 ← *Put the decimal points in line.*
$5\overline{)21^16.^13^35}$

Example 11 1.2 metres of fabric costs £1.56. What is the cost per metre?

1.56 ÷ 1.2 ← **This is not a whole number. To change 1.2 to a whole number multiply by 10.**

1.2 × 10 = 12

1.56 × 10 = 15.6 ← **Do the same to 1.56.**

The division becomes 15.6 ÷ 12

$$\begin{array}{r} 1.\ 3 \\ 12\overline{)15.^36} \end{array}$$

The answer is 1.3 or £1.30.

Results Plus
Examiner's Tip

If the number you are dividing by is not a whole number, change it to a whole number.
Remember to do the same to the number that is divided.

Example 12 Divide 58.2 by 0.03.

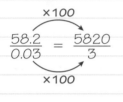

$$\frac{58.2}{0.03} = \frac{5820}{3}$$

← **To make 0.03 a whole number, multiply it by 100. So multiply both 58.2 and 0.03 by 100.**

$$= 3\overline{)5^28^120}\quad \begin{array}{l}1\ 9\ 40\end{array}$$

Exercise 3G

Work out questions 1–3.

1
  a 34.5 ÷ 10   b 3.45 ÷ 10   c 0.345 ÷ 10   d 2071 ÷ 10
  e 2.701 ÷ 10   f 0.2071 ÷ 10   g 65 ÷ 10   h 65 ÷ 100
  i 65 ÷ 1000

2
  a 64.48 ÷ 4   b 3.165 ÷ 5   c 133.56 ÷ 9   d 205.326 ÷ 6
  e 35.189 ÷ 7   f 0.0368 ÷ 8

3
  a 15 ÷ 2   b 23 ÷ 4   c 9 ÷ 8   d 3.5 ÷ 2
  e 14.4 ÷ 12   f 17 ÷ 20   g 310 ÷ 50   h 16.2 ÷ 9

4 Seven people share £107.80 equally. How much does each person get?

5 A 5 kilogram cheese is cut into 8 equal pieces. How much does each piece weigh?

6 Work out
  a 7.75 ÷ 0.5   b 7.92 ÷ 0.6   c 0.84 ÷ 0.04   d 7.7 ÷ 2.2
  e 1.284 ÷ 1.2   f 15.5 ÷ 2.5   g 1.242 ÷ 0.03   h 51.2 ÷ 1.6

3.7 Rounding decimal numbers

⊙ Objective

⊙ You can write numbers to a suitable degree of accuracy.

⊙ Why do this?

You can use rounding when adding up a bill in a restaurant to check that it is roughly correct.

⊙ Get Ready

1. How many decimal places are there in 2.0106?
2. How many digits are there in 2.0106?
3. Work out in your head: **a** 7×8 **b** $15 - 3 \times 4$

🔍 Key Points

⊙ To round a decimal to the nearest whole number, look at the digit in the tenths column (the first decimal place). If it is 5 or more, round the whole number up.

⊙ To round a decimal to one decimal place (1 d.p.), look at the second decimal place. If it is 5 or more, round up the first decimal place. If it is less than 5, leave it and any further decimal places out.

⊙ To round (or correct) to a given number of decimal places (d.p.), count that number of decimal places from the decimal point. Look at the next digit on. If it is 5 or more, you need to round up. Otherwise, leave off this digit and any that follow it.

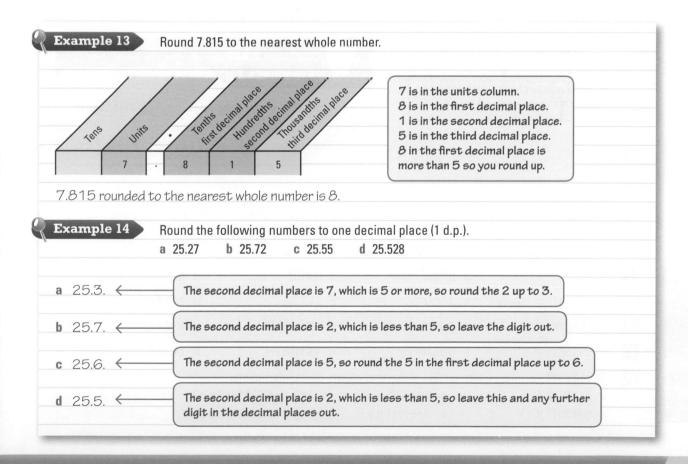

Example 13 Round 7.815 to the nearest whole number.

	Tens	Units	.	Tenths first decimal place	Hundredths second decimal place	Thousandths third decimal place
		7	.	8	1	5

7 is in the units column.
8 is in the first decimal place.
1 is in the second decimal place.
5 is in the third decimal place.
8 in the first decimal place is more than 5 so you round up.

7.815 rounded to the nearest whole number is 8.

Example 14 Round the following numbers to one decimal place (1 d.p.).
   a 25.27   **b** 25.72   **c** 25.55   **d** 25.528

a 25.3. ← The second decimal place is 7, which is 5 or more, so round the 2 up to 3.

b 25.7. ← The second decimal place is 2, which is less than 5, so leave the digit out.

c 25.6. ← The second decimal place is 5, so round the 5 in the first decimal place up to 6.

d 25.5. ← The second decimal place is 2, which is less than 5, so leave this and any further digit in the decimal places out.

Exercise 3H

G

1 Round these numbers to the nearest whole number.

a 7.8	b 13.29	c 14.361	d 5.802
e 10.59	f 19.62	g 0.771	h 20.499
i 0.89	j 100.09	k 19.55	l 1.99

F

2 Round these numbers to one decimal place (1 d.p.).

a 3.6061	b 5.3391	c 0.0901	d 9.347
e 10.6515	f 7.989	g 2.0616	h 0.4999
i 2.45	j 125.67	k 0.05	l 9.890

E

3 Round:

a 13.6 mm to the nearest mm

b 80.09 m to the nearest m

c 0.907 kg to the nearest kg

d £204.49 to the nearest £

e 3.601 lb to the nearest lb

f 0.299 tonne to the nearest tonne

g 10.5001 g to the nearest gram

h 8.066 min to the nearest minute.

Example 15

Round these numbers   i to 3 d.p.   ii to 2 d.p.

a 4.4315   b 7.3962

a   i 4.4315 rounded to 3 d.p. is 4.432.

⟵ In the number 4.4315 the next digit after the 3rd d.p. is 5, so round up and the 1 becomes 2.

   ii 4.4315 rounded to 2 d.p. is 4.43.

⟵ In the number 4.4315 the next digit after the 2nd d.p. is 1, so you can round down and the 3 remains the same.

b   i 7.3962 to 3 d.p. is 7.396

   ii 7.3962 to 2 d.p. is 7.40

The 6 makes the 9 round up to 10 and this changes the 3 to a 4.

Results**Plus**

Examiner's Tip

The final zero is important because 2 d.p. means that two decimal digits need to be shown.

e.g. The number 5.4926 would be written 5.50 to 2 d.p.

Exercise 3I

E

In questions **1** to **4** round the numbers: i to 3 d.p.   ii to 2 d.p.

1	a 4.2264	b 9.7868	c 0.4157	d 0.058 38
2	a 10.5167	b 7.5034	c 21.7295	d 9.088 95
3	a 15.5978	b 0.4081	c 7.2466	d 6.050 77
4	a 29.1582 cm	b 0.054 86 kg	c 13.3785 km	d £5.9976

5 Round each number to the number of decimal places given in brackets.

a 5.6166 (3 d.p.)

b 0.0112 (1 d.p.)

c 0.923 98 (4 d.p.)

d 0.8639 (1 d.p.)

e 9.6619 (1 d.p.)

f 1.0076 (2 d.p.)

3.8 Rounding to 1 significant figure

Objective

- You can write amounts to a suitable degree of accuracy using significant figures.

Why do this?

You can understand an approximate answer more easily. A local newspaper reported that 50 000 people had attended a rock concert. The true number was 47 231 but 50 000 made a better story and was easier to understand.

Get Ready

1. **a** Is 22 nearer to 20 or 30? **b** Is 46 nearer to 40 or 50? **c** Is 155 nearer to 100 or 200?

Key Point

- To write a number to 1 **significant figure** (1 s.f.) look at the place value of the first (highest valued) non-zero digit and round the number to this place value.

Example 16 Write these numbers to 1 significant figure (1 s.f.).
   a 32   **b** 452   **c** 0.0878

a 32 to 1 significant figure is 30.

> The first digit is in the tens column, so you need to round to the nearest ten. 32 to the nearest ten is 30.

b 452 to 1 significant figure is 500.

> The first digit is in the hundreds column, so round to the nearest hundred.

c 0.0878 to 1 significant figure is 0.09.

> The first digit is in the hundredths column, so round to the nearest hundredth.

Exercise 3J

1 Write down these numbers to 1 significant figure (1 s.f.).
   a 41 **b** 709 **c** 287 **d** 0.348 **e** 21 899
   f 0.007 41 **g** 973 **h** 4.6 **i** 13.309 **j** 19.07

2 England won 110 medals in the 2006 Commonwealth Games. Write this number to 1 significant figure (1 s.f.).

3 The number of spectators at a rugby match final was 34 862. Write this number to 1 s.f.

3.9 Rounding to a given number of significant figures

◎ Objective

- You can write numbers to the degree of accuracy that is sensible.

❓ Why do this?

The exact value is not always suitable for ordinary use. For example, at 3 pm on 22 April 2010 the exchange rates were £1 = $1.53889

£1 = 1.15075 euros

In real money you cannot have more accuracy than $1.54 which is 1 dollar and 54 cents.

◈ Get Ready

1. Write these average attendances for football clubs (2007/8 season) to 1 s.f.

Manchester United 75 690   Chelsea 41 395   Real Madrid 76 200

Newcastle United 51 320   Liverpool 43 530   Barcelona 67 560

🔧 Key Point

- To round numbers to a given number of significant figures, you count that number of digits from the first non-zero digit. If the next digit is 5 or more then you round up.

Example 17   Round 642.803   **a** to 1 s.f.   **b** to 2 s.f.   **c** to 3 s.f.   **d** to 4 s.f.   **e** to 5 s.f.

You need the zeros to show the place value of the 6.

1 s.f.	2 s.f.	3 s.f.	.	4 s.f.	5 s.f.	6 s.f.								
6	4	2	.	8	0	3	=	6	0	0				(to 1 s.f.)
6	4	2	.	8	0	3	=	6	4	0				(to 2 s.f.)
6	4	2	.	8	0	3	=	6	4	3				(to 3 s.f.)
6	4	2	.	8	0	3	=	6	4	2	.	8		(to 4 s.f.)
6	4	2	.	8	0	3	=	6	4	2	.	8	0	(to 5 s.f.)

You need the zero to show 5 significant figures.

⚙ Exercise 3K

In questions **1** to **5** round the numbers to: **i** 2 significant figures   **ii** 3 significant figures.

1 **a** 0.061 78   **b** 0.1649   **c** 96.303   **d** 41.475

2 **a** 734.56   **b** 0.079 47   **c** 5.6853   **d** 586.47

3 **a** 0.014 84   **b** 2222.8   **c** 76.249   **d** 0.3798

4 **a** 8.3846   **b** 35.959   **c** 187.418   **d** 0.066 63

5 **a** 218 736   **b** 3 989 375   **c** 307 096   **d** 25 555

6 The exchange rate was £1 = $1.6071 on 29 May 2009. Write £1.6071 to 3 significant figures.

7 The fastest lap time in a motor race Grand Prix was 83.345 seconds. Write this time to 3 significant figures.

3.10 Estimating

⊚ Objective

⊙ You can get a rough idea of the answer by working with each of the numbers to 1 significant figure.

⏃ Why do this?

When you shop for food, having a rough idea of how much you are spending means you can keep to a budget.

⬧ Get Ready

For each of the following, round the numbers to 1 s.f., then add them.

1. $338 + 286$
2. $711 + 479$
3. $0.543 + 0.265$

🌐 Key Point

⊙ To get an **estimate** for an answer, you first round each number to 1 significant figure. Then you can usually do the calculation in your head.

Example 18 Estimate the answer to $289 \times \dfrac{96}{184}$

Rounding each of the numbers to 1 significant figure gives

$300 \times \dfrac{100}{200}$

This works out as 150, which is a suitable estimate.

⚙ Exercise 3L

1. Showing your rounding, work out estimates for:

   a $65 \times \dfrac{57}{31}$

   b $\dfrac{206 \times 311}{154}$

   c $\dfrac{9 \times 31 \times 97}{304}$

   d $\dfrac{200}{12 \times 99}$

   e $\dfrac{498}{11 \times 51}$

   f $\dfrac{103 \times 87}{21 \times 32}$

2. A football grandstand has 48 rows of seats.
   Each row has 102 seats.
   Work out an estimate for the total number of seats.

3. A carton of peaches contains 48 tins.
   Work out an estimate for the total number of tins in 73 cartons.

4. Hazel is buying 28 paving stones.
   Each paving stone costs £4.85.
   Work out an estimate for her total cost.

5. Work out estimates for each of the following calculations.

   a $17.3 \times \dfrac{0.21}{4.1}$

   b $5.67 \times \dfrac{27.8}{0.86}$

   c $\dfrac{873}{23.1} \times 0.476$

D

C

3.11 Manipulating decimals

Objective

- You can use one calculation to find the answer to another.

Why do this?

If you knew that $1.50 was worth £1 then you could use this to work out how much 10p was worth.

Get Ready

1. Work out **a** 20×3 **b** 200×3 **c** 2000×3
2. Work out **a** $300 \div 10$ **b** $30 \div 10$ **c** $3 \div 10$

Key Point

- You can use the answer from one calculation to help you find the answer to a second calculation.

Example 19 Given that $3.8 \times 5.2 = 19.76$, find the values of each of the following.

a 38×5.2 **b** 380×0.52

$38 = 3.8 \times 10$

a $38 \times 5.2 = 3.8 \times 10 \times 5.2$

Rearrange the terms and substitute the known answer.

$= (3.8 \times 5.2) \times 10$

$= 19.76 \times 10$

You can check the answer by estimating.
38×5.2 is roughly 40×5, which is 200.

$= 197.6$

b $380 \times 0.52 = 3.8 \times 100 \times 5.2 \div 10$

$380 = 3.8 \times 100$ and $0.52 = 5.2 \div 10$

$= (3.8 \times 5.2) \times 100 \div 10$

Rearrange the terms and substitute the known answer.

$= 19.76 \times 10$

$= 197.6$

You can check the answer by estimating.
380×0.52 is roughly 400×0.5, which is 200.

Example 20 Given that $\frac{40.8}{8.5} = 4.8$, find the value of each of the following.

a $\frac{408}{8.5}$ **b** $\frac{40.8}{85}$

$408 = 40.8 \times 10$

a $\frac{408}{8.5} = \frac{40.8 \times 10}{8.5}$

Rearrange the terms and substitute the known answer.

$= \left(\frac{40.8}{8.5}\right) \times 10$

You can check the answer by estimating.
$408 \div 8.5$ is roughly $400 \div 8$, which is 50.

$= 4.8 \times 10$

$= 48$

$85 = 8.5 \times 10$

b $\frac{40.8}{85} = \frac{40.8}{8.5 \times 10}$

Rearrange the terms and substitute the known answer. Multiplying the bottom number by 10 is the same as dividing the top number by 10.

$= \left(\frac{40.8}{8.5}\right) \div 10$

$= 4.8 \div 10$

You can check the answer by estimating.
$40.8 \div 85$ is roughly $40 \div 80$, which is 0.5.

$= 0.48$

Exercise 3M

1 Given that $6.4 \times 2.8 = 17.92$, work out

    a 64×28 **b** 640×2.8 **c** 0.64×28

2 Given that $\dfrac{18.3}{1.25} = 14.64$, work out

    a $\dfrac{183}{1.25}$ **b** $\dfrac{1.83}{1.25}$ **c** $\dfrac{0.183}{1.25}$

3 Given that $13.2 \times 5.5 = 72.6$, work out

    a 132×5.5 **b** 1.32×0.55 **c** 0.132×55

4 Given that $\dfrac{30.4}{4.75} = 6.4$, work out

    a $\dfrac{30.4}{47.5}$ **b** $\dfrac{3.04}{4.75}$ **c** $\dfrac{304}{4.75}$

D

Chapter review

- In a decimal number, the decimal point separates the whole number from the part that is smaller than 1.
- To put decimal numbers in order of size, first compare the whole number parts, then the digits in the tenths place, then the digits in the hundredths place, and so on.
- When adding and subtracting decimals, you need the decimal points in line so that the place values match.
- When multiplying, the total number of decimal places in the answer is the same as the sum of the decimal places in the question.
- **Squares** are the result of multiplying any number by itself.
- Finding a **square root** of a number is the opposite (inverse) of squaring.
- **Cubes** come from multiplying any number by itself and then multiplying the result by the original number again.
- Finding a **cube root** is the inverse of cubing.
- To divide decimals, multiply both numbers by 10, 100, 1000 etc. until you are dividing with a whole number.
- To round a decimal to the nearest whole number, look at the digit in the tenths column (the first decimal place). If it is 5 or more, round the whole number up.
- To round a decimal to one decimal place (1 d.p.), look at the second decimal place. If it is 5 or more, round up the first decimal place. If it is less than 5, leave it and any further decimal places out.
- To round (or correct) to a given number of decimal places (d.p.), count that number of decimal places from the decimal point. Look at the next digit on. If it is 5 or more, you need to round up. Otherwise, leave off this digit and any that follow it.
- To write a number to 1 **significant figure**, look at the place value of the first (highest valued) non-zero digit and round the number to this place value.
- To round numbers to a given number of significant figures, you count that number of digits from the first non-zero digit. If the next digit is 5 or more then you round up.
- To get an **estimate** for an answer, you first round each number to 1 significant figure. Then you can usually do the calculation in your head.
- You can use the answer from one calculation to help you find the answer to a second calculation.

Review exercise

1 Put these decimal numbers in order of size. Start with the smallest.

    a 4.85, 5.9, 5.16, 4.09, 5.23

    b 0.34, 0.07, 0.37, 0.021, 0.4

    c 5, 7.23, 5.01, 7.07, 5.007

    d 1.001, 0.23, 1.08, 1.14, 0.06

2 A new cereal gives these weights of vitamins and minerals per 100 g.

Fibre	1.5 g	Iron	0.014 g	Vitamin B6	0.002 g
Thiamin B1	0.0014 g	Riboflavin	0.0015 g	Sodium	0.02 g

    Write down these weights in order. Start with the lowest.

3 Work out

    a 3.4 + 5.1   **b** 12.3 + 6.27   **c** 0.046 + 0.0712

    d 5.68 + 3.093 + 2.3702   **e** 75.3 − 16.9   **f** 20.3 − 4.72

    g 50 − 3.6   **h** 0.03 − 0.0182

4 Diana is packing to go on holiday. The baggage allowance is 20 kg.

    Her suitcase weighs 2.6 kg. In it she packs her clothes weighing 11.3 kg, her shoes weighing 3.7 kg and her toiletries weighing 2.3 kg. Is her packed suitcase within the 20 kg limit?

A03

5 An empty container weighs 27.1 kg.

    When filled, it weighs 238.7 kg.

    What is the weight of the contents?

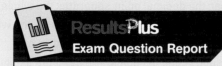

6 Work out:

    a 2.34 × 5   **b** 0.24 × 6

    c 0.3 × 0.4   **d** 25.6 × 1.6

    e 15.3 ÷ 3   **f** 81.4 ÷ 4

86% of students answered part **d** of this question poorly. The most common error using a long multiplication method was to add 1536 and 256.

7 In 2008, 355 024 candidates took Edexcel GCSE Mathematics.

    Write the number of candidates to 2 significant figures.

8

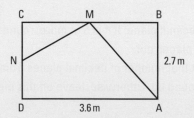

    N is the midpoint of CD. M is the midpoint of CB.

    Work out the difference in length between N to C to M and M to B to A.

9 Copy and complete the numbers in the boxes.

    a 6.24 × 100 = 0.624 × ☐   **b** 5.08 ÷ 10 = 0.508 × ☐

    c 0.455 ÷ 100 = 4.55 ÷ ☐   **d** 1.52 × 1000 = 152 ÷ ☐

10

Rob's Café Price List

Cup of tea	75p	Roll	£1.70
Cup of coffee	85p	Sandwich	£1.35
Can of cola	75p		

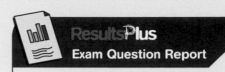

Results Plus
Exam Question Report

82% of students did very well on this type of question. They were awarded method marks for explaining clearly how their answer had been achieved.

Joe buys a can of cola and a roll at Rob's café.
a Work out the total cost.

Susan buys two cups of tea and one sandwich.
b Work out the total cost.

Kim buys a cup of coffee and a roll.
She pays with a £5 note.
c How much change should she get?

Specimen paper 2009

11 Josh has two parcels of weights 2.8 kg and 1.35 kg on a trolley. The greatest total weight the trolley can carry is 5 kg. Work out the largest weight of parcel that Josh could add to the trolley.

12 A car travels 17.2 kilometres on 1 litre of fuel. How far will it travel on 8.5 litres of fuel?

13 Angela has £15.76.
She buys as many bottles of drink costing £1.20 each as she can.
How many does she buy and how much money does she have left?

14 a 51.3 ÷ 0.9   b 0.0412 ÷ 0.4   c 30 ÷ 0.05

15 Round these numbers to 1 decimal place.
a 23.48   b 1.7502   c 0.3479   d 150.03

16 Round these numbers to the number of decimal places given in brackets.
a 7.263 (2)   b 73.0448 (2)   c 0.041 68 (3)   d 0.7208 (3)

17 Round these numbers to the number of significant figures given in brackets.
a 8317 (2)   b 20 056 (3)   c 0.546 72 (1)   d 20.873 (3)

18 Work out an estimate for the total cost of 36 books costing £7.97 each.

19 A packet of 18 slices of bacon costs £5.80.
Work out an estimate for the cost of each slice of bacon.

20 Here are the rates of pay in a company.

Grade	Basic Pay for an hour's work	Overtime pay for an hour's work
Operative	£5.40	£8.10
Technician	£7.50	£11.25
Supervisor	£9.00	£13.50
Driver	£7.20	£10.80

Lily has a part-time job as an operative.
Last week Lily earned basic pay for 24 hours and overtime pay for 3 hours.
a Work out Lily's total pay for last week.
b If Lily had been paid as a technician, work out how much extra pay she would have received.

June 2008 adapted

A03 E A03 A02 A02 A02 F

21 Sam earns £5.95 for each hour that he works.
The table shows the hours he worked one week.
Work out an estimate for the amount of money
that Sam earned that week.

Day	1	2	3	4	5
Hours worked	5	6	6	5	7

22 For each of these calculations,
work out an estimated answer.

a $\dfrac{823 \times 4872}{3261}$

b $\dfrac{3.6 \times 4.5}{9.8}$

c $\dfrac{2.4 \times 7.9}{3.9 \times 2.3}$

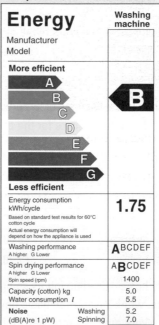

ResultsPlus
Exam Question Report

89% of students answered this type of question
poorly.

23 Using the information that $4.8 \times 34 = 163.2$
write down the value of

a 48×34 b 4.8×3.4 c $163.2 \div 48$ *June 2008*

24 Use the information that $322 \times 48 = 15\,456$ to find the value of

a 3.22×4.8 b 0.322×0.48 c $15\,456 \div 4.8$

25 Sasha works for a company. She gets paid expenses of 40p for each mile she drives during work.
Last year she worked for 48 weeks. Her total expenses for driving for the year were £2116.80.
Work out an estimate for the average number of miles Sasha drove during work each week last year.

June 2009 Specimen Paper

26 Rashid and his 3 friends order 3 pizzas and 2 bottles of cola.
They split the cost equally. How much do they each pay?

Pizza	£2.70
Cola	£1.56

27 Compare the labels from
the Henry Turbowash and
the Henry Ecowash.

a Why do you think that the
Turbowash has a better
energy rating?

b If you do 200 washes a
year, how much energy
(in kWh) do you save by
buying the Turbowash
rather than the Ecowash?
If the electricity charges
are 12p per kWh, how
much do you save per year?

c The Turbowash is £50 more
expensive. Is it better value
for money?

4 FRACTIONS

We use fractions in describing the phases of the moon. The moon in the photo is commonly called a half moon as it looks like half a circle. However, it is more correctly called a quarter moon as when it looks like this it has completed one quarter of an orbit around the Earth and one quarter of the moon's surface is visible from Earth.

◉ Objectives

In this chapter you will:
- learn how to recognise fractions
- learn how to work with fractions
- learn how to convert between fractions and decimals.

◈ Before you start

You need to:
- know your multiplication tables and the rules of basic arithmetic.

4.1 Understanding fractions

◎ Objective

○ You can identify fractions from diagrams or words.

❓ Why do this?

Fractions such as halves and quarters are used in many contexts, and shops often have offers, such as '$\frac{1}{3}$ off'.

⬆ Get Ready

1. Work out the value of:
   a $12 \div 3$ **b** 5×4 **c** $28 \div 4$ **d** $54 \div 9$

🔍 Key Points

◎ Fractions are parts of a unit.
   ⊙ The top number is called the **numerator**.
   ⊙ The bottom number is called the **denominator**.

Example 1   Identify the fraction of the parking spaces that are occupied from this information.
The car park is divided into three spaces.
Two spaces have cars in them.

Two thirds or $\frac{2}{3}$ of the parking spaces are occupied. ⟵ | There are 3 spaces, so 3 is the denominator. 2 spaces are occupied, so 2 is the numerator.

Example 2   John has 150 marbles.
30 of the marbles are made from metal. The remainder are made from glass.
What fraction of John's marbles are

   a metal   **b** glass?

a 30 of the 150 marbles are metal.
The fraction is $\frac{30}{150}$

b $150 - 30 = 120$ marbles are glass.
The fraction is $\frac{120}{150}$

⚙ Exercise 4A

| Questions in this chapter are targeted at the grades indicated. |

G

1 Copy these shapes into a table like the one on the right.
Complete your table.
The first shape has been done for you.

Shape	Fraction shaded	Fraction not shaded
⊘	$\frac{1}{2}$	$\frac{1}{2}$

G

2 Make four copies of this rectangle.
Shade them to show these fractions (one fraction on each copy).

a $\frac{1}{16}$ **b** $\frac{3}{16}$ **c** $\frac{8}{16}$ **d** $\frac{16}{16}$

3 Make three copies of this circle.
Shade them to show these fractions.

a $\frac{1}{6}$ **b** $\frac{3}{6}$ **c** $\frac{1}{2}$

4 There are 28 people on a martial arts course. Thirteen are female and 15 are male.
What fraction of the people are

a male   **b** female?

5 28 competitors took part in a surfing competition on a Saturday, and 47 other competitors took part on Sunday.

a How many surfers were there altogether?

b What fraction of the surfers competed on Sunday?

c What fraction of the surfers competed on Saturday?

4.2 Equivalent fractions

◎ Objective

○ You can find families of fractions that are equal.

⌾ Why do this?

To share a pizza equally with a friend, cut it into six pieces and have three each, or cut it into four pieces and have two each.

◈ Get Ready

1. Copy and complete

a $7 \times ? = 56$ **b** $3 \times ? = 27$ **c** $9 \times ? = 54$

🕐 Key Points

◉ You can make an **equivalent fraction** by multiplying or dividing both numerator and denominator by the same whole number.

◉ You can write a fraction in its **simplest form** by dividing numerator and denominator by the same common factor (see Section 2.1).

🔍 Example 3 ▶ Write the fraction $\frac{18}{24}$ in its simplest form.

$$\frac{18}{24} = \frac{3}{4}$$

⟵ Divide the numerator and denominator by the *same* number.

$\frac{3}{4}$ is equivalent to $\frac{18}{24}$

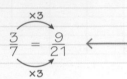

Example 4   Convert $\frac{3}{7}$ to an equivalent fraction with a denominator of 21.

$$\frac{3}{7} = \frac{9}{21}$$

×3

×3

Multiply the numerator and the denominator by the *same* number.

$\frac{3}{7}$ is equivalent to $\frac{9}{21}$

Exercise 4B

1   For each of these diagrams write down at least two equivalent fractions that describe the shaded fraction.

a   b

c   d

2   Copy and complete each set of equivalent fractions.

a $\frac{3}{4} = \frac{\square}{8} = \frac{\square}{12} = \frac{\square}{16} = \frac{\square}{20} = \frac{\square}{24}$
    b $\frac{2}{7} = \frac{\square}{14} = \frac{\square}{21} = \frac{\square}{28} = \frac{\square}{35} = \frac{\square}{42}$

c $\frac{4}{5} = \frac{\square}{10} = \frac{\square}{15} = \frac{\square}{20} = \frac{\square}{25} = \frac{\square}{30}$
    d $\frac{1}{3} = \frac{\square}{9} = \frac{\square}{18} = \frac{\square}{27} = \frac{\square}{36} = \frac{\square}{45}$

3   Copy and complete these equivalent fractions.

a $\frac{1}{6} = \frac{\square}{18}$ b $\frac{3}{7} = \frac{\square}{14}$ c $\frac{3}{8} = \frac{\square}{48}$ d $\frac{4}{7} = \frac{\square}{21}$

e $\frac{5}{6} = \frac{\square}{36}$ f $\frac{2}{3} = \frac{6}{\square}$ g $\frac{4}{9} = \frac{24}{\square}$ h $\frac{5}{7} = \frac{\square}{56}$

i $\frac{9}{10} = \frac{90}{\square}$ j $\frac{7}{12} = \frac{84}{\square}$ k $\frac{7}{8} = \frac{49}{\square}$ l $\frac{2}{9} = \frac{\square}{81}$

4   a   Find a fraction equivalent to $\frac{1}{2}$ and a fraction equivalent to $\frac{1}{3}$ so that the denominators of the two new fractions are the same.

    b   Repeat part **a** for

      i $\frac{2}{5}$ and $\frac{3}{6}$ ii $\frac{1}{10}$ and $\frac{1}{7}$ iii $\frac{1}{4}$ and $\frac{5}{6}$

      iv $\frac{1}{2}$ and $\frac{3}{5}$ v $\frac{2}{3}$ and $\frac{1}{8}$ vi $\frac{3}{4}$ and $\frac{3}{5}$

4.3 Ordering fractions

⊚ Objective

⊚ You can put fractions in order of size.

⊘ Why do this?

To get the best deal when shopping, you need to know if half price is better than one-third off.

⬥ Get Ready

1. Find the lowest common multiple of
   a 5 and 4
   b 10 and 7
   c 2, 4 and 3

⬤ Key Points

⊚ Fractions can be ordered by using equivalent fractions with the same denominator.
⊚ To find a **common denominator**, you need to find a common multiple (see Section 2.1).

Example 5   Write these fractions in order, starting with the largest.

$$\frac{1}{3} \quad \frac{2}{5} \quad \frac{3}{10} \quad \frac{1}{6}$$

Step 1 Decide on a common denominator to use. Find a common multiple of 3, 5, 10 and 6. 30 is the lowest.

$$\frac{1}{3} = \frac{1 \times 10}{3 \times 10} = \frac{10}{30} \qquad \frac{2}{5} = \frac{12}{30} \qquad \frac{3}{10} = \frac{9}{30} \qquad \frac{1}{6} = \frac{5}{30}$$

Step 2 Re-write the fractions as their equivalent fractions using 30 as denominator.

$$\frac{2}{5} \quad \frac{1}{3} \quad \frac{3}{10} \quad \frac{1}{6}$$

Step 3 Compare the equivalent fractions then write the original fractions in order.

⚙ Exercise 4C

1 By writing equivalent fractions, find the smaller fraction in each pair.
   a $\frac{2}{5}$ or $\frac{1}{4}$ b $\frac{2}{4}$ or $\frac{4}{5}$ c $\frac{2}{3}$ or $\frac{3}{4}$ d $\frac{3}{5}$ or $\frac{7}{10}$

2 Which is larger?
   a $\frac{2}{5}$ or $\frac{3}{6}$ b $\frac{1}{10}$ or $\frac{1}{7}$ c $\frac{1}{4}$ or $\frac{5}{6}$ d $\frac{1}{2}$ or $\frac{3}{5}$ e $\frac{2}{3}$ or $\frac{1}{8}$ f $\frac{3}{4}$ or $\frac{3}{5}$

3 Write these fractions in order of size. Put the smallest one first.
   a $\frac{1}{2}, \frac{3}{4}, \frac{2}{3}$ b $\frac{4}{5}, \frac{5}{6}, \frac{7}{15}$ c $\frac{3}{4}, \frac{4}{5}, \frac{1}{2}$ d $\frac{3}{7}, \frac{5}{14}, \frac{1}{2}, \frac{4}{7}$

4 Put these fractions in order of size, starting with the largest.
   $$\frac{2}{5} \quad \frac{1}{2} \quad \frac{7}{8} \quad \frac{3}{4} \quad \frac{2}{10}$$

G

4.4 Improper fractions and mixed numbers

◎ Objective

○ You can use fractions that are bigger than 1.

⑦ Why do this?

Distances on road signs are often displayed as mixed numbers, e.g. 'Oxford $7\frac{1}{2}$ miles'.

⬆ Get Ready

1. Calculate

    a $23 \div 4$ **b** $2 \times 7 + 4$ **c** $4 \times 8 + 3$

🌐 Key Points

◉ A fraction with a numerator that is larger than the denominator is called an **improper fraction**.

◉ An improper fraction can be written as a **mixed number** with a whole number part and a proper fraction part.

🔍 Example 6   Change $\frac{23}{7}$ into a mixed number.

$23 \div 7 = 3$ remainder 2 ⟵   **Step 1** Do the division.

$\quad\quad = 3\frac{2}{7}$ ⟵   **Step 2** As this is 3 whole ones with 2 left over, the mixed number is $3\frac{2}{7}$

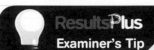
ResultsPlus
Examiner's Tip

Remember: a fraction is a division waiting to be done.

🔍 Example 7   Change $5\frac{3}{4}$ into an improper fraction.

$5\frac{3}{4} = \frac{20}{4} + \frac{3}{4} = \frac{23}{4}$ ⟵   Each whole number is the same as $\frac{4}{4}$ (4 quarters). So, the whole number 5 is $5 \times \frac{4}{4} = \frac{20}{4}$ (20 quarters).

⚙ Exercise 4D

G

1   Change these improper fractions to mixed numbers.

    a $\frac{5}{2}$ **b** $\frac{7}{4}$ **c** $\frac{9}{7}$ **d** $\frac{11}{8}$ **e** $\frac{9}{8}$ **f** $\frac{16}{5}$

    g $\frac{23}{10}$ **h** $\frac{24}{5}$ **i** $\frac{16}{7}$ **j** $\frac{12}{5}$ **k** $\frac{20}{3}$ **l** $\frac{16}{9}$

    m $\frac{39}{4}$ **n** $\frac{27}{5}$ **o** $\frac{26}{9}$ **p** $\frac{17}{10}$

2   Change these mixed numbers to improper fractions.

    a $1\frac{1}{2}$ **b** $5\frac{1}{2}$ **c** $2\frac{3}{4}$ **d** $1\frac{2}{3}$ **e** $3\frac{1}{4}$ **f** $4\frac{2}{5}$

    g $3\frac{7}{10}$ **h** $5\frac{1}{5}$ **i** $7\frac{3}{4}$ **j** $2\frac{1}{4}$ **k** $1\frac{9}{10}$ **l** $9\frac{1}{3}$

    m $2\frac{5}{6}$ **n** $5\frac{3}{8}$ **o** $3\frac{5}{8}$ **p** $1\frac{9}{100}$

4.5 Multiplying fractions

Objective

○ You can multiply fractions by other fractions or by a whole number.

Why do this?

Chefs need to multiply fractions to adjust some recipes to larger quantities.

Get Ready

1. Write down

　a 13×6　　　**b** 7×19　　　**c** 15×7

Key Points

◉ To multiply two fractions, multiply the numerators together and multiply the denominators.

◉ To find a fraction of a quantity, multiply the fraction by the quantity. For example, $\frac{3}{4}$ of £60 $= \frac{3}{4} \times$ £60.

Example 8　Work out

　a $\frac{3}{5} \times \frac{2}{3}$　　　**b** $\frac{4}{7} \times 3$　　　**c** $2\frac{3}{5} \times 1\frac{1}{4}$

> Step 1　Convert any mixed numbers to improper fractions.
> Step 2　Multiply numerators together and denominators together.
> Step 3　Simplify if possible.

a $\frac{3}{5} \times \frac{2}{3} = \frac{3 \times 2}{5 \times 3} = \frac{6}{15} = \frac{2}{5}$　⟵ Divide top and bottom by 3.

b $\frac{4}{7} \times 3$

$= \frac{4}{7} \times \frac{3}{1}$　⟵ First, write the whole number 3 as $\frac{3}{1}$

$= \frac{4 \times 3}{7 \times 1} = \frac{12}{7}$

c $2\frac{3}{5} \times 1\frac{1}{4} = \left(\frac{10}{5} + \frac{3}{5}\right) \times \left(\frac{4}{4} + \frac{1}{4}\right) = \frac{13}{5} \times \frac{5}{4}$

$= \frac{13 \times 5}{5 \times 4} = \frac{65}{20} = \frac{13}{4}$　⟵ Divide top and bottom by 5.

Exercise 4E

1　Work out

　a $\frac{1}{2}$ of £8　　**b** $\frac{1}{5}$ of £25　　**c** $\frac{1}{3}$ of £21　　**d** $\frac{1}{6}$ of £54

　e $\frac{1}{4}$ of 28 cm　　**f** $\frac{3}{4}$ of 28 cm　　**g** $\frac{1}{10}$ of 440 kg　　**h** $\frac{3}{5}$ of 30 kg

2　A machine takes $5\frac{1}{2}$ minutes to produce a special type of container.
　How long would the machine take to produce 15 containers at the same rate?

G

3 Work out

a $\frac{1}{2} \times \frac{3}{4}$ b $\frac{3}{8} \times \frac{1}{4}$ c $\frac{2}{5} \times \frac{4}{5}$ d $\frac{3}{8} \times \frac{3}{4}$

e $\frac{5}{12} \times \frac{1}{3}$ f $\frac{7}{10} \times \frac{3}{4}$ g $\frac{3}{10} \times \frac{3}{5}$ h $\frac{2}{3} \times \frac{2}{3}$

i $\frac{1}{2} \times \frac{3}{8}$ j $\frac{4}{5} \times \frac{2}{3}$ k $\frac{4}{7} \times \frac{1}{3}$ l $\frac{2}{3} \times \frac{2}{5}$

m $\frac{2}{7} \times \frac{1}{5}$ n $\frac{2}{3} \times \frac{5}{7}$ o $\frac{3}{2} \times \frac{3}{4}$ p $\frac{3}{5} \times \frac{1}{3}$

4 Work out

a $\frac{1}{2} \times \frac{4}{5}$ b $\frac{3}{4} \times \frac{4}{5}$ c $\frac{5}{6} \times \frac{3}{5}$ d $\frac{4}{5} \times \frac{3}{10}$

e $\frac{5}{6} \times \frac{3}{4}$ f $\frac{7}{12} \times \frac{3}{14}$ g $\frac{8}{9} \times \frac{3}{10}$ h $\frac{3}{4} \times \frac{16}{21}$

i $\frac{1}{3} \times \frac{6}{7}$ j $\frac{6}{7} \times \frac{5}{12}$ k $\frac{1}{2} \times \frac{4}{10}$ l $\frac{2}{3} \times \frac{1}{4}$

m $\frac{3}{7} \times \frac{2}{6}$ n $\frac{6}{5} \times \frac{1}{3}$ o $5 \times \frac{7}{10}$ p $\frac{9}{10} \times \frac{13}{18}$

5 Work out

a $\frac{1}{2} \times 7$ b $\frac{2}{3} \times 5$ c $6 \times \frac{4}{5}$ d $8 \times \frac{3}{4}$

e $\frac{7}{10} \times 20$ f $9 \times \frac{2}{3}$ g $10 \times \frac{2}{5}$ h $\frac{5}{6} \times 12$

6 Work out

a $\frac{2}{3} \times 1\frac{1}{3}$ b $\frac{2}{5} \times 2\frac{1}{3}$ c $1\frac{1}{2} \times \frac{1}{4}$ d $3\frac{1}{4} \times \frac{1}{2}$

e $\frac{2}{3} \times 4\frac{1}{4}$ f $\frac{5}{6} \times 1\frac{1}{3}$ g $2\frac{1}{2} \times \frac{7}{10}$

7 Work out

a $3\frac{1}{2} \times 1\frac{1}{2}$ b $2\frac{1}{3} \times 2\frac{3}{8}$ c $1\frac{4}{5} \times 2\frac{1}{3}$ d $3\frac{3}{4} \times 1\frac{2}{5}$

e $2\frac{1}{2} \times \frac{1}{4}$ f $1\frac{2}{5} \times 1\frac{1}{3}$ g $6 \times 2\frac{2}{3}$ h $2\frac{1}{7} \times 1\frac{2}{5}$

4.6 Dividing fractions

Objective

- You can divide fractions by other fractions or a whole number.

Why do this?

Doctors calculate the dosage of drugs for children as a fraction of the adult dose, e.g. half a half-teaspoon.

Get Ready

1. Work out

a 8×3 b 2×15 c 10×7

Key Point

- To divide fractions, turn the dividing fraction upside down and multiply the fractions.

Example 9 Work out

$$\text{a } \frac{1}{4} \div \frac{3}{5} \qquad \text{b } \frac{15}{16} \div 5 \qquad \text{c } 3\frac{1}{2} \div 4\frac{3}{4}$$

a $\frac{1}{4} \div \frac{3}{5} = \frac{1}{4} \times \frac{5}{3}$ ← Turn the second fraction upside down and multiply.

$\qquad = \frac{1 \times 5}{4 \times 3} = \frac{5}{12}$

b $\frac{15}{16} \div 5 = \frac{15}{16} \div \frac{5}{1}$

$\qquad = \frac{15}{16} \times \frac{1}{5}$

$\qquad = \frac{15 \times 1}{16 \times 5}$

$\qquad = \frac{15}{80}$

$\qquad = \frac{3}{16}$ ← Divide top and bottom by 5.

Watch Out!

Sometimes students turn the first fraction upside down by mistake. Make sure you turn the second fraction upside down.

c $3\frac{1}{2} \div 4\frac{3}{4} = \frac{7}{2} \div \frac{19}{4}$ ← Change mixed numbers to improper fractions.

$\qquad = \frac{7}{2} \times \frac{4}{19}$ ← Turn the second fraction upside down and multiply.

$\qquad = \frac{28}{38}$

$\qquad = \frac{14}{19}$ ← Simplify the answer.

Exercise 4F

1 Work out

a $\frac{1}{3} \div \frac{1}{4}$ **b** $\frac{1}{4} \div \frac{1}{3}$ **c** $\frac{3}{4} \div \frac{1}{2}$ **d** $\frac{1}{2} \div \frac{7}{10}$

e $\frac{2}{3} \div \frac{1}{5}$ **f** $\frac{5}{8} \div \frac{1}{3}$ **g** $\frac{5}{6} \div \frac{3}{4}$ **h** $\frac{7}{10} \div \frac{4}{5}$

i $\frac{2}{9} \div \frac{1}{2}$ **j** $\frac{2}{5} \div \frac{3}{4}$ **k** $\frac{3}{8} \div \frac{2}{3}$ **l** $\frac{1}{2} \div \frac{1}{4}$

2 Work out

a $8 \div \frac{1}{2}$ **b** $12 \div \frac{3}{4}$ **c** $6 \div \frac{3}{5}$ **d** $8 \div \frac{7}{8}$

e $4 \div \frac{4}{5}$ **f** $1 \div \frac{7}{12}$ **g** $5 \div \frac{1}{3}$ **h** $6 \div \frac{1}{4}$

3 Work out

a $2\frac{1}{2} \div \frac{1}{2}$ **b** $3\frac{1}{4} \div 2\frac{1}{2}$ **c** $3\frac{3}{4} \div 2\frac{1}{4}$ **d** $1\frac{5}{8} \div 3\frac{1}{6}$

e $3\frac{2}{3} \div 7\frac{1}{3}$ **f** $5\frac{1}{2} \div 2\frac{3}{4}$ **g** $1\frac{7}{10} \div 2\frac{7}{10}$ **h** $\frac{7}{8} \div 1\frac{2}{3}$

4 Work out

a $\frac{3}{4} \div 8$ **b** $\frac{5}{6} \div 2$ **c** $\frac{3}{5} \div 6$ **d** $\frac{4}{5} \div 5$

e $1\frac{1}{3} \div 4$ **f** $3\frac{1}{4} \div 6$ **g** $2\frac{5}{6} \div 10$ **h** $2\frac{1}{2} \div 15$

i $\frac{8}{9} \div 4$ **j** $\frac{2}{3} \div 6$ **k** $4\frac{2}{3} \div 4$ **l** $5\frac{1}{4} \div 3$

E

C

4.7 Adding and subtracting fractions

◎ Objective

○ You can add and subtract fractions.

❓ Why do this?

You might need to add or subtract distances involving fractions.

⬆ Get Ready

1. Find the lowest common multiple of these.

   a 8 and 7   **b** 3 and 5   **c** 6 and 3

🔍 Key Points

◉ Fractions can be added or subtracted when they have the same denominator.

◉ To add or subtract fractions that have different denominators you need to find equivalent fractions that have the same denominator (see Section 4.3).

Example 10 ▶ Work out $\frac{5}{8} + \frac{3}{7}$

Find equivalent fractions to these that have the same denominator.

$$\frac{5}{8} \overset{\times 7}{=} \frac{35}{56}$$

$$\frac{3}{7} \overset{\times 8}{=} \frac{24}{56}$$

Notice that $7 \times 8 = 56$.
56 is the lowest common multiple of 7 and 8.

$$\frac{35}{56} + \frac{24}{56} = \frac{59}{56}$$

Same denominator – now you can add the numerators.

So $\frac{5}{8} + \frac{3}{7} = \frac{59}{56}$

Example 11 ▶ Work out $2\frac{1}{4} + 3\frac{1}{5}$

$2 + 3 = 5$ ◀ These are mixed numbers. First add the whole numbers.

$\frac{1}{4} + \frac{1}{5} = ?$ ◀ Then add the fractions.

Results Plus
Watch Out!

Make sure you don't add the denominators.

$$\frac{1}{4} \overset{\times 5}{=} \frac{5}{20}$$

$$\frac{1}{5} \overset{\times 4}{=} \frac{4}{20}$$

Convert both denominators to 20 because $4 \times 5 = 20$

$$\frac{5}{20} + \frac{4}{20} = \frac{9}{20}$$

Same denominator – now you can add the numerators.

So $2\frac{1}{4} + 3\frac{1}{5} = 5\frac{9}{20}$ ◀ Now put the whole numbers and fractions back together.

Exercise 4G

1 Work out

a $\frac{3}{8} + \frac{4}{8}$ b $\frac{2}{9} + \frac{5}{9}$ c $\frac{5}{12} + \frac{7}{12}$ d $\frac{5}{18} + \frac{11}{18}$

e $\frac{1}{8} + \frac{3}{8}$ f $\frac{2}{7} + \frac{4}{7}$ g $\frac{2}{5} + \frac{3}{5}$ h $\frac{9}{10} + \frac{7}{10}$

i $\frac{7}{9} + 2\frac{4}{9}$ j $\frac{5}{6} + 1\frac{5}{6}$ k $\frac{3}{4} + \frac{3}{4} + \frac{1}{4}$ l $\frac{3}{8} + \frac{5}{8} + \frac{7}{8}$

2 Work out

a $\frac{1}{2} + \frac{1}{4}$ b $\frac{1}{4} + \frac{3}{8}$ c $\frac{1}{2} + \frac{7}{8}$ d $\frac{2}{3} + \frac{1}{6}$

e $\frac{5}{6} + \frac{1}{3}$ f $\frac{2}{5} + \frac{3}{10}$ g $\frac{7}{12} + \frac{3}{4}$ h $\frac{3}{4} + \frac{7}{20}$

3 Work out

a $\frac{1}{2} + \frac{7}{8}$ b $\frac{3}{4} + \frac{1}{10}$ c $\frac{4}{9} + \frac{5}{12}$ d $\frac{7}{8} + \frac{9}{10}$

e $\frac{3}{10} + \frac{4}{15}$ f $\frac{5}{6} + \frac{1}{4}$ g $\frac{3}{8} + \frac{7}{12}$ h $\frac{1}{6} + \frac{8}{9}$

i $\frac{5}{8} + \frac{1}{4}$ j $\frac{1}{6} + \frac{5}{8}$ k $\frac{3}{8} + \frac{11}{16}$

4 Work out

a $\frac{1}{2} + \frac{1}{3}$ b $\frac{2}{5} + \frac{1}{6}$ c $\frac{5}{8} + \frac{1}{5}$ d $\frac{3}{4} + \frac{1}{9}$

e $\frac{5}{6} + \frac{3}{7}$ f $\frac{9}{10} + \frac{2}{7}$ g $\frac{2}{3} + \frac{7}{10}$ h $\frac{3}{5} + \frac{2}{7}$

i $\frac{1}{5} + \frac{3}{8}$ j $\frac{1}{5} + \frac{1}{6}$ k $\frac{2}{3} + \frac{2}{7}$

5 Work out

a $1\frac{1}{2} + 2\frac{1}{8}$ b $2\frac{3}{4} + 3\frac{7}{8}$ c $1\frac{3}{4} + 2\frac{5}{16}$ d $\frac{3}{4} + 3\frac{5}{8}$

e $2\frac{9}{16} + 1\frac{5}{8}$ f $1\frac{3}{10} + 1\frac{2}{3}$ g $3\frac{1}{6} + \frac{2}{7}$ h $2\frac{5}{6} + 1\frac{1}{7}$

i $3\frac{2}{5} + 2\frac{7}{15}$ j $1\frac{2}{3} + 1\frac{2}{9}$

6 Jo cycled $2\frac{3}{4}$ miles to one village and then a further $4\frac{1}{3}$ miles to her home. What is the total distance Jo travelled?

7 Work out

a $3\frac{1}{4} + 2\frac{1}{2}$ b $2\frac{1}{2} + \frac{2}{3}$ c $1\frac{1}{4} + 2\frac{7}{8}$ d $3\frac{1}{3} + 5\frac{3}{4}$

e $3\frac{5}{16} + 1\frac{7}{8}$ f $2\frac{11}{12} + \frac{3}{4}$ g $\frac{5}{6} + 6\frac{1}{3}$ h $2\frac{2}{3} + 4\frac{3}{5}$

Example 12   Work out $4\frac{2}{5} - 1\frac{1}{2}$

Step 1 Change to a common denominator.

$4\frac{2}{5} - 1\frac{1}{2} = 4\frac{4}{10} - 1\frac{5}{10}$ ⟵   10 is a common denominator for 5 and 2.

Step 2 You cannot take away the fraction part as $\frac{5}{10}$ is bigger than $\frac{4}{10}$. So use 1 from $4\frac{4}{10}$ and change it into $\frac{10}{10}$ so that $4\frac{4}{10}$ becomes $3\frac{14}{10}$

Step 3 You can do the whole numbers and the fractions separately.

$3\frac{14}{10} - 1\frac{5}{10} = 2\frac{9}{10}$ ⟵ $\left(3 - 1 = 2 \text{ and } \frac{14}{10} - \frac{5}{10} = \frac{9}{10}\right)$

Exercise 4H

E

1 Work out

a $\frac{5}{11} - \frac{3}{11}$ b $\frac{7}{9} - \frac{5}{9}$ c $\frac{7}{8} - \frac{1}{8}$ d $\frac{7}{12} - \frac{5}{12}$

e $\frac{3}{4} - \frac{1}{4}$ f $\frac{5}{8} - \frac{3}{8}$ g $\frac{15}{16} - \frac{7}{16}$ h $\frac{6}{7} - \frac{3}{7}$

2 $\frac{2}{5}$ of the students at Hay College wear contact lenses. What fraction of the students do not wear them?

D

3 Work out

a $\frac{1}{2} - \frac{1}{4}$ b $\frac{7}{8} - \frac{3}{4}$ c $\frac{5}{8} - \frac{1}{2}$ d $\frac{3}{4} - \frac{1}{8}$

e $\frac{5}{6} - \frac{1}{3}$ f $\frac{7}{12} - \frac{1}{3}$ g $\frac{9}{10} - \frac{2}{5}$ h $\frac{1}{4} - \frac{1}{20}$

i $\frac{1}{2} - \frac{3}{8}$ j $\frac{7}{8} - \frac{1}{2}$ k $\frac{11}{12} - \frac{3}{4}$

4 Work out

a $\frac{2}{3} - \frac{1}{2}$ b $\frac{5}{8} - \frac{1}{3}$ c $\frac{1}{5} - \frac{1}{6}$ d $\frac{3}{5} - \frac{1}{6}$

e $\frac{4}{5} - \frac{2}{3}$ f $\frac{3}{4} - \frac{3}{5}$ g $\frac{7}{10} - \frac{1}{3}$ h $\frac{9}{10} - \frac{3}{4}$

i $5\frac{1}{4} - \frac{1}{10}$ j $7\frac{1}{2} - \frac{1}{3}$

5 In a school, $\frac{7}{16}$ of the students are girls. What fraction of the students are boys?

C

6 Work out

a $4\frac{5}{8} - 2\frac{1}{4}$ b $6\frac{1}{2} - 5\frac{1}{4}$ c $9\frac{1}{2} - 7\frac{3}{10}$ d $4 - 1\frac{3}{10}$

e $4\frac{4}{5} - 3\frac{9}{10}$ f $1\frac{2}{3} - \frac{11}{12}$ g $5\frac{3}{4} - 2\frac{19}{20}$ h $4\frac{7}{8} - 1\frac{2}{3}$

i $5\frac{7}{9} - 3\frac{1}{3}$ j $3\frac{4}{5} - \frac{3}{8}$ k $7\frac{4}{7} - 4\frac{2}{5}$

4.8 Converting between fractions and decimals

◎ Objectives

○ You can convert a fraction into a decimal and a decimal into a fraction.

○ You can recognise that some fractions are recurring decimals.

❓ Why do this?

Food labels show the amount of protein, fat etc in the food we eat, and what fractions these are of your recommended daily amount.

⬦ Get Ready

1. Do these divisions.

a $3 \div 8$ b $1 \div 2$ c $4 \div 100$

Key Points

- ◉ All fractions can be written as decimals.
- ◉ You can convert a fraction into a decimal by doing the division.
- ◉ You can convert a decimal into a fraction by looking at the place value of its smallest digit.
- ◉ You should be familiar with the equivalent fractions and decimals in this table.

Decimal	0.01	0.1	0.25	0.5	0.75
Fraction	$\frac{1}{100}$	$\frac{1}{10}$	$\frac{1}{4}$	$\frac{1}{2}$	$\frac{3}{4}$

- ◉ Fractions that have an exact decimal equivalent are called **terminating decimals**.
- ◉ Fractions that have a decimal equivalent that repeats itself are called **recurring decimals**.

Example 13 ▶ Convert these fractions into decimals. **a** $\frac{3}{4}$ **b** $\frac{1}{3}$ **c** $\frac{3}{11}$

a $0.\,7\,5$
   $4\overline{)3.^30^20}$

$\frac{3}{4} = 0.75$

b $0.\,3\,3\,3$
   $3\overline{)1.^10^10^10}$

$\frac{1}{3} = 0.\dot{3}$

c $0.\,2\,7\,2\,7\,2\,7$
   $11\overline{)3.^30^80^30^80^30^80}$

$\frac{3}{11} = 0.\dot{2}\dot{7}$ ◀ Place dots over the 2 and 7 to show that they recur.

Example 14 ▶ Convert these decimals into fractions. **a** 0.3   **b** 0.709

a The smallest digit of 0.3 is 3 tenths.
$0.3 = 3$ tenths $= \frac{3}{10}$

b The smallest digit of 0.709 is 9 thousandths.
$0.709 = 709$ thousandths $= \frac{709}{1000}$

Put the number into a place value table.

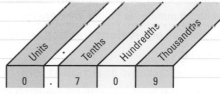

Units	.	Tenths	Hundredths	Thousandths
0	.	7	0	9

Exercise 4I

1 Convert these fractions into decimals. Show your working.

a $\frac{3}{5}$	**b** $\frac{1}{2}$	**c** $\frac{7}{10}$	**d** $\frac{7}{20}$
e $\frac{4}{25}$	**f** $\frac{3}{50}$	**g** $\frac{7}{8}$	**h** $\frac{9}{20}$
i $\frac{19}{25}$	**j** $\frac{5}{16}$	**k** $\frac{1}{8}$	**l** $\frac{27}{50}$
m $\frac{9}{100}$	**n** $\frac{13}{200}$	**o** $\frac{2}{3}$	**p** $\frac{19}{20}$

G

2 Convert these decimals into fractions.

a 0.3	**b** 0.37	**c** 0.93	**d** 0.137
e 0.293	**f** 0.7	**g** 0.59	**h** 0.003
i 0.00003	**j** 0.0013	**k** 0.77	**l** 0.077
m 0.39	**n** 0.0041	**o** 0.019	**p** 0.031

3 Write these fractions as decimals.

a $\frac{4}{5}$	**b** $\frac{3}{4}$	**c** $1\frac{1}{8}$	**d** $\frac{19}{100}$
e $3\frac{3}{5}$	**f** $\frac{13}{25}$	**g** $\frac{5}{8}$	**h** $3\frac{17}{40}$
i $\frac{7}{50}$	**j** $4\frac{3}{16}$	**k** $3\frac{3}{20}$	**l** $4\frac{5}{16}$
m $\frac{7}{1000}$	**n** $1\frac{7}{25}$	**o** $15\frac{15}{16}$	**p** $2\frac{7}{20}$

4 Write these decimals as fractions in their simplest form.

a 0.48	**b** 0.25	**c** 1.7	**d** 3.406
e 4.003	**f** 2.025	**g** 0.049	**h** 4.875
i 3.75	**j** 10.101	**k** 0.625	**l** 2.512
m 0.8125	**n** 14.14	**o** 9.1875	**p** 60.065

Chapter review

- Fractions are parts of a unit.
  - The top number is called the **numerator**.
  - The bottom number is called the **denominator**.
- You can make an **equivalent fraction** by multiplying or dividing both numerator and denominator by the same whole number.
- You can write a fraction in its **simplest form** by dividing numerator and denominator by the same common factor.
- You can order fractions by using equivalent fractions with the same denominator.
- To find a **common denominator**, you need to find a common multiple.
- A fraction whose numerator is larger than its denominator is called an **improper fraction**.
- An improper fraction can be written as a **mixed number** with a whole number part and a proper fraction part.
- To multiply two fractions, multiply the numerators together and multiply the denominators.
- To find a fraction of a quantity, multiply the fraction by the quantity.
- To divide fractions, turn the dividing fraction upside down and multiply the fractions.
- Fractions can be added or subtracted when they have the same denominator.
- To add or subtract fractions that have different denominators you need to find equivalent fractions that have the same denominator.
- You can convert a fraction into a decimal by doing the division.
- You can convert a decimal to a fraction by looking at the place value of its smallest digit.
- You should be familiar with the most common equivalent fractions and decimals.
- Fractions that have an exact decimal equivalent are called **terminating decimals**.
- Fractions that have a decimal equivalent that repeats itself are called **recurring decimals**.

Review exercise

1 In each case write down the fraction of the shape that has been shaded.

a   b    c   d

2 Write down three fractions equivalent to each of the following.

a $\frac{4}{5}$ b $\frac{2}{7}$ c $\frac{18}{24}$ d $\frac{16}{20}$ e $\frac{30}{100}$ f $\frac{18}{27}$

3 Convert these fractions into decimals.

a $\frac{1}{4}$ b $\frac{3}{8}$ c $\frac{7}{10}$ d $\frac{3}{5}$ e $\frac{24}{200}$ f $\frac{37}{50}$

4 Convert these decimals into fractions.

a 0.34   b 0.125   c 0.3   d 0.025   e 0.15   f 3.1

5 There are 600 counters in a bag. 90 of the counters are yellow.
Work out 90 as a fraction of 600.
Give your answer in its simplest form.

May 2009

6 On an average day, William spends 9 hours asleep, 8 hours at school, 2 hours watching television, 1 hour travelling and 4 hours on other things.
Work out the fraction of time spent

a asleep   b at school   c watching television
d travelling   e on other things.

7 In each case, write down which fraction is larger. You must show your working.

a $\frac{2}{3}$ or $\frac{3}{5}$ b $\frac{3}{4}$ or $\frac{4}{5}$ c $\frac{7}{10}$ or $\frac{3}{4}$ d $\frac{4}{9}$ or $\frac{11}{25}$

8 Put these fractions in order of size. Start with the largest.

$\frac{3}{10}$ $\frac{1}{3}$ $\frac{2}{7}$ $\frac{4}{15}$ $\frac{29}{100}$

9 Here are the fractions $\frac{3}{4}$ and $\frac{4}{5}$.
Which is the larger fraction? You must show working to explain your answer.
You may copy and use the grids to help with your explanation.

June 2007

10 Put these fractions and decimals in order. Start with the smallest.

$\frac{7}{10}$ $\frac{3}{4}$ 0.6 $\frac{450}{1000}$

E | **11** | Work out

  a $\frac{3}{5} \times \frac{2}{9}$ b $\frac{7}{15} \times \frac{5}{21}$ c $\frac{9}{16} \div \frac{3}{4}$

12 Which one of these fractions is equal to 0.28?

  $\frac{28}{10}$ $\frac{1}{28}$ $\frac{2}{8}$ $\frac{7}{25}$ $\frac{28}{1}$

D | **13** | Work out

  a $\frac{5}{12} + \frac{1}{4}$ b $\frac{7}{15} + \frac{4}{45}$ c $\frac{3}{5} + \frac{3}{8}$ d $\frac{7}{12} + \frac{4}{9}$ e $\frac{7}{9} - \frac{1}{3}$

A02 | **14** | The diagram shows a rectangle.
Work out the distance all the way around the rectangle.

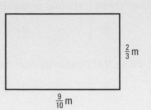

$\frac{2}{3}$ m

$\frac{9}{10}$ m

A03 | **15** | Carats are used to measure the purity of gold jewellery. Purity is measured in 24ths.
24 carat gold is pure gold but is not often used in jewellery as is it very expensive and very soft.
12 carat gold is $\frac{1}{2}$ gold. It is an alloy of which $\frac{1}{2}$ the weight is gold and $\frac{1}{2}$ the weight is made up of another metal.

  a Anisha buys an 18 carat gold bracelet.
    Explain what this means in terms of the amount of gold in the bracelet.

  b Glen buys an 8 carat ring. What fraction of gold does the ring contain?
    Give your answer in its simplest form.

  c Kelly has a 12 carat gold bangle weighing 60 g. What weight of gold is in the bangle?

  d Raj buys a 10 carat necklace that weighs 120 g. What is the weight of gold in the necklace?

  e Carol's watch strap weighs 144 g. The weight of gold in the strap amounts to 60 g. What carat is this?

C | **16** | a $2\frac{1}{2} \times 3\frac{2}{5}$ b $3\frac{3}{8} \div 2\frac{4}{7}$ c $1\frac{2}{5} \div 1\frac{11}{14}$

17 Work out $2\frac{2}{3} \times 1\frac{1}{4}$
Give your answer in its simplest form.

Results Plus
Exam Question Report

82% of students answered this question poorly.
Many students did turn the mixed numbers
into improper fractions, but wrote them with a
common denominator of 12.

June 2007

18 Work out

  a $1\frac{3}{4} - \frac{7}{16}$ b $2\frac{1}{4} - \frac{1}{2}$ c $4\frac{3}{8} - 1\frac{1}{2}$ d $5\frac{1}{5} - 2\frac{7}{15}$

19 George won some money in a lottery.
He gave $\frac{2}{5}$ to his wife and $\frac{1}{3}$ to his daughter.
What fraction did he have left?

20 The diagram represents a part of a machine. In order to fit the machine, the part must be between $6\frac{1}{16}$ cm and $6\frac{3}{16}$ cm long.

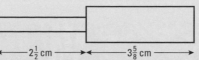

Diagram **NOT** accurately drawn

$\longleftarrow 2\frac{1}{2}$ cm $\longrightarrow$ $\longleftarrow 3\frac{5}{8}$ cm $\longrightarrow$

Will the part fit the machine? You must explain your answer.

21 A full glass of water holds $\frac{1}{6}$ of a bottle of water.
How many glasses of water can be filled from $2\frac{1}{2}$ bottles of water?

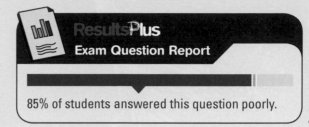

ResultsPlus
Exam Question Report

85% of students answered this question poorly.

June 2007

Relative humidity is a measure of how much water is held by air at a given temperature as a percentage of the maximum amount of water that air could hold. Under a rainforest canopy, on the forest floor, relative humidity is an amazing 95%. The continent with the lowest humidity is Antarctica where humidity is just 0.03%. This is because water vapour freezes in the extreme low temperatures so very little remains in the air.

Objectives

In this chapter you will:
- convert between percentages, fractions and decimals
- put lists of percentages, fractions and decimals in order of size
- work out a percentage of a quantity
- find quantities after a percentage increase or decrease.

Before you start

You need to be able to:
- simplify a fraction
- order fractions
- multiply fractions and decimals
- convert between fractions and decimals.

5.1 Converting between percentages, fractions and decimals and ordering them

Objectives

- You can convert between percentages, fractions and decimals.
- You can write a list of percentages, fractions and decimals in order of size.

Why do this?

You need to be able to convert between percentages, fractions and decimals to make sure that you are getting the best discounts in a shop.

Get Ready

1. What fraction of the shape is shaded?

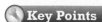

2. Write $\frac{40}{100}$ in its simplest form.

3. Write $\frac{27}{100}$ as a decimal.

Key Points

- Per cent means 'out of 100'.
- 40 per cent means 40 out of 100 or $\frac{40}{100}$.
- 40 per cent is written as 40%.

This large square is divided into 100 small squares.
40 of the 100 small squares are shaded.
40% of the large square is shaded.

- You can write percentages as decimals or fractions.
- You can write decimals or fractions as percentages.
- You can put a list with fractions, decimals and percentages in order of size by changing them to the same type of number.

Example 1 What percentage of the shape is shaded?

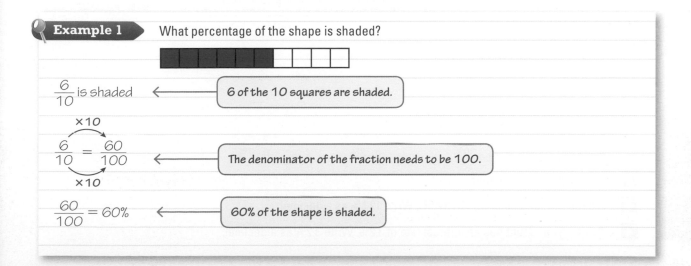

$\frac{6}{10}$ is shaded ← 6 of the 10 squares are shaded.

$\overset{\times 10}{\frac{6}{10}} = \underset{\times 10}{\frac{60}{100}}$ ← The denominator of the fraction needs to be 100.

$\frac{60}{100} = 60\%$ ← 60% of the shape is shaded.

G

Questions in this chapter are targeted at the grades indicated.

1 What percentage of each shape is shaded?

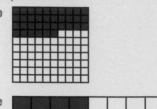

2 For each shape in question **1** write down the percentage of the shape that is not shaded.

3 a Copy this shape and shade 70% of it.

b Copy this shape and shade 40% of it.

4 60% of the houses in a street have a satellite dish.
What percentage of the houses do not have a satellite dish?

5 Louise has some flowers. 25% of the flowers are yellow. What percentage of the flowers are not yellow?

6 73% of bingo players are female. What percentage of bingo players are male?

7 On a wall the tiles are white or green or yellow. 15% of the tiles are green. 20% of the tiles are yellow.
What percentage of the tiles are white?

Example 2 Write 37% as a decimal.

$37\% = \dfrac{37}{100}$ ← Write the percentage as a fraction.

$\dfrac{37}{100} = 37 \div 100 = 0.37$ ← Change the fraction into a decimal.

so $37\% = 0.37$

Exercise 5B

G

1 Write these percentages as decimals.
a 50%   b 45%   c 62%   d 95%
e 29%   f 30%   g 3%   h 7%

2 Write 125% as a decimal.

F

3 A shop reduced its prices by 12.5%. Write 12.5% as a decimal.

4 A savings account has an interest rate of 3.2%. Write 3.2% as a decimal.

Example 3   Write 65% as a fraction in its simplest form.

$65\% = \dfrac{65}{100}$ ← Write the percentage as a fraction.

$$\dfrac{65}{100} \overset{\div 5}{\underset{\div 5}{=}} \dfrac{13}{20}$$ ← Simplify the fraction by dividing both 65 and 100 by a common factor (see Section 2.1).

so $65\% = \dfrac{13}{20}$ ← $\frac{13}{20}$ cannot be simplified.

Exercise 5C

1   Write these percentages as fractions in their simplest form.
- **a** 60%
- **b** 75%
- **c** 35%
- **d** 90%
- **e** 5%
- **f** 80%
- **g** 84%
- **h** 32%

2   64% of the spectators at a football match were male.
Write down the fraction of the spectators that were male. Give your fraction in its simplest form.

3   24% of students cycled to school. What fraction of students cycled to school?
Give your fraction in its simplest form.

4   A jacket is made from 55% silk and 45% linen. The shopkeeper wants to put on the label the fraction of silk the jacket is made of. Write 55% as a fraction in its simplest form.

5   Write these percentages as fractions in their simplest form.
- **a** 12.5%
- **b** 2.5%
- **c** 37.5%
- **d** $17\frac{1}{2}\%$

Example 4   Write these numbers in order of size. Start with the smallest number.

$\frac{3}{8}$ 0.4   35%

$\frac{3}{8} = 0.375$ ← Change $\frac{3}{8}$ into a decimal.

0.4 ← 0.4 is already a decimal.

$35\% = 0.35$ ← Change 35% into a decimal.

0.35   0.375   0.4 ← Write the decimals in order of size (see Section 3.2).

35% $\frac{3}{8}$ 0.4 ← Write each number in its original form.

ResultsPlus
Examiner's Tip

Remember to show your working out so that an examiner can follow your reasoning.

Exercise 5D

G

1 a Write 23% as a decimal.

  b Write $\frac{1}{4}$ as a decimal.

  c Which is bigger, 23% or $\frac{1}{4}$?

2 a Write 74% as a decimal.

  b Write $\frac{7}{10}$ as a decimal.

  c Which is bigger, 74% or $\frac{7}{10}$?

3 Write each list in order of size, starting with the smallest number.

  a $\frac{1}{2}$ 48%   0.45   b 55% $\frac{6}{10}$ 0.53

  c 0.7 $\frac{3}{4}$ 68%   d 27% $\frac{3}{10}$ 0.2

4 Which is bigger, 15% or $\frac{7}{40}$?

F

5 Write each list in order of size, starting with the smallest number.

  a 0.4 $\frac{1}{2}$ 45%   30% $\frac{1}{3}$

  b 0.12 $\frac{1}{20}$ 15% $\frac{1}{5}$ 10%

  c $\frac{2}{3}$ 68%   0.63 $\frac{13}{20}$ 0.6

  d 70% $\frac{27}{40}$ 0.65 $\frac{3}{5}$ 62%

Mixed exercise 5E

G

1 What percentage of the shape is shaded?

2 15% of the cars in a car park are red. What percentage of the cars are not red?

3

  a What fraction of the shape is shaded?

  b What percentage of the shape is shaded?

  c Copy the shape and shade in more squares so that 80% of the shape is shaded.

4 Write 70% as a decimal.

5 Write 37% as a fraction.

6 Write 28% as fraction. Give your answer in its simplest form.

7   Write these numbers in order of size. Start with the smallest number.

$\frac{3}{8}$ 35%   0.3 $\frac{1}{3}$ 0.25

8   Two students did a test. Sam gained $\frac{2}{5}$ of the total marks. Ryan gained 45% of the total marks. Who did better? Explain your answer.

5.2 Finding percentages of quantities

Objectives

- You can work out a percentage of a quantity.
- You can use percentages in real-life problems.

Why do this?

This is useful if you need to work out your exam marks as a percentage of the total to see if you have reached the 75% pass mark.

Get Ready

1. Work out 32 ÷ 4.   2. Work out 80 ÷ 5.   3. Work out 45 ÷ 10.

Key Points

- You should know these percentages and their fraction and decimal equivalents.

Percentage	1%	10%	25%	50%	75%
Decimal	0.01	0.1	0.25	0.5	0.75
Fraction	$\frac{1}{100}$	$\frac{1}{10}$	$\frac{1}{4}$	$\frac{1}{2}$	$\frac{3}{4}$

- If a percentage can be written as a simple fraction it is easy to work out a percentage of a quantity without using a calculator. For example:
  - to work out 50% of a quantity you work out $\frac{1}{2}$ of it
  - to work out 25% of a quantity you work out $\frac{1}{4}$ of it.
- To work out a percentage of a quantity using fractions you should:
  - write the percentage as a fraction, and then
  - multiply the fraction by the quantity.
- To find a percentage of a quantity using decimals you should:
  - write the percentage as a decimal, and then
  - multiply the decimal by the quantity.

Example 5   Work out 30% of 50. Do not use a calculator.

Method 1

$10\% = \frac{1}{10}$ ← 10% is equivalent to $\frac{1}{10}$.

$\frac{1}{10}$ of 50 = 50 ÷ 10 = 5 ← To find $\frac{1}{10}$ of 50, divide 50 by 10.

10% of 50 = 5

so 30% of 50 = 3 × 5 ← 30% is 3 lots of 10%.

= 15

Method 2

$\frac{30}{100} \times 50$ ← Replace of with times.

30% is $\frac{30}{100}$.

30 × 0.5 ← Divide by bottom $\frac{50}{100}$ = 0.5.

= 15 ← Multiply by top.

Exercise 5F

1 Work out
   a 50% of £24   **b** 50% of 80 kg   **c** 25% of 32 m   **d** 50% of 76p
   e 25% of £200   **f** 75% of 12 cm   **g** 75% of $40   **h** 25% of £92

2 Work out
   a 10% of £60   **b** 10% of 70 km   **c** 20% of 70 km   **d** 30% of £120
   e 20% of £80   **f** 15% of 20 kg   **g** 70% of 300 ml **h** 35% of £40

3 Simon's salary last year was £35 400. He saved 10% of his salary. Simon wants to buy a car costing £3650. Has he saved enough?

4 A packet of breakfast cereal contains 750 g of cereal plus '20% extra free'.
Work out how much extra cereal the packet contains.

5 Jamal earns £28 000 in one year. He gets £1000 tax free. On the remainder he pays income tax at 20%. Work out how much income tax he pays in that year.

6 Hannah earns £6.80 per hour. She is given a pay rise of 5%. Work out how much extra she gets per hour.

7 The price of a new sofa is £480. Leah pays a deposit of 15% of the price.
Work out the deposit she pays.

8 Naomi works in a shop. Each week she is paid £150 plus 5% of her weekly sales over £500. Last week, Naomi's weekly sales totalled £1200. Work out how much Naomi was paid last week.

9 The normal cost of a suit is £120. In a sale the cost of the suit is reduced by 35%.
Work out how much the cost of the suit is reduced by in the sale.

10 Rahma pays income tax. She pays 20% on the first £37 400 of her income and 40% on income over £37 400. Rahma's income last year was £59 400. Work out how much income tax she paid last year.

Example 6

The normal price of a television is £200.
Harry is given a discount of 20%.
Work out the discount that Harry is given.

$20\% = 0.2$ ←   *Write 20% as a decimal.*

$0.2 \times 200 = 40$ ← $2 \times 200 = 400$, so $0.2 \times 200 = 40$

Harry is given a discount of £40.

Exercise 5G

1 Work out
   a 10% of £40   **b** 80% of 45 kg   **c** 3% of £270   **d** 90% of 800 km

5.3 Finding the new amount after a percentage increase or decrease

Objective

⊙ You can find quantities after a percentage increase or decrease.

Why do this?

You can work out the cost of something reduced in the sales.

Get Ready

1. Work out   **a** 10×1.5 **b** 150×1.2 **c** 10×0.65 **d** 3000×0.9

Key Points

◉ There are two methods that can be used to increase an amount by a percentage:
   ⊙ you can find the percentage of that number and then add this to the starting number
   ⊙ you can use a multiplier.
◉ There are two methods that can be used to decrease an amount by a percentage:
   ⊙ you can find the percentage of that number and then subtract this from the starting number
   ⊙ you can use a multiplier.

Example 7   A packet contains 500 g of cereal plus 20% extra cereal free.
Work out the weight of the cereal in the packet.

$100\% + 20\% = 120\%$ ← The packet contains 120% of the usual amount of cereal.

$120\% = 1.2$ ← 1.2 is called the multiplier.

120% of $500 = 1.2 \times 500$ ← $5 \times 12 = 60$, so $1.2 \times 500 = 600$
$= 600\,g$

Example 8   In a sale all normal prices are reduced by 40%.
The normal price of a pair of jeans is £40.
Work out the sale price of the jeans.

$100\% - 40\% = 60\%$ ← The jeans are 60% of their normal price.

$60\% = 0.6$ ← 0.6 is the multiplier.

60% of $40 = 0.6 \times 40$ ← $6 \times 4 = 24$, so $0.6 \times 40 = 24$
$= £24$

Exercise 5H

1 Write down the single number you can multiply by to work out an increase of:
   a 10%   **b** 20%   **c** 50%   **d** 3%

2 **a** Increase £300 by 20%   **b** Increase 90 kg by 30%
    c Increase 40 km by 25%   **d** Increase £1200 by 100%

3 Write down the single number you can multiply by to work out a decrease of:
   a 10%   **b** 20%   **c** 50%   **d** 3%

4 **a** Decrease £400 by 10%   **b** Decrease 200 kg by 20%
    c Decrease 70 m by 30%   **d** Decrease £1500 by 80%

Chapter review

- Per cent means 'out of 100'.
- You can write percentages as decimals or fractions.
- You can write decimals or fractions as percentages.
- You can put a list with fractions, decimals and percentages in order of size by changing them to the same type of number.
- You should know these percentages and their fraction and decimal equivalents.

Percentage	1%	10%	25%	50%	75%
Decimal	0.01	0.1	0.25	0.5	0.75
Fraction	$\frac{1}{100}$	$\frac{1}{10}$	$\frac{1}{4}$	$\frac{1}{2}$	$\frac{3}{4}$

- If a percentage can be written as a simple fraction it is easy to work out a percentage of a quantity without using a calculator.
- To work out a percentage of a quantity using a fraction you should:
  - write the percentage as a fraction, and then
  - multiply the fraction by the quantity.
- To find a percentage of a quantity using decimals you should:
  - write the percentage as a decimal, and then
  - multiply the decimal by the quantity.
- There are two methods that can be used to increase an amount by a percentage:
  - you can find the percentage of that number and then add this to the starting number
  - you can use a multiplier.
- There are two methods that can be used to decrease an amount by a percentage:
  - you can find the percentage of that number and then subtract this from the starting number
  - you can use a multiplier.

Review exercise

1 **a** Write 10% as a decimal.
    b Write 4% as a decimal.
    c Write 26% as a fraction.
       Give your answer in its simplest form.

June 2007

2 The weight of a coin is 25% nickel and 75% copper.
   a Write 25% as a decimal.
   b Write 25% as a fraction.
   Give your answer in its simplest form.

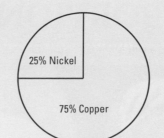

25% Nickel

75% Copper

3 **a** Write 60% as a fraction. Give your fraction in its simplest form.
   b 55% of the students in a school are female.
   What percentage of students are male? *Specimen 2006*

4 **a** Work out 50% of £60.   **b** Work out 25% of 20 metres. *May 2009*

5 **a** Write 92% as a decimal.
   b Write 3% as a fraction.
   c Work out 5% of 400 grams. *May 2008*

6 The weight of a coin is 8 grams. 25% of the weight is nickel and 75% of the weight is copper. A bag contains 10 coins. Work out the weight of
   a copper   **b** nickel. *June 2006*

7 A test had 60 questions. Saima answered 75% of the questions correctly. How many questions did she answer correctly?

8 There are 240 children at a disco. 35% of the children are boys. How many of the children are boys?

9 VAT is charged at a rate of $17\frac{1}{2}$%. Work out how much VAT will be charged on a laptop costing £300.

10 In a maths test, Mojda scored 27 out of 40. In the same test Emma scored 70%.
   Who got the better score? Explain your answer.

11 The price of a holiday is £670. Pip receives a 10% discount for booking online. Work out the discount she receives.

12 Last year, a football team won 40% of all its games.
   The team won 14 of the first 30 games.
   The team played 10 more games.
   How many of these 10 games did the team win?

13 The normal price of a pack of croissants is £1.80. The normal price is reduced by 20%
   Work out the price after the reduction.

14 A store reduced all normal prices by 10% in a two-day-sale. Work out the sale price of:
   a a drill with a normal price of £70
   b a lawnmower with a normal price of £180
   c a tin of paint with a normal price of £14

15 The population of a town is 68 000. In ten years' time the population is expected to have increased by 20%. What is the population expected to be in ten years' time?

16 A concert ticket costs £40 plus a booking charge of 15%. Work out the cost of a concert ticket.

G

F

E

A02
A03
D

A03

To understand bicycle gears, you need to work with ratios. Gears work by changing the distance that you move forward each time you pedal. The lowest gear of a bicycle might use a front cog with 22 teeth and a back cog with 33 teeth. That's a ratio of 0.67 : 1. So for each pedal stroke, the wheels do 0.67 of a turn. The highest gear might have a front chain wheel with 44 teeth and a back chain wheel with 11 teeth, giving a ratio of 4 : 1. So the wheel turns four times for every pedal stroke.

⊙ Objectives

In this chapter you will:
- write a ratio in various forms
- use a ratio to write down a fraction
- use equivalent ratios to find unknown quantities
- solve problems using scales of maps and scale drawings and problems involving proportion
- divide a quantity in a given ratio and use the unitary method to find quantities.

◈ Before you start

You need to be able to:
- multiply and divide by an integer
- simplify a fraction
- find equivalent fractions
- convert between metric units.

6.1 Introducing ratio

⊙ Objectives

- ⊙ You can write down a ratio.
- ⊙ You can use a ratio to write down a fraction.
- ⊙ You can write a ratio in its simplest form.
- ⊙ You can write a ratio in the form $1 : n$.

⊘ Why do this?

Scales on maps and drawings are often written as ratios. You can work out the distance between two places on a map by using the map's scale.

◈ Get Ready

1. What fraction of this rectangle is blue?

2. Write $\frac{9}{12}$ in its simplest form.

3. Write $\frac{80}{100}$ in its simplest form.

Key Points

- ● **Ratios** are used to compare quantities.

  The ratio of green triangles to white triangles is $3 : 2$.
  The ratio of white triangles to green triangles is $2 : 3$.

- ● Ratios can be simplified like fractions.

  To simplify a ratio you divide each of its numbers by a common factor. Common factors appeared in Section 2.1.

  This necklace has 6 yellow beads and 3 red beads.
  The ratio of yellow beads to red beads is $6 : 3$.
  Both 6 and 3 can be divided by 3 to give the ratio $2 : 1$.
  This means that for every 2 yellow beads there is 1 red bead.

- ● When a ratio cannot be simplified it is in its simplest form.
- ● It is sometimes useful to write ratios in the form $1 : n$.

  The number n is written as a whole number or a decimal.
  When one of the numbers in a ratio is 1, the ratio is in **unitary** form.

- ● To work out real-life ratios you may need to convert between units of measure (see Chapter 17).

🔍 Example 1

Write down the ratio of yellow counters to blue counters.

The ratio is $5 : 4$. ⟵

> 5 comes first as 'yellow counters' is written before 'blue counters'.

 Questions in this chapter are targeted at the grades indicated.

E

1 For each pattern of tiles, write down the ratio of the number of white tiles to the number of red tiles.

a  b c

2 Adam is 16 years old. Sarah is 13 years old.

a What is the ratio of Adam's age to Sarah's age?

b What is the ratio of Sarah's age to Adam's age?

3

a What is the ratio of the number of circles to the number of squares?

b What is the ratio of the number of circles to the number of triangles?

c What is the ratio of the number of triangles to the number of squares?

d What is the ratio of the number of circles to the number of squares to the number of triangles?

Example 2 A box contains blue cubes and red cubes in the ratio 3 : 2.
What fraction of these cubes are blue?

← For every 3 blue cubes there are 2 red cubes.

$3 + 2 = 5$

$\frac{3}{5}$ of the cubes are blue.

Exercise 6B

E

1 In a fitness centre, the ratio of the number of men to the number of women is 1 : 2.
What fraction of these people are men?

2 A vase contains red roses and white roses in the ratio 2 : 3.

a What fraction of these roses are red?

b What fraction of these roses are white?

3 A box of chocolates contains milk chocolates and dark chocolates in the ratio 5 : 3.

a What fraction of these chocolates are milk chocolates?

b What fraction of these chocolates are dark chocolates?

D

4 A box contains blue pens, red pens and black pens in the ratio 7 : 1 : 2.

   a What fraction of these pens are red?

   b What fraction of these pens are blue?

5 On a farm, $\frac{1}{2}$ of the horses are female.

   What is the ratio of female horses to male horses on this farm?

6 On the farm, $\frac{1}{3}$ of the pigs are female.

   What is the ratio of female pigs to male pigs on this farm?

Example 3 Write the ratio 15 : 20 in its simplest form.

$$15 : 20$$

$\div 5$ ( ) $\div 5$ ← Divide both 15 and 20 by 5.

$$= 3 : 4$$

3 : 4 is the simplest form.

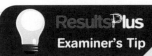

ResultsPlus
Examiner's Tip

A question will often ask you to write your ratio in its simplest form. Remember to do this!

Exercise 6C

1 Write these ratios in their simplest form.

   a 5 : 10   **b** 4 : 12   **c** 21 : 15   **d** 30 : 45

   e 80 : 60   **f** 24 : 72   **g** 160 : 300   **h** 600 : 2400

2 In a fruit bowl there are 8 oranges and 12 apples.

   Write down the ratio of the number of oranges to the number of apples.

   Give your ratio in its simplest form.

3 There are 200 boys and 150 girls in a school.

   Write down the ratio of the number of boys to the number of girls.

   Give your ratio in its simplest form.

4 In a bird aviary there are 20 finches, 36 canaries and 24 budgies.

   Write down the ratio of the number of finches to the number of canaries to the number of budgies.

   Give your ratio in its simplest form.

5 In a class there are 28 students. Twelve of the students are boys.

   Write down the ratio of the number of boys to the number of girls.

   Give your ratio in its simplest form.

E

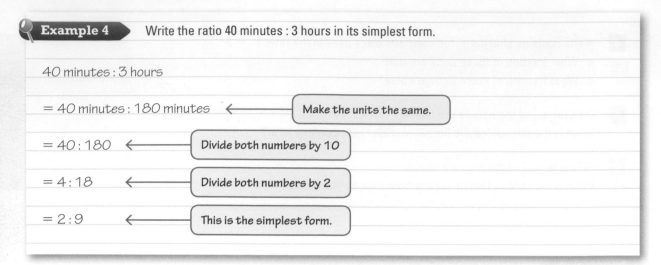

Example 4   Write the ratio 40 minutes : 3 hours in its simplest form.

40 minutes : 3 hours

= 40 minutes : 180 minutes ←   Make the units the same.

= 40 : 180 ←   Divide both numbers by 10

= 4 : 18 ←   Divide both numbers by 2

= 2 : 9 ←   This is the simplest form.

Exercise 6D

E

1   Write these ratios in their simplest form.

    a 20 minutes : 1 hour   **b** 40p : £1   **c** 30 cm : 1 m   **d** 1 day : 6 hours

2   Write these ratios in their simplest form.

    a 600 g : 2 kg   **b** 4 cm : 6 mm   **c** 2 hours : 45 minutes   **d** 75 cm : 3 m

3   A glass contains 450 ml of water and a bottle contains 1 l of water.
    Write down the ratio of the amount of water in the glass to the amount of water in the bottle.
    Give your ratio in its simplest form.

4   A small bag of sugar weighs 375 g and a large bag of sugar weighs 2 kg.
    Write down the ratio of the weight of the small bag to the weight of the large bag.
    Give your ratio in its simplest form.

5   A table has a length of 1.2 m and a width of 40 cm.

    a Find, in its simplest form, the ratio of the length of the table to the width of the table.

    b Find, in its simplest form, the ratio of the width of the table to the length of the table.

Example 5   Write the ratio 6 : 9 in the form 1 : n.

6 : 9

÷6 (   ) ÷6 ←   Divide by 6 to make the first number into 1.

= 1 : 1.5

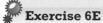

Exercise 6E

1 Write these ratios in the form $1 : n$.
  a 2 : 5   **b** 5 : 12   **c** 10 : 3   **d** 200 : 250

2 The length of a model car is 20 cm. The length of the real car is 360 cm.
Write down the ratio of the length of the model car to the length of the real car.
Give your answer in the form $1 : n$.

3 In a school there are 120 teachers and 1740 students. The headteacher wants to display on his website the ratio of the number of teachers to the number of students in the form $1 : n$. Work out this ratio.

4 Write these ratios in the form $1 : n$.
  a 20p : £1   **b** 50 g : 1 kg   **c** 4 cm : 8 mm   **d** 30 minutes : 3 hours

Mixed exercise 6F

1 Write, in its simplest form, the ratio of the number of red squares to the number of white squares.

2 A necklace has yellow beads and green beads only.
The ratio of the number of yellow beads to the number of green beads is 4 : 5.
  a What fraction of the beads are yellow?
  b What fraction of the beads are green?

3

Ingredient	Weight in grams
Protein	6
Carbohydrate	16
Fat	4

Write, in its simplest form, the ratio of:
  a the weight of carbohydrate to the weight of protein
  b the weight of fat to the weight of carbohydrate
  c the weight of protein to the weight of carbohydrate to the weight of fat.

4 Write these ratios in their simplest form.
  a 2 hours : 40 min   **b** 800 m : 2 km

5 A recipe uses 150 g of margarine and 1 kg of fruit.
Write down the ratio of the weight of margarine to the weight of fruit.
Give your ratio in its simplest form.

6 Write the ratio 24 : 36
  a in its simplest form   **b** in the form $1 : n$.

D

A03

7 On a bus there are 12 adults and 27 children.
Write the ratio of the number of adults to the number of children in the form $1 : n$.

8 An art gallery issues this guidance to schools planning to visit the gallery.

> The recommended adult/pupil ratio is:
>
> For Years 1 to 3, a minimum of 1 adult to every 5 pupils
>
> For Years 4 to 9, a minimum of 1 adult to 10 pupils
>
> For Years 10 onwards, a minimum of 1 adult to 15 pupils

A primary school is planning a visit to the art gallery.
The table shows information about the pupils going on the visit.

Year	Number of pupils
2	7
3	13
4	12
5	14

Work out the minimum number of adults that need to go on the visit.

6.2 Solving ratio problems

◎ Objectives

- You can use equivalent ratios to find unknown quantities.
- You can solve problems using scales of maps and scale drawings.

❔ Why do this?

You could use ratios if you were trying to mix a specific colour of paint for decorating your room that the shop did not stock.

⬦ Get Ready

1. $\frac{1}{5} = \frac{?}{20}$
2. Which of these fractions are equivalent to $\frac{1}{4}$? $\frac{2}{6}$ $\frac{2}{8}$ $\frac{3}{9}$ $\frac{3}{12}$
3. Change 3.65 centimetres to metres.

⬙ Key Points

- The ratio $15 : 20$ simplifies to the ratio $3 : 4$.
  $15 : 20$ and $3 : 4$ are **equivalent ratios**.
- If you know the ratio of two quantities and you know one of the quantities, you can use equivalent ratios to find the other quantity.
- Maps have a scale to tell us how a distance on the map relates to the real distance.
  A scale of $1 : 25\,000$, for example, means that 1 cm on the map represents a real distance of $25\,000$ cm.
- Someone making a scale model or producing a scale drawing will also need to use a scale.

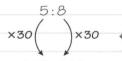 **Example 6**

To make puff pastry, Dylan mixes margarine and flour in the ratio 5 : 8 by weight.
He uses 150 g of margarine. How much flour should he use?

$$5:8$$

×30 ( ) ×30 ⟵ The amount of margarine is multiplied by 30 so multiply the amount of flour by 30.

$$150:240$$

He should use 240 g of flour.

Exercise 6G

D

1   James makes mortar by mixing cement and sand in the ratio 1 : 5.
    He uses 4 buckets of cement. Work out how many buckets of sand he uses.

2   Margaret makes porridge by mixing oats and water in the ratio 1 : 2.
    Work out the number of cups of water she uses for:
    a 2 cups of oats   b 3 cups of oats   c 10 cups of oats.

3   An alloy contains iron and aluminium in the ratio 4 to 1 by weight.
    a If there is 24 kg of iron, work out the weight of aluminium.
    b If there is 15 kg of aluminium, work out the weight of iron.

4   In a recipe for making pastry the ratio of the weight of flour to the weight of margarine is 3 : 2.
    a Work out the weight of margarine needed for:
       i 60 g of flour   ii 300 g of flour   iii 450 g of flour.
    b Work out the weight of flour needed for:
       i 60 g of margarine   ii 100 g of margarine   iii 250 g of margarine.

5   Sidra is making a fruit drink.
    She mixes orange juice, pineapple juice and syrup in the ratio 3 : 4 : 1.
    a If she uses 600 m*l* of orange juice, how much syrup will she need?
    b If she uses 1 *l* of pineapple juice, how much orange juice will she need?

6   In the 2008 Olympic Games, the ratio of the number of bronze medals won by Great Britain to the
    number of bronze medals won by Japan was 3 : 2. Great Britain won 5 more bronze medals than Japan.
    How many bronze medals did Great Britain win?

7   Paul makes green paint by mixing 2 parts of yellow paint with 3 parts of blue paint.
    Paul has 500 m*l* of yellow paint and 1 litre of blue paint.
    What is the maximum amount of green paint that Paul can make?

A03

Example 7   The scale of a map is 1 : 20 000.
Work out the real distance, in kilometres, that 9 cm on the map represents.

1 : 20 000   ← 1 cm on the map represents a real distance of 20 000 cm.
9 × 20 000 = 180 000   ←
   Multiply the length on the map by 20 000.

The real distance is 180 000 cm.
180 000 ÷ 100 = 1800 m   ←
1800 ÷ 1000 = 1.8 km   Change 180 000 cm to kilometres.

9 cm on the map represents a real distance of 1.8 km.

Exercise 6H

D

1   Alex uses a scale of 1 : 50 to draw a plan of his bedroom.
On the plan the length of the bedroom is 8 cm.
Work out the real length of the bedroom.

2   Shannon makes a scale model of a house. She uses a scale of 1 : 12.
The height of the model house is 60 cm.
Work out the height of the real house.

3   A model of a ship is made using a scale of 1 : 600.
The length of the model ship is 40 cm.
Work out the length of the real ship.

4   The length of a car is 3 metres.
Asif makes a model of the car. He uses a scale of 1 : 12.
Work out the length, in centimetres, of the model car.

5   A company makes model cars using a scale of 1 : 18.
   a   Work out the length of the real car if the length of a model car is 20 cm.
   b   Work out the length of a model car if the length of the real car is 4.68 m.

6   The scale of a map is 1 : 100 000.
On the map the distance between two towns is 6 cm.
Work out the real distance between the two towns.

7   The scale of a map is 1 : 50 000.
On the map the length of a railway tunnel is 3.5 cm.
Work out the real length of the railway tunnel.

8   The scale of a map is 1 : 200 000.
The real distance between two towns is 24 km.
Work out the distance between the towns on the map.

6.3 Sharing in a given ratio

◎ **Objective**

◉ You can divide a quantity in a given ratio.

Why do this?

Builders use ratios to make sure they have the exact quantities of ingredients needed for mixing concrete.

Get Ready

1. In a necklace the ratio of blue beads to yellow beads is 3 : 4.
   What fraction of these beads are yellow?
2. Write the ratio 16 : 24 in its simplest form.
3. Write the ratio 12 : 180 in the form 1 : n.

Key Point

◉ Sometimes we want to divide a quantity in a certain ratio.
   Suppose Sam and Hannah buy a box of chocolates costing £3.00.
   If Sam paid £2.00 and Hannah paid £1.00 they might decide to share the chocolates in the ratio 2 : 1.

Example 8

Heidi and Kirsty share £75 in the ratio 2 : 3.
Work out how much money each girl receives.

$2 + 3 = 5$ ← Work out the total number of shares.

$75 ÷ 5 = 15$ ← Work out the size of one share.

$15 × 2 = 30$ ← Heidi receives 2 shares.

$15 × 3 = 45$ ← Kirsty receives 3 shares.

Heidi gets £30 and Kirsty gets £45.

ResultsPlus
Watch Out!

Some students divide the total amount by the numbers in the ratio. Make sure you work out the number of shares first.

Exercise 6I

1 Share £80 in the ratio 1 : 4.

2 Share £24 in the ratio 3 : 5.

3 Share £45 in the ratio 2 : 3.

4 Alex and Ben share 40 sweets in the ratio 3 : 1.
   Work out how many sweets each receives.

D

D

5 At school the technician is going to make some brass. Brass is made from copper and nickel in the ratio 17 : 3.
Work out how much copper and how much nickel he will need to make 800 g of brass.

6 The ratio of boys to girls in a class is 4 : 5.
There are 27 students in the class. Work out the number of girls in the class.

C

7 Share £40 in the ratio 1 : 3 : 4.

8 Rebecca is making shortbread. She uses flour, sugar and butter in the ratio 3 : 1 : 2.
Work out how much of each ingredient she needs to make 900 g of shortbread.

9 Three boys shared £30 in the ratio 5 : 3 : 2.
William received the smallest amount. Work out how much William received.

10 Masud made some compost. He mixed soil, manure and leaf mould in the ratio 3 : 2 : 1.
Masud made 120 litres of compost. Work out how much manure he used.

Mixed exercise 6J

D

1 A model of a helicopter is made using a scale of 1 : 72.
The length of the model helicopter is 20 cm.
Work out the length, in metres, of the real helicopter.

2 Tom is making cakes. In a recipe the ratio of the weight of margarine to the weight of caster sugar is 3 : 1.
Work out the weight of caster sugar Tom needs if the recipe uses 300 g of margarine.

3 The ratio of Martin's height to Tom's height is 7 to 8.
Martin's height is 140 cm. What is Tom's height?

4 In a school choir the ratio of the number of boys to the number of girls is 2 : 5.
There is a total of 35 boys and girls in the choir.
Work out the number of girls in the choir.

5 The scale of a map is 1 : 300 000.
On the map the distance between two towns is 4.5 cm.
Work out the real distance, in kilometres, between the towns.

6 Ashley and Farjad share £42 in the ratio 3 : 4.
Work out how much more money Farjad receives than Ashley.

C

7 An alloy contains copper, manganese and nickel in the ratio 14 : 5 : 1 by weight.
The weight of the copper is 70 kg.
Work out the weight of the manganese and the weight of the nickel.

8 Matt needs 3 lengths of wood for a shelf display in the ratio 5 : 2 : 3.
The plank of wood the shelves are cut from is 1200 cm long.
Work out the length of each piece of wood for the 3 shelves.

6.4 Solving ratio and proportion problems using the unitary method

Objectives

- You can use the unitary method to find quantities.
- You can solve problems involving proportion.

Why do this?

Chefs use proportion to make sure they use the correct quantity of each ingredient.

Get Ready

1. If two cups of tea cost £2.70, what is the cost of one cup of tea?
2. If five pens cost 85p, what is the cost of one pen?
3. If four pizzas cost £7.80, what is the cost of one pizza?

Key Points

- If two quantities increase or decrease at the same rate they are in **direct proportion**.
- If 1 pound buys 2 dollars, then 2 pounds will buy 4 dollars.
  When the number of pounds is doubled, the number of dollars is also doubled.
  The number of dollars is **proportional** to the number of pounds.

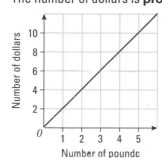

As the number of dollars is proportional to the number of pounds, the graph is a straight line passing through the origin.

- When two quantities are in direct proportion the ratio between them says the same.
- There are many examples of proportion in everyday life.
  The cost of a bag of tomatoes is directly proportional to the weight of the tomatoes.
  The cost of buying some petrol is proportional to the number of litres of petrol you put in your tank.
- You can use the unitary method to solve problems involving direct proportion.
  In the unitary method you always find the value of one item (or unit) first.
  For example, if three snack bars cost 96p, you can work out the cost of one snack bar.
  Once you know that one snack bar costs 32p you can then work out the cost of any number of snack bars.

Example 9

Seven bricks weigh 21 kg.

Work out the weight of 10 of these bricks.

$21 \div 7 = 3$ ← First work out the weight of one brick.

$3 \times 10 = 30$ ← Multiply the weight of one brick by 10.

10 bricks weigh 30 kg.

Exercise 6K

D

1 Five pens cost 75p. Work out the cost of eight of these pens.

2 Three identical stamps cost 96p. Work out the cost of seven of these stamps.

3 Eight cinema tickets cost £46. What would 10 of these tickets cost?

4 Fifteen identical pipes laid end to end make a length of 120 metres.
   What length will 12 of the pipes make if they are laid end to end?

5 The cost of 5 metres of ribbon is £6.25. Daniel wants to buy 3 metres of ribbon.
   Work out the cost of 3 metres of ribbon.

6 Three 2.5-litre tins of paint cost £44.94. Work out the cost of five of the 2.5-litre tins of paint.

7 100 g of cheese contains 14 g of carbohydrates.
   Work out the weight of carbohydrates in 125 g of cheese.

8 300 sheets of paper have a total thickness of 2.7 cm.
   Charlotte's paper feeder for her printer holds 4.5 cm of paper.
   How many sheets will she need to completely fill the paper feeder?

Example 10 This is a list of ingredients needed to make fruit crumble for four people.

| 350 g fruit | 100 g flour | 50 g margarine | 50 g sugar |

Work out the amount of margarine needed to make fruit crumble for 12 people.

$12 \div 4 = 3$

> 3 times as much of each ingredient is needed.

> ÷ 4 then × 12 is the same as × 3.

$50 \times 3 = 150$

> Multiply the amount of margarine by 3.

150 g of margarine is needed.

Exercise 6L

D

1 This is a list of ingredients needed to make 50 cheese straws.

   100 g flour

   50 g margarine

   75 g cheese

   Connor wants to make 100 cheese straws. Work out the amount of each ingredient Connor needs.

2 This is a recipe for making Quiche Lorraine for four people.

   100 g pastry

   100 g bacon

   75 g cheese

   2 eggs

   150 ml milk

   Rhys is making Quiche Lorraine for 6 people.
   Work out the amount of each ingredient he needs.

3 This is a list of ingredients needed to make 20 almond biscuits.

175 g flour
75 g caster sugar
50 g ground almonds
150 g margarine

a Work out the amount of flour needed to make 40 almond biscuits.

b Work out the amount of margarine needed to make 10 almond biscuits.

c Work out the amount of ground almonds needed to make 30 almond biscuits.

Example 11

a Martin went to Spain. He changed £400 into euros.
The exchange rate was £1 = 1.08 euros. How many euros did he receive?

b When Martin came home he changed 63 euros into pounds.
The new exchange rate was £1 = 1.05 euros. How many pounds did he receive?

a 400 × 1.08 = 432 ⟵ | Multiply the number of pounds by 1.08. |
Martin received 432 euros.

b 63 ÷ 1.05 = 60 ⟵ | Divide the number of euros by 1.05. |
Martin received £60.

Exercise 6M

1 Hannah went on holiday to France. She changed £200 into euros.
The exchange rate was £1 = 1.08 euros. Work out how many euros Hannah received.

2 Matas changed £600 into Russian roubles. The exchange rate was £1 = 44.95 roubles.
Work out how many roubles Matas received.

3 Suha is going to the USA. The exchange rate is £1 = $1.42
   a Change £300 into dollars.
   b Change $355 into pounds.

4 Paolo bought a railway ticket for €84 in Italy. The exchange rate was £1 = €1.12.
Work out the cost of the ticket in pounds.

5 Danny paid 74 francs for a meal in Switzerland. The exchange rate was £1 = 1.85 francs.
Work out the cost of the meal in pounds.

D

Mixed exercise 6N

D

1 The weight of 10 identical coins is 250 g. Work out the weight of 12 of these coins.

2 Matthew was paid £75 for 12 hours' work in a shop.
At the same rate, how much would he be paid for 7 hours' work?

3 The cost of three rolls of wallpaper is £25.50. Mandeep needs five rolls of wallpaper to wallpaper her dining room. Work out the total cost.

4 £1 = 1.18 euros
   a Change £400 into euros.   **b** Change 236 euros into pounds.

5 This is a recipe for making sponge pudding for six people.
   100 g margarine
   100 g caster sugar
   2 eggs
   225 g flour
   30 ml milk
   Work out the amount of each ingredient needed to make sponge pudding for 15 people.

6 Aliyah came back from a holiday in Australia. She changed $164 into pounds.
The exchange rate was £1 = $2.05. Work out how many pounds she received.

A03

7 This is a recipe for making fruit crumble for four people.
   350 g fruit
   50 g margarine
   100 g flour
   50 g sugar
   Hazel has only 175 g of flour.
   She has plenty of each of the other ingredients.
   Work out how many people she can make fruit crumble for.

A02
A03

8 Calum is on holiday in Switzerland.
He buys a pair of sunglasses.
He can pay either 80 francs or 55 euros.
Is it cheaper for Calum to pay 80 francs or to pay 55 euros?
Explain your answer.

Exchange rates
£1 = 1.12 euros
£1 = 1.68 francs

Chapter review

● **Ratios** are used to compare quantities.
● Ratios can be simplified like fractions.
   To simplify a ratio you divide each of its numbers by a common factor.
● When a ratio cannot be simplified it is in its simplest form.

- It is sometimes useful to write ratios in the form $1 : n$.

  The number n is written as a whole number or a decimal.

  When one of the numbers in a ratio is 1, the ratio is in **unitary** form.

- The ratio 15 : 20 simplifies to the ratio 3 : 4.

  15 : 20 and 3 : 4 are **equivalent ratios**.

- If you know the ratio of two quantities and you know one of the quantities, you can use equivalent ratios to find the other quantity.

- Maps have a scale to tell us how a distance on the map relates to the real distance.

- Someone making a scale model or producing a scale drawing will also need to use a scale.

- Sometimes we want to divide a quantity in a certain ratio.

- If two quantities increase or decrease at the same rate they are in **direct proportion**.

- When two quantities are in direct proportion the ratio between them stays the same.

- You can use the unitary method to solve problems involving direct proportion.

  In the unitary method you always find the value of one item (or unit) first.

Review exercise

1 There are some sweets in a bag.

8 of the sweets are toffees. 12 of the sweets are mints.

Write down the ratio of the number of toffees to the number of mints.

Give your ratio in its simplest form.

June 2009

2 A coin is made from copper and nickel.

84% of its weight is copper.

16% of its weight is nickel.

Find the ratio of the weight of copper to the weight of nickel.

Give your answer in its simplest form.

June 2008

3 The distance from Ailing to Beeford is 2 km. The distance from Ceetown to Deeton is 800 metres.

Write as a ratio

Distance from Ailing to Beeford : Distance from Ceetown to Deeton.

Give your answer in its simplest form.

4 There are some oranges and apples in a box.

The total number of oranges and apples is 54.

The ratio of the number of oranges to the number of apples is 1 : 5.

Work out the number of apples in the box.

June 2009

5 A garage sells British cars and foreign cars.

The ratio of the number of British cars sold to the number of foreign cars sold is 2 : 7.

The garage sells 45 cars in one week.

Work out the number of British cars the garage sold that week.

June 2008

6 Alice builds a model of a house. She uses a scale of 1 : 20.

The height of the real house is 10 metres.

a Work out the height of the model.

The width of the model is 80 cm.

b Work out the width of the real house.

E

D

7 There are 600 counters in a bag.

90 of the 600 are yellow. 180 of the 600 are red.

The rest of the counters in the bag are blue or green.

There are twice as many blue counters as green counters.

Work out the number of green counters in the bag.

May 2009

8 Here is a list of ingredients for making fudge for 6 people.

Fudge

Ingredients for 6 people

600 g of sugar

12 g of butter

480 g of condensed milk

90 ml of milk

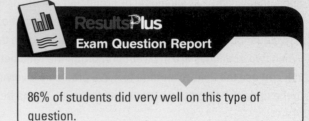

Results**Plus**

Exam Question Report

86% of students did very well on this type of question.

Work out how much of each ingredient is needed to make fudge for 9 people.

Nov 2006

9 Ron went to Spain.

He changed £200 into euros (€).

The exchange rate was £1 = €1.40.

a How many euros did he get?

When he came home he changed €10.64 back into pounds.

The exchange rate was now £1 = €1.33.

b How many pounds did he receive?

June 2006

10 Bob lays 200 bricks in one hour.

He always works at the same speed.

He starts work at 9 am.

Bob takes 15 minutes for morning break and 30 minutes for lunch break.

Bob has to lay 960 bricks.

Work out the time at which he will finish laying bricks.

June 2006 adapted

11 The exchange rate between pounds (£) and euros (€) is £1 = €1.08 in London and €1 = 88p in Paris.

Will has £1200 to change into euros. Should he do it in London or Paris?

12 Mr Brown makes some compost.

He mixes soil, manure and leaf mould in the ratio 3 : 1 : 1.

Mr Brown makes 75 litres of compost.

How many litres of soil does he use?

Nov 2006

One of the most famous formulae you may come across is Einstein's $e = mc^2$ from his theory of relativity. Einstein's brain was removed after his death and researchers in Canada compared it with the brains of 91 people of average intelligence to try to discover the secret of his outstanding intelligence. They found that the area of Einstein's brain that is responsible for mathematical thought and spatial awareness was much larger and his brain was 15% wider than the others.

⊙ Objectives

In this chapter you will:
- use letters instead of numbers
- learn the difference between variables, terms and expressions
- collect like terms
- multiply and divide variables and write them in their simplest form
- expand and factorise brackets
- learn the difference between expressions, equations and formulae.

⟨⟩ Before you start

You need to know:
- $d + d = 2d$
- $2g = 2 \times g$
- $5p - 2p = 3p$
- $4p - p = 3p$

7.1 Using letters to represent numbers

Objective

○ You can use letters instead of numbers.

Why do this?

You can use algebra to make a rule that will work for lots of different situations. The rule for time taken to get to a place when travelling on a motorway works for all towns and all motorways.

Get Ready

1. 3 apples and  2 apples = 5 apples

$a + a + a + a + a =$

2. 3 apples and 🍌🍌 2 bananas will stay as 3 apples and 2 bananas.

$a + a + a + b + b =$

Key Point

◉ Letters can be used instead of numbers to fit all situations.

Example 1

Find these.

　　　　a $3a + 5a =$　　**b** $6b - b =$

a $3a + 5a = 8a$

b $6b - b = 5b$

> a is the same as $1a$ so
> $a + a = 2a$　and　$a + a + a = 3a$
> Remember that 6 bananas take away
> 1 banana would be 5 bananas!

Exercise 7A

Questions in this chapter are targeted at the grades indicated.

Find these.

1　$a + a + a + a =$

2　$a + a + a + a + a + a =$

3　$p + p + p =$

4　$6x - x =$

5　$9j - 3j =$

F

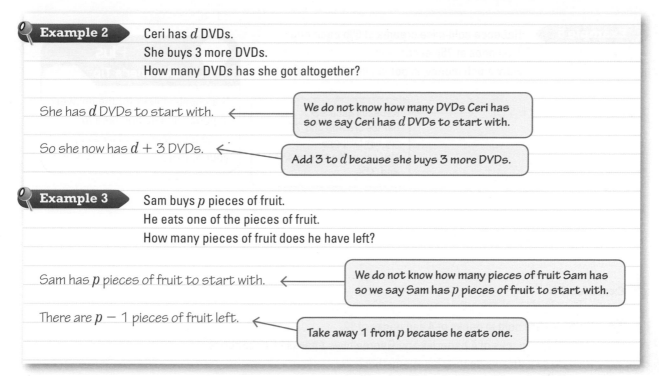

Example 2

Ceri has d DVDs.
She buys 3 more DVDs.
How many DVDs has she got altogether?

She has d DVDs to start with. ← We do not know how many DVDs Ceri has so we say Ceri has d DVDs to start with.

So she now has $d + 3$ DVDs. ← Add 3 to d because she buys 3 more DVDs.

Example 3

Sam buys p pieces of fruit.
He eats one of the pieces of fruit.
How many pieces of fruit does he have left?

Sam has p pieces of fruit to start with. ← We do not know how many pieces of fruit Sam has so we say Sam has p pieces of fruit to start with.

There are $p - 1$ pieces of fruit left. ← Take away 1 from p because he eats one.

Exercise 7B

1 Use algebra to write:
   a 3 more than p
   b x plus 4
   c q take away 5
   d 5 less than g
   e 4 more than h
   f k minus 6
   g j with 6 taken away
   h a plus 3
   i y minus 4
   j m with 3 taken away
   k p with 6 added
   l l together with h.

2 James had c CDs. He buys 12 more. How many CDs does James have now?

3 Terri had a apples. She eats 3 of them. How many apples does she have now?

4 Rashmi had d downloads on his MP3 player. He downloads 12 more.
   How many downloads has he altogether?

5 Hajra has g computer games. She sells 7 on the internet. How many computer games has she got now?

6 Helen and Robin go shopping. Robin buys x T-shirts and Helen buys y T-shirts.
   How many T-shirts do they have altogether?

Example 4

Rachel sells eggs.
She sells eggs in boxes of 6 or in boxes of 12.
One day she sells s boxes of 6 eggs and t boxes of 12 eggs.
How many eggs does she sell altogether?

You can write this as $6s$ and $12t$ so
Rachel sold $6s + 12t$ eggs altogether.
← If Rachel sells s boxes with 6 eggs in them there will be $6 \times s$ eggs in those s boxes and $12 \times t$ eggs in the t boxes of 12 eggs.

Example 5

Rebecca sold c ice creams at 99p each and
l lollipops at 75p each.
How much money, in pence, did she receive?

The total is $99c + 75l$. ←

Rebecca gets $99 \times c$ for
her ice creams and $75 \times l$
for her lollipops.
You can write this as $99c$
and $75l$.

Results**Plus**

Examiner's Tip

If a question asks for the answer
in pence you do not need to
write p or pence and mess up
your algebra.

Exercise 7C

E

1. Batteries are sold in packs of 4 and packs of 12.
   Harry buys f packs of 4 batteries and George buys t packs of 12 batteries.
   How many batteries do they buy altogether?

2. Naomi buys eggs in boxes of 4 and boxes of 10.
   One day she bought f boxes of 4 eggs and t boxes of 10 eggs.
   How many eggs did she buy altogether?

3. Moshe collects stickers in packs of 4 and packs of 9.
   One month he collected a packs of 4 stickers and b packs of 9 stickers.
   How many stickers did he collect altogether?

4. Richard sells rings. The gold rings cost £50 each and the silver rings cost £20 each.
   One day he sold g gold rings and s silver rings. How much money, in £, did he receive?

5. Jane packs boxes of chocolates.
   Plain chocolate boxes take 5 minutes to pack, milk chocolate boxes take 3 minutes and mixed boxes
   take 4 minutes.
   One day she packed p plain boxes, m milk boxes and n mixed boxes of chocolates.
   How much time did she spend packing chocolates on that day?

6. Nimer bakes ordinary cakes and iced cakes.
   Iced cakes are baked in batches of 6 and ordinary cakes are baked in batches of 12.
   One day, he baked x batches of iced cakes and y batches of ordinary cakes.
   Write down the total number of cakes he baked.

7. Lucy packs pencils in boxes of 12 and pens in boxes of 6.
   One week she packed x boxes of pencils and y boxes of pens.
   How many pens and pencils did she pack in total?

8. Cheryl sells flowers. She makes a profit of 90p on roses and a profit of 10p on daffodils.
   One day she sold r roses and d daffodils. How much profit, in pence, did she make on that day?

7.2 Understanding variables, terms and expressions

⊙ Objective

⊙ You know the difference between a variable, a term and an expression.

⑦ Why do this?

You need to understand mathematical words to be able to understand questions in exams.

⬥ Get Ready

1. There were s tickets for the stands at a rugby match and b tickets for the boxes.
   How many tickets were there altogether?
2. There were h members of a hockey club at the start of the season. Twelve left at the end of the season.
   How many were still in the club at the end of the season?
3. Helena has r rabbits and g guinea pigs. How many pets does she have altogether?

🔍 Key Points

⊙ Variables, **expressions** and **terms** are the building blocks of **algebra**.
⊙ A variable is something that can change, e.g. speed, and is shown using a letter, e.g. a, b or c.
⊙ A term is a multiple of a letter that denotes a variable, for example $5a$, $6b$, c.
⊙ An expression is a collection of terms or variables, e.g. $5a + 6b - c$.

Example 6   Write down   **a** the letters that are variables   **b** the terms in this expression $2c + 5d$.

a c and d are the variables. ⟵ | c and d can change value. |

b $2c$ and $5d$ are the terms.

⚙ Exercise 7D

1   Write down the letters that are the variables in these expressions.
   a $3a + b$ **b** $x + 4y$ **c** $5a - 4t$ **d** $x - y$ **e** $2t - 5d$
   f $2a - 5s$ **g** $4b - 6$ **h** $9g + 6$ **i** $5t + 7$ **j** $2a + 5b$

2   Write down the terms in these expressions.
   a $3a + 4b$ **b** $x + 4y$ **c** $5a - 4t$ **d** $x - y$ **e** $2t - 5d$
   f $2a - 5s + 8$ **g** $4b - 6h$ **h** $9g + 6r + 4$ **i** $5t + 7s - 3$ **j** $2a + 5b$

3   Write down the variables in these terms and expressions.
   a $3a$ **b** $4y$ **c** $4t$ **d** x **e** $5d$
   f $2a - 5s$ **g** $4b - 6$ **h** $9g + 6$ **i** $5t + 7$ **j** $2a + 5b$

4   Use some of the terms in question 3 to make five new expressions.

5   Use some of the variables you identified in question 1 to make five new terms.

F

7.3 Collecting like terms

Objectives

- You can collect like terms when there is only one variable.
- You can collect like terms when there is more than one variable.
- You can collect like terms when there are numbers as well.
- You can collect like terms when there are powers and numbers.

Why do this?

Collecting like terms makes them easier to deal with.

Get Ready

1. Find these.

   a 5 apples + 3 apples   **b** 10 bananas − 5 bananas − 1 pear   **c** 2 apples + 5 bananas + 4 apples

Key Points

- Terms that use the same variable or letter or arrangement of letters are called **like terms**.
  a is the same as $1a$ so $a + a = 2a$ and $a + a + a = 3a$.
  Don't forget $2a - a$ is $2a - 1a$ or a.
  - a and $3a$ are like terms.
  - $5p$ and $8p$ are also like terms.
- You can add and subtract like terms to **simplify** expressions.
- Sometimes algebraic expressions have more than one term. You can make them simpler by collecting like terms together.
- Numbers will often be included as well as variables and terms. You treat these in exactly the same way as any other term.
- There may also be terms where variables are combined such as x^2, x^3 and ab.
- When you collect like terms you have to keep these more complicated terms together as well.
- It is also possible to have negative values when you collect like terms.

Example 7

Add or subtract these like terms to simplify the expressions.

   a $5p + 7p + p$ **b** $7a - 3a$

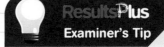
ResultsPlus
Examiner's Tip

'Simplify' means collect like terms.

a $5p + 7p + p = 13p$
b $7a - 3a = 4a$

Exercise 7E

Simplify

F

1 $t + t + t + t + t$	**2** $c + c$
3 $x + x + x + x$	**4** $a + a + a$
5 $y + y + y + y + y + y$	**6** $a + a + a + a + a + a + a + a$

Exercise 7F

Simplify

1	$3a + 2a$	**2**	$5p - 3p$	**3**	$6s + 2s$
4	$6x - 3x$	**5**	$4b + 2b$	**6**	$8k - 3k$
7	$4a + 2a + 3a$	**8**	$5x + x$	**9**	$5b + 3b + b$
10	$2p - p$	**11**	$2n + 3n + 5n$	**12**	$2p + 5p - 3p$
13	$7x + 2x - 5x$				

Example 8   Simplify $2a + 7b + 5a - 2b$.

$$2a + 7b + 5a - 2b$$

$2a + 5a + 7b - 2b$

Collect the terms in a together and collect the terms in b together.

$= 7a + 5b$

Combine the a terms together and the b terms together.

Results**Plus**

Watch Out!

Do not try to combine the as and the bs together when you are adding and subtracting.
They are different variables.

Exercise 7G

Simplify

1	$3a + 4b + 4a + 5b$	**2**	$6m + 5n + 3m + 2n$
3	$2p + 5q + 5p + 3q$	**4**	$7e + 2f - 5e$
5	$6g + 2h - 4g + 5h$	**6**	$8p - 6p + 7r - 2r$
7	$6j + 5k - 3j - 2k$	**8**	$7m + 8n - m$
9	$3a + 7b + 5a - 2b$	**10**	$7a + 8b - 4a - 5b$
11	$6m + 5n - 3m - 2n$	**12**	$2p + 5q + 5p - 3q$
13	$7e + 2f - e - f$	**14**	$6g + 8h - 4g - 5h$
15	$8p - 6r + 7r - 2p$	**16**	$6j + 5k + 3j - 2k$
17	$m + 8n - m$	**18**	$8a + 7b - 5a - 5b$
19	$8p - 2j - 5p + 5j$	**20**	$6p + 5t - 6p - t$
21	$6x + 4x - 3x - x$	**22**	$5y + 4b - 5y - 4b$
23	$k + 5g - k - g$	**24**	$4m - 6m + 2m + m$

Example 9 Simplify $2a + 7b + 9 + 5a - 2b - 6$. ← $2a + 7b + 9 + 5a - 2b - 6$

$2a + 5a + 7b - 2b + 9 - 6$

Collect the terms in a and in b together as before, then deal with the numbers on their own.

$= 7a + 5b + 3$

Combine the a terms together and the b terms together.

Exercise 7H

Simplify

E

1 $3a + 4 + 4a + 5$

2 $6m + 5 + 3m + 2$

3 $2 + 5p + 5 + 3q$

4 $7e + 2 - 5e$

5 $6 + 2h - 4 + 5h$

6 $6g + 8 - 4 - 5g$

7 $6j + 5 - 3j - 2$

8 $7m + 8 - m - 7$

D

9 $3a + 7b - 6 + 5a - 2b + 7$

10 $7a + 8 - 4a - 5 + c$

11 $6m + 5n - 3 + 12$

12 $2p + 9q - 8 + p - 3q + 10$

13 $5e + 2f - e - f + 2$

14 $8p + 6 - 5p + 7r - 3r - 2$

15 $8p - 6r - 7 + 7r - 7p + 7$

Example 10 Simplify $2a^2 + 5ab + 3a^2 - 3ab$.

$2a^2 + 3a^2 + 5ab - 3ab$ ← The like terms here are those in a^2 and those in ab.
$= 5a^2 + 2ab$

Exercise 7I

Simplify

E

1 $x^2 + x^2$

2 $3y^2 + 2y^2$

3 $7a^2 - 5a^2$

D

4 $3a^2 + 4b^3 + 4a^2 + 5b^3$

5 $6m^2 + 5n + 3m^2 + 2n$

6 $2p^3 + 5pq + 5p^3 - 3pq$

7 $7ef + 2ef - 5ef$

8 $6g^2 + 2h^3 - 4g^2 + 5h^3$

9 $8pq - 6pq + 7r^3 - 2r^2$

10 $6jk + 5jk - 3jk - 2jk$

11 $7m^3 + 8n - m^3$

12 $3a^2 + 7b^2 + 4a^2 - 3b^2$

13 $7a^3 + 8b^2 - 5a^3 - 5b^2$

14 $6m + 5n^2 - 4m - 2n^2$

15 $2pq + 5pq + 5p^3 - 3q^2$

16 $7e^3 + 2f^2 - e^3 - f^2$

17 $6gh + 8h^3 - 5gh - 6h^3$

18 $8pqr - 6pqr + 7pqr - 2pqr$

Example 11 Simplify $2a + 7 - 5a - 2$.

$$2a + 7 - 5a - 2$$

$2a - 5a + 7 - 2$ ← Collect the terms in a together and collect the number terms together.

$= -3a + 5$ ← Combine the a terms together and the number terms together.

Don't forget $2 - 5$ is -3. So $2a - 5a = -3a$.

Exercise 7J

Simplify

1 $3a - 6a$

2 $6m + 3n - 8m + 2n$

3 $2p + 2q - 5p - 6q$

4 $2e - 7e + 4 - 5$

5 $6g + 2h - 4g - 5h$

6 $8p^2 - 6p^2 + 3r^3 - 8r^3$

7 $3j + 5k - 3j - 8k$

8 $7m^3 - 8n - 8m^3$

9 $3a - 7b - 5a + 2b$

10 $7ab + ab - 9ab$

11 $6m + 5 - 8m - 7$

12 $2p + 5 - 5p - 11$

13 $7e + f - 6e - 2f$

14 $4g^2 + 3 - 6g^2 - 5$

15 $8p - 6r + 2 + 8r - 12p$

16 $6j + 5k - 8j - 2k$

17 $m - 8n - m - 5$

18 $8a - 3b - 5a - 5b$

19 $2p - 2j - 5p - 5j$

20 $6p - 5 - 6p - 3$

21 $x + 4 - 3 - 5x$

22 $5y^3 + 4 - 5y^3 - 4$

23 $k^2 + 5g^3 - k^2 - 8g^3$

24 $4mn - 6m^2 - 12mn + 8m^2$

D

7.4 Multiplying with numbers and letters

◎ Objective

◉ You can multiply variables and write them in their simplest form.

⑦ Why do this?

Multiplying with letters and numbers means you can solve problems such as how much carpet you would need for your bedroom.

⬆ Get Ready

1. Simplify

a $2 \times 6 \times 3 \times 4$

b $4 \times 2 \times 5 \times 2$

c $2 \times 5 \times 2 \times 5 \times 8$

🔍 Key Points

- $2a$ can also be written as 2 lots of a or $2 \times a$.
- When you multiply terms and variables in algebra you can combine them by writing them next to each other.

  So $\quad p \times q \quad$ is written as $\quad pq$

  and $\quad a \times a \quad$ is written as $\quad aa \quad$ or $\quad a^2 \quad$ (a squared)

  while $\quad a \times a \times a \quad$ is written as $\quad aaa \quad$ or $\quad a^3 \quad$ (a cubed).

Example 12 Simplify $p \times q \times r$.

$= pqr \quad \longleftarrow \quad$ Combine the p, q and r into pqr

⚙️ Exercise 7K

Simplify

1 $\quad a \times b$

2 $\quad x \times y$

3 $\quad b \times b$

4 $\quad d \times d \times d$

5 $\quad r \times s \times t$

6 $\quad a \times b \times c$

7 $\quad g \times g \times g$

8 $\quad 2 \times e \times f$

9 $\quad 3 \times j \times k$

10 $\quad h \times h$

11 $\quad 5 \times s \times s$

12 $\quad 6 \times t \times t \times t$

13 $\quad r \times t$

14 $\quad x \times y \times t$

15 $\quad 3 \times m \times n$

16 $\quad 7 \times a \times b \times c$

Example 13 Simplify $2p \times 5t$.

$= 2 \times 5 \times p \times t \quad \longleftarrow \quad$ Combine the numbers 2×5.
Combine the variables $p \times t$.

$= 10pt \quad \longleftarrow \quad$ $2 \times 5 = 10 \qquad p \times t = pt$.

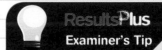

Examiner's Tip

You combine the letters when you multiply.

Example 14 Simplify $2a \times 3a \times 4a$.

$= 2 \times 3 \times 4 \times a \times a \times a \quad \longleftarrow \quad$ Combine the numbers $2 \times 3 \times 4$.
Combine the variables $a \times a \times a$.

$= 24 \times aaa \quad \longleftarrow \quad$ $2 \times 3 \times 4 = 24$

$= 24a^3 \quad \longleftarrow \quad$ $a \times a \times a = aaa = a^3$

Exercise 7L

Simplify

1	$2a \times 3b$	**2**	$4x \times 5y$	**3**	$2b \times 3b$
4	$2d \times 3d \times 2d$	**5**	$5r \times 7s$	**6**	$6b \times 2c$
7	$5g \times 3g$	**8**	$2e \times 7f$	**9**	$3j \times 8k$
10	$4h \times 5h$	**11**	$5s \times 5s$	**12**	$3t \times 2t \times 2t$
13	$6r \times 2t$	**14**	$5x \times 7y$	**15**	$3m \times 6n$
16	$5a \times 2b \times 3c$	**17**	$2g \times 2g$	**18**	$7h \times 7h$
19	$2x \times 2x \times 2x$	**20**	$5n \times 5n$	**21**	$2f \times 3g \times 5h$
22	$4j \times 6k$	**23**	$6h \times 6i$	**24**	$2a \times 2a \times b$

E

D

7.5 Dividing with numbers and letters

⊙ Objective

⊙ You can divide variables and write them in their simplest form.

⊘ Why do this?

Simplifying expressions makes them easier to work with.

⬆ Get Ready

1. Write these fractions in their simplest form.

a $\frac{12}{4}$ **b** $\frac{12}{8}$ **c** $\frac{9}{12}$

🌐 Key Point

⊙ Dividing algebraic expressions is like **cancelling** fractions.

Example 15

Simplify   **a** $10a \div 2$ **b** $\dfrac{20ab}{5b}$

a $10a \div 2$ ← $10a \div 2$ is the same as $\dfrac{{}^{5}10a}{{}^{}2_{1}}$

You can cancel the 2 into the 10 just as you would with fractions.

$= 5a$ ←   This leaves you with the answer $5a$ on the top.

b $\dfrac{20ab}{5b}$ ← $\dfrac{20ab}{5b}$ is the same as $\dfrac{{}^{4}20 \times a \times \cancel{b}}{{}_{1}5 \times \cancel{b}}$

You can cancel the 5 into the 20 and cancel the b.

$= 4a$ ←   This leaves you with the answer $4a$ on the top.

Exercise 7M

Simplify

1 $12pq \div 3q$ **2** $3p \div 3$ **3** $4h \div h$ **4** $12n \div 3$

5 $4t \div 2$ **6** $15x \div x$ **7** $24k \div 4$ **8** $12ab \div 6b$

9 $\dfrac{8x}{4}$ **10** $\dfrac{12p}{4p}$ **11** $\dfrac{30xy}{6y}$ **12** $\dfrac{8pq}{pq}$

13 $\dfrac{8pqr}{4qr}$ **14** $\dfrac{8xy}{4xy}$ **15** $\dfrac{24abc}{6ab}$ **16** $\dfrac{8xy}{4xy}$

7.6 Expanding single brackets

◉ Objective

○ You can expand a single bracket.

⬆ Get Ready

How many small blocks are there in these rectangles?

1.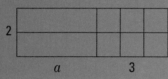

2.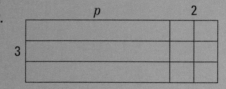

Key Point

◉ To **expand** a bracket, multiply everything inside the bracket by what is outside.
   In question 1 above, the number of blocks could be written as either $2(a + 3)$ or $2 \times a + 2 \times 3$.
   This means that $2(a + 3) = 2 \times a + 2 \times 3$.
   The brackets have been expanded.
   This section gives similar examples.

Example 16   Expand $2(a + 5)$.

$= 2 \times (a + 5)$ ← $2(a + 5)$ means $2 \times (a + 5)$

$= 2 \times a + 2 \times 5$ ← You multiply everything inside the bracket by the 2.

$= 2a + 10$ You multiply the a by the 2 and the 5 by the 2.
$2 \times a + 2 \times 5$
The + sign follows down.

Results Plus
Examiner's Tip

When you see the word 'expand' it means you multiply out the bracket. You need to multiply everything inside the bracket by what is outside.

Exercise 7N

Expand

1	$2(a + 4)$	**2**	$3(b + 2)$	**3**	$4(c + 6)$	**4**	$5(a - 4)$
5	$3(b - 5)$	**6**	$5(x + 3)$	**7**	$2(y - 2)$	**8**	$6(n + 2)$
9	$3(5 + g)$	**10**	$2(5 - x)$	**11**	$3(2 - y)$	**12**	$5(4 - h)$
13	$10(a + 5)$	**14**	$3(g + 7)$	**15**	$4(s - 5)$	**16**	$3(7 - w)$

D

Example 17   Expand $2(3a + 6)$.

$= 2 \times (3a + 6)$ ← | $2(3a + 6)$ means $2 \times (3a + 6)$ |

$= 2 \times 3a + 2 \times 6$ ← | You multiply everything inside the bracket by the 2. |

$= 6a + 12$ ← | You multiply the $3a$ by the 2 and the 6 by the 2.
$2 \times 3a + 2 \times 6$
The + sign follows down. |

Exercise 7O

Expand

1	$2(3a + 4)$	**2**	$3(5b + 4)$	**3**	$4(5c + 6)$	**4**	$3(2a - 5)$
5	$3(5b - 7)$	**6**	$5(2x + 5)$	**7**	$2(3y - 4)$	**8**	$6(2n + 7)$
9	$3(5 + 3g)$	**10**	$2(5 - 2x)$	**11**	$3(2 - 5y)$	**12**	$5(4 - 3h)$
13	$10(4a + 3)$	**14**	$3(5g + 7)$	**15**	$4(3s - 5)$	**16**	$3(7 - 4w)$

D

Example 18   Expand $a(2a + 5)$.

$a \times (2a + 5)$ ← | $a(2a + 5)$ means $a \times (2a + 5)$.
You multiply everything inside the bracket by the a. |

$= a \times 2a + a \times 5$ ← | You multiply the $2a$ by the a and the 5 by the a. |

$= 2aa + 5a$ ← | $2a \times a + a \times 5$ |

$= 2a^2 + 5a$ ← | Don't forget $a \times a = aa = a^2$
The + sign follows down. |

Exercise 7P

Expand

D

1 $a(a + 4)$

2 $b(b + 2)$

3 $a(c + 6)$

4 $a(2a - 4)$

5 $b(b - 5)$

6 $x(x + 3)$

7 $y(2y - 2)$

8 $n(3n + 2)$

9 $g(5 + g)$

10 $x(5 - 2x)$

11 $y(2 - 3y)$

12 $h(4 - 5h)$

13 $2a(a + b)$

14 $2g(g + 7)$

15 $4s(s + t)$

16 $3w(7 - w)$

17 $5p(2p + 3)$

18 $5x(3x - y)$

19 $2h(g + 3h)$

20 $5p(4 - 2p)$

7.7 Factorising

Objective

○ You can factorise an expression using a single bracket.

Get Ready

1. Expand

　　a $3(7p - 4) + 5(10 - 3p)$　　**b** $2(3 - 3s) + 4(5 - 2s)$　　**c** $10(4 + 2g) + 3(5 - 5g)$

Key Points

◉ Multiplying out a bracket is called expanding a bracket.

◉ Putting in a bracket is called **factorising**.

Example 19

Factorise $2a + 10$.

> 'Factorise' means take out factors that are in both terms.

$2 \times a + 2 \times 5$ ←

> $2a = 2 \times a$　　2 and a are factors of $2a$.
> $10 = 2 \times 5$　　2 and 5 are factors of 10.

$2(a + 5)$ ←

> You take the 2 and put it outside the bracket. That leaves the a and the $+ 5$ inside the bracket.

ResultsPlus
Examiner's Tip

When you see the word 'factorise' it means you put the bracket back in.

Exercise 7Q

Factorise

1	$2a + 6$	**2**	$2n + 8$	**3**	$2a - 12$	**4**	$3k + 6$
5	$3f - 9$	**6**	$5p - 10$	**7**	$5r + 20$	**8**	$3x - 12$
9	$7w + 14$	**10**	$3m - 15$	**11**	$4q + 8$	**12**	$2s + 2$
13	$5a - 25$	**14**	$6x + 30$	**15**	$8p - 40$	**16**	$5y - 5$

D

Example 20   Factorise $x^2 - 3x$.

$= x \times x - 3 \times x$ ⟵ $x^2 = x \times x$ $3x = 3 \times x$

$= x(x - 3)$ ⟵   The x is in both terms so you take the x and put it outside the bracket. That leaves the other x and the -3 inside the bracket.

Example 21   Factorise $y^2 + y$.

$= y \times y + 1 \times y$ ⟵ $y^2 = y \times y$ $y = 1 \times y$

$= y(y + 1)$ ⟵   The y is in both terms so you take the y and put it outside the bracket. That leaves the other y and the $+1$ inside the bracket.

Exercise 7R

Factorise

1	$a^2 + 2a$	**2**	$a^2 + 8a$	**3**	$y^3 + 2y$	**4**	$j^3 - 3j$
5	$s^2 - 9s$	**6**	$x^3 - 5x$	**7**	$p^2 + 6p$	**8**	$a^2 - a$
9	$p^3 + p$	**10**	$m^3 - m^2$	**11**	$c^3 + 8c$	**12**	$2a + a^2$
13	$x^3 - 2x^2$	**14**	$x^2 + 7x$	**15**	$p^3 - p$	**16**	$y^2 - 5y$

D

Example 22   Factorise completely $3x^2 - 6x$.

$= 3 \times x \times x - 3 \times 2 \times x$ ⟵ $3x^2 = 3 \times x \times x$ $6x = 3 \times 2 \times x$

$= 3x(x - 2)$ ⟵   The 3 and the x are in both terms so you take the $3x$ and put it outside the bracket. That leaves the other x and the -2 inside the bracket.

ResultsPlus
Examiner's Tip

When you see 'factorise completely' it means there is more than one factor to take outside the bracket.

C

Exercise 7S

Factorise completely

1	$6a^2 + 2a$	**2**	$3a^2 + 9a$	**3**	$4y^3 + 2y$	**4**	$6j^3 - 3j$
5	$3s^2 - 9s$	**6**	$10x^3 - 5x$	**7**	$3p^2 + 6p$	**8**	$4a^2 - 2a$
9	$5p^3 + 10p$	**10**	$6m^3 - 3m^2$	**11**	$4c^3 + 8c$	**12**	$12a + 6a^2$
13	$6x^3 - 2x^2$	**14**	$5x^2 + 30x$	**15**	$8p^3 - 4p$	**16**	$25y^2 - 5y$

7.8 Understanding expressions, equations and formulae

◎ Objective

○ You know the difference between an expression, an equation and a formula.

❓ Why do this?

A formula will allow you to work out how long a journey will take at a given speed, or how much your savings will earn in interest over a given number of years.

⬆ Get Ready

1. Write three examples of expressions and identify the terms and variables in your expressions.

🌐 Key Points

◉ An expression is a collection of terms or variables.
  $2x + 2y$ is an expression.
◉ **Equations** and **formulae** have equals signs in them.
◉ An equation is where it is possible to find one or more numerical values for a variable.
  $2x + 1 = 7$ is an example of an equation. The value of x is 3.
◉ A formula is where one variable is equal to an expression in a different variable.
  $P = 2l + 2w$ gives the perimeter of a rectangle of length l and width w. This is an example of a formula.

🔍 Example 23 Draw arrows to link the equation, expression and formula in this list.

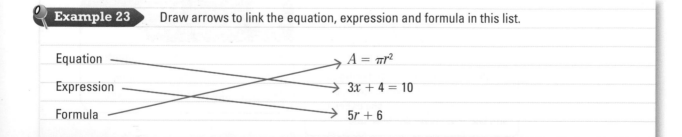

Equation ⟶ $A = \pi r^2$

Expression ⟶ $3x + 4 = 10$

Formula ⟶ $5r + 6$

Exercise 7T

State whether each of the following is an equation, expression or formula.

1 $y = mx + c$

2 $mx + c$

3 $3y + 1 = 10$

4 $2p + 6 = 5$

5 $F = ma$

6 $F + 3 = 10$

7 $6P + 2G$

8 $C = \pi D$

9 $a^2 + b^2 = c^2$

10 $C = \frac{9}{5}(F - 32)$

D

7.9 Replacing letters with numbers

⊙ Objective

⊙ You can replace letters with numbers in expressions.

◈ Why do this?

You might want to calculate the area of a square using the equation area $= l^2$ when $l = 3$ cm.

◈ Get Ready

1. Work out

  a $(2 \times 3) + 5$

  b $(5 \times 2) - (3 \times 3)$

  c $(2 \times 5) + (4 \times 3) - 8$

Example 24   Find the value of   **a** $3p$ when $p = 5$

  b $5x$ when $x = 4$

  c $2a + b$ when $a = 3$ and $b = 5$

  d $3p - 2q$ when $p = 2$ and $q = 5$

  e $2(x + y)$ when $x = 4$ and $y = 3$.

a $3p = 3 \times 5 = 15$ ⟵

> $3p$ means $3 \times p$ or $p + p + p$
> If $p = 5$ then $3p = 5 + 5 + 5$ or 3×5

b $5x = 5 \times 4 = 20$ ⟵

> $5x$ means $5 \times x$ or $x + x + x + x + x$
> If $x = 4$ then $5x = 5 \times 4$

c $2a + b = 2 \times 3 + 5 = 6 + 5 = 11$

d $3p - 2q = 3 \times 2 - 2 \times 5 = 6 - 10 = -4$

e $2(x + y) = 2(4 + 3) = 2 \times 7 = 14$

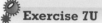
Exercise 7U

Find the value of these expressions when $a = 2$, $b = 5$ and $c = 3$.

1 $a + a$	**2** $b + b + b$	**3** $c + c + c + c$	**4** $b - a$
5 $3a$	**6** $5c$	**7** $4b$	**8** $5a$
9 $2a + b$	**10** $3b + c$	**11** $5c - a$	**12** $2b - a$
13 $5c + 2a$	**14** $2b - c$	**15** $4b - 5a$	**16** $4c - 3a$
17 $6a + 3b$	**18** $a + b + c$	**19** $2a + 4b - 8c$	**20** $5c - 4c$

Exercise 7V

Find the value of these expressions when $p = 5$, $q = 3$ and $r = -2$.

1 $p + p$	**2** $r + r + r$	**3** $q + q + q + q$	**4** $p - r$
5 $3r$	**6** $5q$	**7** $4p$	**8** $5r + p$
9 $5p + q$	**10** $3q + r$	**11** $5q - p$	**12** $2r - q$
13 $5p + 2q$	**14** $2q - r$	**15** $2p - 5r$	**16** $r - 3q$
17 $2p - 3r$	**18** $p + q + r$	**19** $2p + 4q - r$	**20** $2p - 5q$

Exercise 7W

Find the value of these expressions when $a = 2$, $b = 5$ and $c = 3$.

1 $2(a + b)$	**2** $3(b + c)$	**3** $4(a + c)$	**4** $5(b - a)$
5 $3(a + 2b)$	**6** $5(a + 2c)$	**7** $4(c - b)$	**8** $5(2a - b)$
9 $2(a + 2b)$	**10** $3(4b + 2c)$	**11** $5(c - a)$	**12** $2(3b - 2a)$
13 $5(c + 2a)$	**14** $2(2b - 3c)$	**15** $4(5b - 3a)$	**16** $4(2b - 3c)$
17 $6(a + 3b)$	**18** $2(a + b + c)$	**19** $2(a - 4b)$	**20** $5(c - a - b)$

Chapter review

- Letters can be used instead of numbers to fit all situations.
- Variables, **expressions** and **terms** are the building blocks of **algebra**.
- A variable is something that can change, e.g. speed, and is shown using a letter, e.g. a, b or c.
- A term is a multiple of a letter that denotes a variable, for example $5a$, $6b$, c.
- An expression is a collection of terms or variables, e.g. $5a + 6b - c$.
- Terms that use the same variable or letter or arrangement of letters are called **like terms**.

- You can add and subtract like terms to **simplify** expressions.
- Sometimes algebraic expressions have more than one term. You can make them simpler by collecting like terms together.
- Numbers will often be included as well as variables and terms. You treat these in exactly the same way as any other term.
- There may also be terms where variables are combined such as x^2, x^3 and ab.
- When you collect like terms you have to keep these more complicated terms together as well.
- It is also possible to have negative values when you collect like terms.
- $2a$ can also be written as 2 lots of a or $2 \times a$.
- When you multiply terms and variables in algebra you can combine them by writing them next to each other.
- When you divide algebraic expressions you do it like **cancelling** fractions.
- To expand a bracket, multiply everything inside the bracket by what is outside.
- Multiplying out a bracket is called **expanding** a bracket.
- Putting in a bracket is called **factorising**.
- An expression is a collection of terms or variables.
- **Equations** and **formulae** have equals signs in them.
- An equation is where it is possible to find one or more numerical values for a variable.
- A formula is where one variable is equal to an expression in a different variable.

Review exercise

1 Callum has £3 more than Luke. Becky has twice as much as Callum.
Write down an expression for the total amount in pounds Callum, Luke and Becky have altogether.

2 Here is a rod. Its length is x and its width is y.

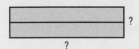

Two of these rods are put with their widths alongside.

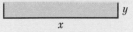

a Copy and complete the diagram. Give your answer in its simplest form.

Three of these rods are put together as shown in the diagram.

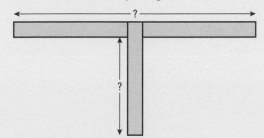

b Copy and complete the diagram.
Give your answer in its simplest form.

D

A03

3 In this set of rectangles, each expression in a rectangle is obtained by adding the two expressions immediately underneath.

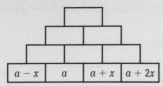

$a - x$ | a | $a + x$ | $a + 2x$

What expression should go in the top square?

A03

4 **a** Simplify $13x - 17y - 3x - 3y$.
   b Work out the value of $13 \times 99 - 17 \times 39 - 3 \times 99 - 3 \times 39$.

5 Simplify
   a $c + c + c$ **b** $e + f + e + f + e$ **c** $2a + 3a$
   d $2xy + 3xy - xy$ **e** $3a + 5b - a + 2b + 8$ *June 2006*

6 Simplify
   a $5bc + 2bc - 4bc$ **b** $4x + 3y - 2x + 2y$ **c** $m \times m \times m$ **d** $3n \times 2p$ *Nov 2008*

7 Factorise $x^2 + 4x$ *June 2006*

8 Expand $4(3a - 7)$ *May 2008*

8 ALGEBRA 2

In this chapter, we will look at writing numbers using powers. For example, one trillion is written in long form as 1 000 000 000 000 but can be written using powers as 10^{12}. This form of writing large numbers is very useful in science. The distance from the Earth to the Sun and back is approximately 10^8 miles.

Objectives

In this chapter you will:
- calculate with powers
- write expressions as a single power of the same number
- use powers in algebra
- multiply out brackets in algebra
- factorise expressions.

Before you start

You need to know how to:
- use letters instead of numbers
- collect like terms
- multiply and divide variables and write them in their simplest form.

8.1 Calculating with powers

Objectives

- You can work out the value of numbers raised to a power.
- You can write numbers using index notation.
- You can work out values of expressions given in index notation.

Why do this?

It is easy to make mistakes when working with very large numbers with lots of zeros, or with repeated multiples. Using powers reduces mistakes.

Get Ready

1. Work out the following.

   a 2×2 b $5 \times 5 \times 5$ c $8 \times 8 \times 8$ d $10 \times 10 \times 10 \times 10 \times 10 \times 10$

Key Points

- The 2 in 5^2 is called a **power** or an **index**. It tells you how many times the given number must be multiplied by itself. The plural of index is indices.
- The 5 in 5^2 is called the **base**.
- You can solve equations such as $3^x = 81$ by working out how many times the base has to be multiplied by itself to give the answer.

Example 1   Work out the value of the following.

   a 3^2 b 2 to the power of 5   c $2^3 \times 3^2$

a $3^2 = 3 \times 3 = 9$ ← 3 to the power of 2 is usually called 3 squared.

b 2 to the power of $5 = 2^5$
$2^5 = 2 \times 2 \times 2 \times 2 \times 2 = 32$ ← Write as a power. 2^5 means 2 multiplied by itself 5 times.

c $2^3 \times 3^2$
$2^3 = 2 \times 2 \times 2 = 8$ ← Work out 2^3 first.
$3^2 = 3 \times 3 = 9$ ← Work out 3^2.
$2^3 \times 3^2 = 8 \times 9$ ← Use the two values above.
$\qquad = 72$

Example 2   Rewrite these expressions using index notation.

   a $3 \times 3 \times 3 \times 3$ b $4 \times 4 \times 5 \times 5 \times 5$ c $6 \times 6 \times 6 \times 6 \times 6$

a $3 \times 3 \times 3 \times 3 = 3^4$

b $4 \times 4 \times 5 \times 5 \times 5 = 4^2 \times 5^3$ ← Replace 4×4 with 4^2 and $5 \times 5 \times 5$ with 5^3.

c $6 \times 6 \times 6 \times 6 \times 6 = 6^5$ ← The index is 5 because 5 lots of 6 are multiplied together.

Example 3 Find the value of x.

  a $5^x = 25$ b $4^x = 64$

a $5 \times 5 = 25$
So $5^2 = 25$ ← Write as a power. Compare powers.
$x = 2$

b $4 \times 4 \times 4 = 64$
So $4^3 = 64$ ← Write as a power. Compare powers.
$x = 3$

Exercise 8A

Questions in this chapter are targeted at the grades indicated.

E

1 Find the value of
  a 2^5 b 4 to the power 4   c 1^6
  d 10 to the power 4   e 5^4 f 6 to the power 5

2 Write these using index notation.
  a $2 \times 2 \times 2 \times 2$ b $4 \times 4 \times 4 \times 4 \times 4$
  c $1 \times 1 \times 1 \times 1 \times 1 \times 1$ d $8 \times 8 \times 8$
  e $3 \times 3 \times 8 \times 8 \times 8 \times 8$ f $4 \times 4 \times 4 \times 4 \times 2 \times 2 \times 2$

3 Work out the value of
  a 2^4 b 3^5 c 6^3 d 5^2
  e 8^3 f $2^4 \times 9^3$ g $2^6 \times 4^5$ h $5^3 \times 3^4$
  i $2^7 \times 3^5$ j $4^3 \times 4^1$

4 Copy and complete the table for powers of 10.

Power of 10	Index	Value	Value in words
	3		One thousand
10^2		100	
		1 000 000	One million
	1	10	
10^5	5		

5 Work out the value of
  a $4^3 \times 10^2$ b 4×10^2 c 6×10^3
  d $10^2 \div 5^2$ e $10^3 \div 2^3$ f $4^3 \div 2^2$

6 Find x when
  a $5^x = 125$ b $3^x = 81$ c $2^x = 64$ d $10^x = 10\,000$
  e $9^x = 81$ f $3^x = 27$ g $2^x = 16$ h $7^x = 49$

D

8.2 Writing expressions as a single power of the same number

Objective

○ You can use rules of indices to simplify expressions.

Why do this?

Using rules of indices saves time and makes calculations easier.

Get Ready

1. Work out the value of
   a 2^2
   b 4^3
   c 7^4
   d 10^4

Key Points

⦿ To multiply powers of the same number, add the indices.
  e.g. $3^2 \times 3^3 = 3^{2+3} = 3^5$

⦿ To divide powers of the same number, subtract the indices.
  e.g. $5^6 \div 5^3 = 5^{6-3} = 5^3$

⦿ Any number raised to the power of 1 is equal to the number itself.
  e.g. $4^1 = 4$

⦿ To raise a power of a number to a further power, multiply the powers (or indices).
  e.g. $(10^3)^2 = 10^{3 \times 2} = 10^6$

Example 4 Simplify these expressions by writing them as a single power of the number.
   a $2^3 \times 2^4$ b $5^8 \div 5^3$ c $(8^2)^5$

a $2^3 \times 2^4 = 2^{3+4}$ ← Add the powers.
   $= 2^7$

b $5^8 \div 5^3 = 5^{8-3}$ ← Subtract the powers.
   $= 5^5$

c $(8^2)^5 = 8^{2 \times 5}$ ← $(8^2)^5 = (8 \times 8) \times (8 \times 8) \times (8 \times 8) \times (8 \times 8) \times (8 \times 8)$
   $= 8^{10}$ $= 8^{10}$

   Multiply the powers.

Exercise 8B

Simplify these expressions by writing as a single power of the number.

1 a $6^8 \times 6^3$ b $8^3 \times 8^5$ c $2^4 \times 2^2$

2 a $4^3 \div 4^2$ b $6^6 \div 6^3$ c $7^5 \div 7$

3 a $4^2 \times 4^3$ b $5^3 \div 5$ c $3^9 \div 3^8$

C

C

4 a $5^6 \times 5^4 \times 5^3$ **b** $2^3 \times 2^7 \times 2$

5 a $10^2 \times 10^2 \times 10$ **b** $9^4 \div 9^4$

6 a $6^3 \times 6^7 \times 6$ **b** $5^2 \times 5^2 \times 5^2$

7 a $3^5 \times 3 \times 3^2$ **b** $4^7 \times \dfrac{4^5}{4^6}$

8 a $\dfrac{6^8}{6^2} \times 6^3$ **b** $5^8 \times \dfrac{5^4}{5^7}$ **c** $\dfrac{4^9}{4^2} \times 4^5$

9 a $(5^2)^3$ **b** $(7^4)^2$

8.3 Using powers in algebra to simplify expressions

⊙ Objectives

⊙ You can multiply powers of the same letter.
⊙ You can divide powers of the same letter.
⊙ You can raise a power of a letter to a further power.

⑦ Why do this?

Using powers makes calculating easier.

⬆ Get Ready

1. Simplify
   a $7^2 \times 7^5$ **b** $8^5 \div 8^3$ **c** $(5^3)^2$

🔍 Key Points

⊙ In the expression x^n, the number n is called the power or index.
⊙ $x^m \times x^n = x^{m+n}$
⊙ $x^m \div x^n = x^{m-n}$
⊙ $(x^m)^n = x^{m \times n}$
⊙ Any letter raised to the power of 1 is equal to the letter itself, e.g. $x^1 = x$.
⊙ Any letter raised to the power 0 is equal to 1, e.g. $x^0 = 1$.

Example 5   Simplify   **a** $x^5 \times x^3$ **b** $y^7 \div y^4$ **c** $a^2 \times a^3 \times a^5$ **d** $(x^3)^2$
                              e $3x^2 \times 4x^3$ **f** $10x^6 \div 5x^3$ **g** $(3a^2)^4$

a $x^5 \times x^3 = x^{5+3}$ ← Add the powers.
    $= x^8$

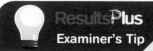
Results Plus
Examiner's Tip

Show your working.
The mark is scored here.

b $y^7 \div y^4 = y^{7-4}$ ← Subtract the powers.
    $= y^3$

c $a^2 \times a^3 \times a^5 = a^{2+3+5}$
                    $= a^{10}$

117

d $(x^3)^2 = x^{3 \times 2}$ ← Multiply the powers.

$= x^6$

e $3x^2 \times 4x^3 = 3 \times 4 \times x^2 \times x^3$ ← Write the numbers and letters together.

$= 12 \times x^{2+3}$

$= 12x^5$ ← Multiply the numbers 3 and 4.

Add the powers 2 and 3.

f $10x^6 \div 5x^3 = \frac{10}{5} \times x^6 \div x^3$ ← Divide the numbers 10 and 5.

$= 2 \times x^{6-3}$ ← Subtract the powers.

$= 2x^3$

g $(3a^2)^4 = 3^4 \times (a^2)^4$ ← Both 3 and a^2 are raised to the power of 4. $3^4 = 3 \times 3 \times 3 \times 3 = 81$

$= 81 \times a^{2 \times 4}$

$= 81a^8$ ← Multiply the powers.

Exercise 8C

Simplify the following.

1 **a** $x^8 \times x^2$ **b** $y^3 \times y^8$ **c** $x^9 \times x^5$

2 **a** $a^5 \times a^3$ **b** $b^3 \times b^3$ **c** $d^7 \times d^4$

3 **a** $p^5 \div p^2$ **b** $q^{12} \div q^2$ **c** $t^8 \div t^4$

4 **a** $j^9 \div j^3$ **b** $k^5 \div k^4$ **c** $n^{25} \div n^{23}$

5 **a** $x^5 \times x^2 \times x^2$ **b** $y^2 \times y^4 \times y^3$ **c** $z^3 \times z^5 \times z^2$

6 **a** $3x^2 \times 2x^3$ **b** $5y^9 \times 3y^{20}$ **c** $6z^8 \times 4z^2$

7 **a** $12p^8 \div 4p^3$ **b** $15q^5 \div 3q^3$ **c** $6r^5 \div 3r^2$

8 **a** $(d^3)^4$ **b** $(e^5)^2$ **c** $(f^3)^3$ **d** $(g^7)^9$

9 **a** $(g^6)^4$ **b** $(h^2)^2$ **c** $(k^4)^0$ **d** $(m^0)^{56}$

10 **a** $(3d^2)^7$ **b** $(4e)^3$ **c** $(3f^{129})^0$

11 **a** $a^4 \times \dfrac{a^5}{a^9}$ **b** $b^7 \times \dfrac{b}{b^4}$ **c** $c^3 \times \dfrac{c^4}{c^2} \times c^5$

12 **a** $4d^9 \times 2d$ **b** $8e^8 \div 4e^4$ **c** $(4f^2)^2$

8.4 Understanding order of operations

Objective

- You can work out the value of numerical expressions.

Why do this?

When making a cake, you need to know what order to add the ingredients in. The same is true of a calculation such as $3 \times 4 + 2 \times 5$. It is important that the operations are carried out in the correct order or the answer will be wrong.

Get Ready

1. Work out

  a $(6 + 3) \times (2 - 1)$ **b** $6 + (3 \times 2) - 1$ **c** $((6 + 3) \times 2) - 1$

Key Points

- **BIDMAS** gives the order in which **operations** should be carried out.
- Remember that B I D M A S stands for

  B rackets   If there are brackets, work out the value of the expression inside the brackets first.

  I ndices   Indices include square roots, cube roots and powers.

  D ivide   If there are no brackets, do dividing and multiplying before adding and subtracting, no

  M ultiply   matter where they come in the expression.

  A dd

  S ubtract   If an expression has only adding and subtracting then work it out from left to right.

Example 6   Work out $(3 \times 2) - 1$

$(3 \times 2) - 1 = 6 - 1$ ← Work out the Brackets first.
$= 5$

Example 7   Work out $3 + 2 \times 5 - 1$

$3 + 2 \times 5 - 1 = 3 + 10 - 1$ ← There is no Bracket or Divide, so start with Multiply, then Add, then Subtract.
$= 13 - 1$
$= 12$

Example 8   Work out $(10 + 2)^2 - 5 \times 3^2$

$(10 + 2)^2 - 5 \times 3^2 = 12^2 - 5 \times 3^2$ ← Brackets first, then Indices, Multiply, and finally Subtract.
$= 144 - 5 \times 9$
$= 144 - 45$
$= 99$

Exercise 8D

1 Use BIDMAS to help you find the value of these expressions.

   a $5 + (3 + 1)$ **b** $5 - (3 + 1)$

   c $5 \times (2 + 3)$ **d** $5 \times 2 + 3$

   e $3 \times (4 + 3)$ **f** $5 \times 4 + 3$

   g $20 \div 4 + 1$ **h** $20 \div (4 + 1)$

   i $6 + 4 \div 2$ **j** $(6 + 4) \div 2$

   k $24 \div (6 - 2)$ **l** $24 \div 6 - 2$

   m $7 - (4 + 2)$ **n** $7 - 4 + 2$

   o $((15 - 5) \times 4) \div ((2 + 3) \times 2)$

2 Make these expressions correct by replacing the • with $+$ or $-$ or $\times$ or $\div$ and using brackets if you need to. The first one is done for you.

   a $4 • 5 = 9$ becomes $4 + 5 = 9$ **b** $4 • 5 = 20$

   c $2 • 3 • 4 = 20$ **d** $3 • 2 • 5 = 5$

   e $5 • 2 • 3 = 9$ **f** $4 • 2 • 8 = 10$

   g $5 • 4 • 5 • 2 = 27$ **h** $5 • 4 • 5 • 2 = 23$

3 Work out

   a $(3 + 4)^2$ **b** $3^2 + 4^2$

   c $3 \times (4 + 5)^2$ **d** $3 \times 4^2 + 3 \times 5^2$

   e $2 \times (4 + 2)^2$ **f** $2^3 + 3^2$

   g $2 \times (3^2 + 2)$ **h** $\dfrac{(2 + 5)^2}{3^2 - 2}$

   i $\dfrac{5^2 - 2^2}{3}$ **j** $4^2 - 2^4$

   k $2^5 - 5^2$ **l** $4^3 - 8^2$

8.5 Multiplying out brackets in algebra

⊙ Objectives

- You can add expressions with brackets.
- You can subtract expressions with brackets.

❓ Why do this?

Using brackets and collecting like terms helps you get the correct answer.

⬦ Get Ready

1. Expand   **a** $6(5 + 3)$ **b** $3(4 \times 4 - 5 \times 3)$ **c** $(5 \times 4)(2 \times 4 - 2 \times 1)$

⬥ Key Points

- Expanding brackets means multiplying each term inside the brackets by the term outside the brackets.
- To simplify an expression with brackets, expand the brackets and collect like terms.

Example 9 Simplify   **a** $2(3x - y) + 5(y - 2x)$
b $3x(2x + y) - 2x(5y - 1)$

a $2(3x - y) + 5(y - 2x)$ ← 2 times $(3x - y)$ plus
5 times $(y - 2x)$
Multiply each pair of terms.

$= 2 \times (3x - y) + 5 \times (y - 2x)$

$= 2 \times 3x - 2 \times y + 5 \times y - 5 \times 2x$

Remember $+ + = +$
$+ - = -$
$- + = -$
$- - = +$

$= 6x - 2y + 5y - 10x$

$= 3y - 4x$ ← Collect the terms.

b $3x(2x + y) - 2x(5y - 1)$

$= 3x \times 2x + 3x \times y - 2x \times 5y - 2x \times -1$

$= 6x^2 + 3xy - 10xy + 2x$ ← $x \times x = x^2$

$= 6x^2 - 7xy + 2x$ ← Collect the terms.

Exercise 8E

Expand and simplify.

1 $3(x + 2) + 2(x + 4)$ **2** $4(2x - 1) + 3(4x + 7)$

3 $5(3x + 2) + 4(2x + 1)$ **4** $7(3 - 2x) + 3(2x - 3)$

5 $6(4 - 2x) - 3(5 + 3x)$ **6** $4(3 - 2x) + 3(1 - 5x)$

7 $2(3x - 5y) + 3(2x - 4y)$ **8** $5(6y + 2x) - 4(3x + 2y)$

9 $3(2x - 3y) - 2(5x + 6y)$ **10** $3(2x + 3y) - 5(x + y)$

11 $4(3y - 2) - 5(y - 2)$ **12** $2(3x + 6) - 3(2x - 5)$

13 $4(3 - 2x) - 3(5 - 3x)$ **14** $2(3 - y - 2x) - 3(4x - 3y)$

15 $3(2x - 3y) + 5(3x - 2y)$ **16** $5(3y - 5x) - 2(x - 3y)$

17 $(4x - 3y) + 2(3x - 2y)$ **18** $7(3x - 5y) - (x - 3y)$

19 $x(2y + 1) + 2x(3y + 1)$ **20** $2x(3y + 1) + y(2x + 1)$

21 $2y(3x - 2) + 3x(2 - 3y)$ **22** $4x(2y - 5x) + 2y(x - y)$

C

8.6 Factorising expressions

Objectives

- You can factorise expressions by taking out a single factor.
- You can factorise an expression by taking out multiple factors.

Why do this?

If you can see a common number or factor it can make calculations easier. Buying two burgers and two fries $(2b + 2f)$ is the same as buying $2(b + f)$.

Get Ready

1. Find the highest common factors of these.
   a 16 and 24   b 48 and 20   c ab and abc

Key Points

- Factorising is the reverse process to expanding brackets.
- To factorise an expression, find the common factor of the terms in the expression and write the common factor outside a bracket. Then complete the bracket with an expression which, when multiplied by the common factor, gives the original expression.

Example 10   Factorise   a $3x^2 + 5x$ b $10a^2 - 15ab$

a $3x^2 + 5x = x(3x + 5)$

> Take x outside the bracket. It is a common factor of $3x^2 + 5x$.

b $10a^2 - 15ab$
   $= 5a(2a - 3b)$

> The common factor may be both a number and a letter.
> Take $5a$ outside the bracket. It is a common factor of both $10a^2$ and $15ab$.

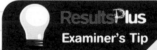

ResultsPlus
Examiner's Tip

Always check your answer by expanding.

Exercise 8F

Factorise each of the expressions in questions 1–6.

D

1 a $2x + 6$ b $6y + 2$ c $15b - 5$
   d $4r - 2$ e $3x + 5xy$ f $12x + 8y$
   g $12x - 16$ h $9 - 3x$ i $9 + 15g$

C

2 a $3x^2 + 4x$ b $5y^2 - 3y$ c $2a^2 + a$
   d $5b^2 - 2b$ e $7c - 3c^2$ f $d^2 + 3d$
   g $6m^2 - m$ h $4xy + 3x$ i $n^3 - 8n^2$

3
  a $8x^2 + 4x$ **b** $6p^2 + 3p$ **c** $6x^2 - 3x$
  d $3b^2 - 9b$ **e** $12a + 3a^2$ **f** $15c - 10c^2$
  g $21x^4 + 14x^3$ **h** $16y^3 - 12y^2$ **i** $6d^4 - 4d^2$

4
  a $ax^2 + ax$ **b** $pr^2 - pr$ **c** $ab^2 - ab$
  d $qr^2 + q^2$ **e** $a^2x + ax^2$ **f** $b^2y - by^2$
  g $6a^3 - 9a^2$ **h** $8x^3 - 4x^4$ **i** $18x^3 + 12x^5$

5
  a $12a^2b + 18ab^2$ **b** $4x^2y - 2xy^2$ **c** $4a^2b + 8ab^2 + 12ab$
  d $4x^2y + 6xy^2 - 2xy$ **e** $12ax^2 + 6a^2x - 3ax$ **f** $a^2bc + ab^2c + abc^2$

6
  a $5x + 20$ **b** $12y - 10$ **c** $3x^2 + 5x$
  d $4y - 3y^2$ **e** $8a + 6a^2$ **f** $12b^2 - 8b$
  g $cy^2 + cy$ **h** $3dx^2 - 6dx$ **i** $9c^2d + 15cd^2$

Chapter review

- The 2 in 5^2 is called a **power** or an **index**. It tells you how many times the given number must be multiplied by itself. The plural of index is indices.
- The 5 in 5^2 is called the **base**.
- You can solve equations such as $3^x = 81$ by working out how many times the base has to be multiplied by itself to give the answer.
- To multiply powers of the same number, add the indices.
- To divide powers of the same number, subtract the indices.
- Any number raised to the power of 1 is equal to the number itself.
- To raise a power of a number to a further power, multiply the powers (or indices).
- In the expression x^n, the number n is called the power or index.
- $x^m \times x^n = x^{m+n}$
- $x^m \div x^n = x^{m-n}$
- $(x^m)^n = x^{m \times n}$
- Any letter raised to the power of 1 is equal to the letter itself, e.g. $x^1 = x$.
- Any letter raised to the power 0 is equal to 1, e.g. $x^0 = 1$.
- **BIDMAS** gives the order in which **operations** should be carried out.
- Remember that (B I D M A S) stands for

  B rackets   If there are brackets, work out the value of the expression inside the brackets first.

  I ndices   Indices include square roots, cube roots and powers.

  D ivide   If there are no brackets, do dividing and multiplying before adding and subtracting, no

  M ultiply   matter where they come in the expression.

  A dd

  S ubtract   If an expression has only adding and subtracting then work it out from left to right.

- Expanding brackets means multiplying each term inside the brackets by the term outside the brackets.
- To simplify an expression with brackets, expand the brackets and collect like terms.
- Factorising is the reverse process to expanding brackets.
- To factorise an expression, find the common factor of the terms in the expression and write the common factor outside a bracket. Then complete the bracket with an expression which, when multiplied by the common factor, gives the original expression.

Review exercise

E

1 Write down these using index notation.
   a $6 \times 6 \times 6$ b 11×11 c $2 \times 2 \times 2 \times 2 \times 2 \times 2$

2 Write down the value of
   a 5^4 b 2^7 c 10^3 d 10^5

3 Work out the value of
   a $3^2 \times 4^2$ b $2^4 \times 7^2$ c 4×10^2 d 3×10^4

4 Rewrite these expressions using index notation.
   a $2 \times 2 \times 3 \times 3 \times 3$ b $5 \times 5 \times 7 \times 7$ c $4 \times 4 \times 8 \times 8 \times 8 \times 8$ d $6 \times 6 \times 6 \times 2 \times 2 \times 2$

5 Work out
   a 8^3 b 10^4 c 5^3 d $2^4 \times 3^2$ e $5^2 \times 2^5$

D

6 Find x when
   a $3^x = 243$ b $2^x = 32$ c $10^x = 1000$

7 Factorise
   a $5x + 15y$ b $15p - 9q$ c $cd + ce$

8 Jake thinks of a number, squares it, multiplies his answer by 2 and gets 72.
   What number did Jake think of?

9 Work out $6^5 \div (2^4 \times 3^5)$.

A02

10 In this set of squares, each number in a square is obtained by
   multiplying the two numbers immediately underneath.
   What number should go in the top square?
   Give your answer as a power of 2.

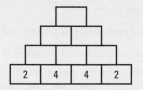

11 Factorise $5p - 20$
   March 2007

C

12 Simplify
   a $2^3 \times 2^4$ b $5^3 \times 5^2$ c 3×3^4 d $7^5 \div 7^2$
   e $9^8 \div 9^4$ f $8^3 \div 8$ g $7^2 \times \dfrac{7^4}{7^3}$ h $6^4 \div \dfrac{6}{6^2}$

13 Simplify
   a $x^6 \times x^3$ b $x^8 \div x^5$ c $(x^3)^5$ d $x^5 \div x^4$
   e $x \times x^4$ f $(x^6)^2$ g $x^8 \div x^8$ h $x^7 \div x$
   i $x^2 \times x^6 \times x^3$ j $x^8 \times x$ k $x^6 \times \dfrac{x^4}{x^7}$ l $x^3 \times \dfrac{x^7}{x^4} \times x^5$

14 Simplify

a $4x^3 \times x^5$ b $3x^2 \times 5x^6$ c $7x \times 3x^4$ d $8x^9 \div 2x^5$

e $24x^6 \div 3x$ f $36x^9 \div 4x^8$ g $(x^5)^2$ h $(x^3)^3$

15 Simplify

a $a^3 \times a^4$ b $3x^2y \times 5xy^3$

16 Expand and simplify

a $3a(b - 2a) + 2b(3a - 2b)$ b $4p(2q + 3p) + 3p(2p + q)$

c $5c(3c + 2d) - 2c(c - d)$ d $a(a + b) + b(a + b)$

e $3a(b + c) + 2b(a + c) - c(2a + 3b)$ f $2a(b - 2c) - 3b(2a + 3c)$

17 Factorise

a $x^2 - 7x$ b $t^2 + at$ c $bx^2 - x$ d $3p^2 + py$ e $aq^2 - at$

18 a Factorise $x^2 - 5x$.

b Work out the value of $105^2 - 5 \times 105$.

Nov 2007 adapted

A03

C

Neptune was the first planet to be discovered by mathematical prediction. It was found by looking at the number patterns of the other planets in the Solar System, and its position was correctly predicted to within a degree. Two scientists were eventually jointly credited with the discovery, one British and one French.

Objectives

In this chapter you will:
- continue number patterns using the four rules of number
- continue patterns using pictures and give the rule for continuing the pattern
- use number machines to produce a number pattern and write down the rule
- complete a table of values using a number machine and write down the rule
- use the first difference to find the nth term and use the nth term to find any number in a sequence
- identify whether or not a number is in a sequence.

Before you start

You need to:
- recognise simple patterns and simple number differences
- know about odd numbers
- know about even numbers
- know that multiples are members of a multiplication table, e.g. 3, 6, 9, 12 are multiples of 3
- know the properties of simple shapes.

9.1 Sequences

Objectives

- You can continue number patterns by adding or subtracting, or multiplying or dividing by a number.
- You can give the term to term rules for continuing number patterns.
- You can continue patterns using pictures and give the rule for continuing the pattern.

Why do this?

You may use sequences when learning a dance routine or a note sequence when playing a musical instrument.

Get Ready

Even numbers form a pattern   2, 4, 6, 8, 10, 12, ...   They go up in twos.
Odd numbers also form a pattern   1, 3, 5, 7, 9, 11, ...   These also go up in twos.

1. Write down all the even numbers up to 20.
2. Write down all the odd numbers up to 20.
3. Check that you have written all the numbers from 1 to 20.

Key Points

- A **sequence** is a pattern of numbers or shapes that follows a rule.
- Number patterns can be continued by adding, subtracting, multiplying and dividing.
- Patterns with pictures can be continued by finding the rule for continuing the pattern.
- The numbers in a number pattern are called terms.
- The **term to term rule** for a number pattern means you can say how you find a term from the one before it.

Continuing patterns by adding

Example 1

a Write down the next two numbers in this number pattern.

    2   6   10   14

b What is the rule you use to find the next number in the number pattern?

c Find the 10th number in this pattern.

a   14 + 4 = 18   2   6   10   14   18
    18 + 4 = 22   +4   +4   +4   +4

b   To get the next number you add 4 each time.

c   The 10th number in the pattern is 38.

2   6   10   14   18   22   26   30   34   38
Carry on the number pattern until you get to the 10th number in the pattern.

Exercise 9A

Questions in this chapter are targeted at the grades indicated.

F

1 Find the two missing numbers in these number patterns.
For each pattern, write down the term to term rule.

a 3, 6, 9, __, __, 18, 21 b 3, 7, 11, __, __, 23, 27
c 5, 10, 15, 20, __, __, 35, 40 d 2, 7, 12, 17, __, __, 32, 37
e 1, 4, 7, 10, __, __, 19, 22 f 5, 7, 9, 11, __, __, 17, 19
g 3, 8, 13, 18, __, __, 33, 38 h 4, 7, 10, 13, __, __, 22, 25
i 2, 6, 10, 14, __, __, 26, 30 j 10, 20, 30, __, __, 60, 70

2 a Write down the next two numbers in these sequences.

  i 1, 5, 9, 13, 17, … ii 2, 5, 8, 11, 14, …
  iii 3, 7, 11, 15, 19, … iv 4, 8, 12, 16, 20, …
  v 5, 8, 11, 14, 17, … vi 5, 11, 17, 23, …
  vii 2, 6, 10, 14, 18, … viii 1, 7, 13, 19, 25, …
  ix 3, 11, 19, 27, 35, … x 5, 9, 13, 17, 21, …

b Write down the rule you used to find the missing numbers in each sequence.

> **ResultsPlus**
> **Examiner's Tip**
>
> … means that the sequence carries on.

3 Find the 10th number of each of the number patterns in questions 1 and 2.

E

4 Jenny saves £2 each week in her piggy bank.
Here is the pattern of how her money grows.

Week	1	2	3	4	5
Money in piggy bank (£)	2	4	6		

a Copy and complete the table.

b Jenny is saving for a present for her Mum's birthday that costs £20.
How many weeks will this take?

A03

Continuing patterns by subtracting

Key Point

 Number patterns can be continued by subtracting the same number from each term.

Example 2

a Write down the next two numbers in this number pattern.

60   54   48   42   36

b What is the rule you use to find the next number in the number pattern?

c Find the 8th number in this pattern.

a $36 - 6 = 30$
$30 - 6 = 24$

60   54   48   42   36
     -6 -6 -6 -6

> To get to the next number you take away 6.
> Take away 6 from 36 to get 30 then
> take away 6 from 30 to get 24.

b To get the next number you subtract 6 each time.

c The 8th number in the pattern is 18.

> 60   54   48   42   36   30   24   18
> Carry on the number pattern until you get to the 8th number in the pattern.

Exercise 9B

1 Find the two missing numbers in these number patterns.
Write down the rule for each number pattern.

a 20, 18, 16, 14, __, __, 8
b 17, 15, 13, 11, __, __, 5
c 55, 50, 45, 40, __, __, 25
d 42, 37, 32, 27, __, __, 12
e 22, 19, 16, 13, __, __, 4
f 19, 17, 15, 13, __, __, 7
g 45, 38, 31, 24, __, __, 3
h 25, 22, 19, 16, __, __, 7
i 29, 25, 21, 17, __, __, 5
j 80, 70, 60, __, __, 30

2 a Write down the next two numbers in these sequences.

   i 41, 37, 33, 29, …   **ii** 27, 24, 21, 18, …
  iii 59, 55, 51, 47, …   **iv** 34, 31, 28, 25, …
   v 30, 27, 24, 21, …   **vi** 61, 55, 49, 43, …
 vii 22, 20, 18, 16, … **viii** 51, 46, 41, 36, …
  ix 64, 57, 50, 43, …   **x** 8, 6, 4, 2, 0, −2, …

b Write down the rule you used to find the missing numbers
in each sequence.

3 Find the 10th number of each of the number patterns in questions 1 and 2.

4 Abdul's mother gives him £20 each week to buy his lunch.
His lunch costs him £3 each day.
Here is the pattern of how he spends his money.

Day	M	Tu	W	Th	F
Money left at end of day (£)	17	14			

How much money will Abdul have left at the end of the week?

Continuing number patterns by multiplying

Example 3

a Write down the next two numbers in this number pattern.

1   3   9   27   81

b What is the rule you use to find the next number in the number pattern?

c Find the 8th number in this pattern.

a $81 \times 3 = 243$
$243 \times 3 = 729$

1   3   9   27   81

×3 ×3 ×3 ×3

To get to the next number you multiply by 3. Multiply 81 by 3 to get 243, then multiply 243 by 3 to get 729.

b To get the next number you multiply by 3 each time.

c The 8th number in the pattern is 2187.

1 3 9 27 81 243 729 2187
Carry on the number pattern until you get to the 8th number in the pattern.

Exercise 9C

1 Find the missing numbers in these number patterns.
For each pattern, write down the rule.

a 1, 2, 4, 8, __, __, 64

b 1, 4, 16, 64, __, 1024

c 1, 5, __, 125, __, 3125

d 1, 10, 100, __, __, 100 000

e 3, 6, 12, 24, __, __, 192

f 2, 6, 18, __, __, 486

g 2, 8, 32, __, __, 2048

h 2, 20, 200, 2000, __, __, 2 000 000

i 2, 10, 50, __, __, 6250

j 3, 15, 75, __, 1875

2 **a** Write down the next two numbers in these sequences.

i 2, 4, 8, 16, …

ii 3, 9, 27, 81, …

iii 4, 16, 64, 256, …

iv 5, 25, 125, 625, …

v 5, 10, 20, 40, …

vi 4, 12, 36, 108, …

vii 10, 30, 90, 270, …

viii 5, 50, 500, 5000, …

ix 10, 20, 40, 80, …

x 6, 36, 216, 1296, …

b Write down the rule you used to find the missing numbers in each sequence.

3 Find the 10th number of each of the number patterns in questions 1 and 2.

4 The number of rabbits in a particular colony doubled every month for 10 months.
The table shows the beginning of the pattern.

Month	1	2	3	4	5
Number of rabbits	2	4	8		

a Copy and complete the table.

b How many rabbits were in the colony in month 10?

Continuing number patterns by dividing

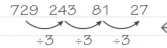 **Example 4**

a Write down the next two numbers in this number pattern.

729   243   81   27

b What is the rule you use to find the next number in the number pattern?

c Find the 8th number in this pattern.

a $27 \div 3 = 9$
$9 \div 3 = 3$

729 243   81   27

$\div 3$ $\div 3$ $\div 3$

> To get to the next number you divide by 3. Divide 27 by 3 to get 9 then divide 9 by 3 to get 3.

b To get the next number you divide by 3 each time.

c The 8th number in the pattern is $1 \div 3 = \frac{1}{3}$

> 729   243   81   27   9   3   1 $\frac{1}{3}$
> Carry on the number pattern until you get to the 8th number in the pattern.
> $1 \div 3 = \frac{1}{3}$

Exercise 9D

1 Find the missing numbers in these number patterns.
Write down the rule for each number pattern.

a 64, 32, 16, 8, __, __, 1

b 1024, 256, 64, __, 4

c 3125, 625, 125, __, __, 1

d 100 000, 10 000, 1000, __, __, 1

e 192, 96, 48, 24, __, __, 3

f 486, 162, 54, 18, __, 2

g 1024, 512, 256, 128, __, __, 16

h 300 000, 30 000, 3000, __, __, 3

i 6250, 1250, 250, __, __, ?

j 2000, 200, 20, __, __, 0.02

2 a Write down the next two numbers in these sequences.

i   64, 32, 16, 8, …

ii   243, 81, 27, 9, …

iii   128, 64, 32, 16, …

iv   625, 125, 25, 5, …

v   80, 40, 20, 10, …

vi   972, 324, 108, 36, …

vii   2430, 810, 270, 90, …

viii   50 000, 5000, 500, 50, …

ix   160, 80, 40, 20, …

x   1296, 216, 36, 6, …

b Write down the rule you used to find the missing number in each sequence.

3 Find the 8th number of each of the number patterns in questions 1 and 2.

4 The number of radioactive atoms in a radioactive isotope halves every 10 years.
The table shows the beginning of the pattern.

Years	0	10	20	30	40
Number of atoms	2560	1280	640		

a Copy and complete the table.

b How many radioactive atoms were in the isotope in year 100?

F

E

AO3

Continuing patterns in pictures

Example 5

a Copy and complete the table for the number of matches used to make each member of the pattern.

Pattern number	1	2	3	4	5	6	7
Number of matches used	4	7	10				

b Write down the rule to get the next number in the pattern.

c How many matches are there in pattern number 10?

a

Pattern number	1	2	3	4	5	6	7
Number of matches used	4	7	10	13	16	19	22

> Count the number of matches in each pattern and write down the number of matches used.

$$10 + 3 = 13 \qquad 13 + 3 = 16$$
$$16 + 3 = 19 \qquad 19 + 3 = 22$$

b Add 3 to the previous number.

c Pattern number 10 has 31 matches. ←
> Continue the patterns.
> 22   25   28   31

Exercise 9E

E

1 For these patterns:
  i draw the next two patterns
  ii write down the rule in words to find the next pattern
  iii use your rule to find the 10th term.

a
```
X    XX    XXX
X    XX    XXX
```

b
```
X      XXX      XXXXX
XX     XXXX     XXXXXX
```

c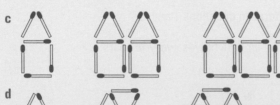

d

e
```
     X       XX      XXX      XXXX
X    XX      XXX     XXXX     XXXXX
```

2 **a** Write down the number of matches in each of these patterns.

Pattern 1

Pattern 2

Pattern 3

b Draw the next two patterns.
c Write down the rule in words to continue the pattern.
d Use your rule to find the number of matches needed for pattern number 10.

3 Repeat question 2 with the hexagon shape shown below.

 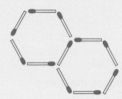 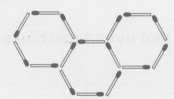

9.2 Using input and output machines to investigate number patterns

Objectives

○ You can use number machines to produce a number pattern and write down the rule (term number to term).

○ You can complete a table of values using a number machine and write down the rule (term number to term).

○ You can find missing values in a table of values and use the term number to term rule.

Why do this?

When baking, you take your ingredients, mix them together and bake them in the oven, and you end up with a cake.

Input ⟶ Action ⟶ Output

This process applies to anything from baking a cake to making a motor car.

Get Ready

1. Put the following numbers into this number machine and write down the answers.

a 5   **b** 3   **c** 11   **d** 17

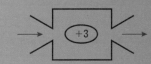

2. Draw a number machine for the process −6 and use it to find the answer when the following numbers are put into it.

a 10   **b** 6   **c** 2

Key Points

⦿ In this number pattern   3   7   11   15   19   23   ...
    Term 1 Term 2 Term 3 Term 4

3, 7, 11, 15, 19, 23, ... are the terms.

⦿ The term number tells you the position of each term in the pattern.
In the sequence 3, 7, 11, 15, 19, ... term 1 is 3, term 2 is 7, etc.

⦿ You can use number machines to produce a number pattern and write down the rule
(term number to term rule).

⦿ You can complete a table of values using a number machine and write down the rule (term number to term).

⦿ Sometimes you can put two number machines together to make a sequence.

One-stage input and output machines

Example 6   This number machine has been used to produce the terms
of a pattern.

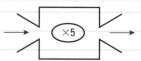

a Complete the term numbers and terms in this table of values for the number machine.

Term number	Term
1	5
2	
3	
4	

b What is the rule for working out the term from the term number?

c Write down the rule for finding the next term from the term before it.

a

Term number	Term
1	5
2	10
3	15
4	20

Term number	Term
1	5
2	10
3	15
4	20

+5
+5
+5

> The rule for the number machine is multiply the term
> number by 5 so the terms will be 5, 10, 15, 20.

b Multiply the term number by 5.   ← | To get to the term from the term number you multiply by 5. |

c Add 5. ← | To get to the next term from the term before it you have to
add 5 since the pattern is 5, 10, 15, 20, ... |

Exercise 9F

For each of these questions:
a copy and complete the table of values for the number machine
b write down the rule for finding the term from the term number
c write down the rule for finding the next term from the term before it.

1

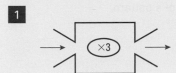

Term number	Term
1	3
2	6
3	
4	

2

Term number	Term
1	7
2	14
3	
4	

3

Term number	Term
1	4
2	8
3	
4	

4

Term number	Term
1	2
2	4
3	
4	

5

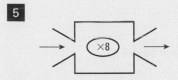

Term number	Term
1	8
2	16
3	
4	

6

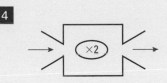

Term number	Term
1	10
2	20
3	
4	

7

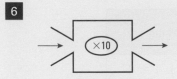

Term number	Term
1	12
2	24
3	
4	

8

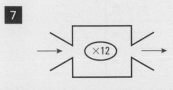

Term number	Term
1	50
2	100
3	
4	

F

Two-stage input and output machines

Example 7

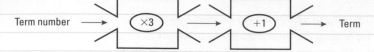

Term number → ×3 → +1 → Term

This number machine has been used to produce the terms of a pattern.

> **ResultsPlus**
> **Examiner's Tip**
>
> Use the rule from the number machine on the term number to get to the term. You feed the result of the first machine into the second machine.

a Complete the terms in a table of values for the number machine.

Term number	Term
1	4
2	
3	
4	

b What is the rule for working out the term from the term number?

c Write down the rule for finding the next term from the term before it.

a

Term number	Term
1	4
2	7
3	10
4	13

Term number	Term	Working
1	4	1 × 3 + 1 = 4
2	7	2 × 3 + 1 = 7
3	10	3 × 3 + 1 = 10
4	13	4 × 3 + 1 = 13

+3
+3
+3

b Multiply by 3 and add 1. ← To get to the term from the term number you ×3 and +1.

c Add 3. ← To get to the next term from the term before it you have to add 3 since the pattern is 4, 7, 10, 13, …

Exercise 9G

For each of these questions:
a copy and complete the table of values for the number machine
b write down the rule for finding the term from the term number
c write down the rule for finding the next term from the term before it.

E

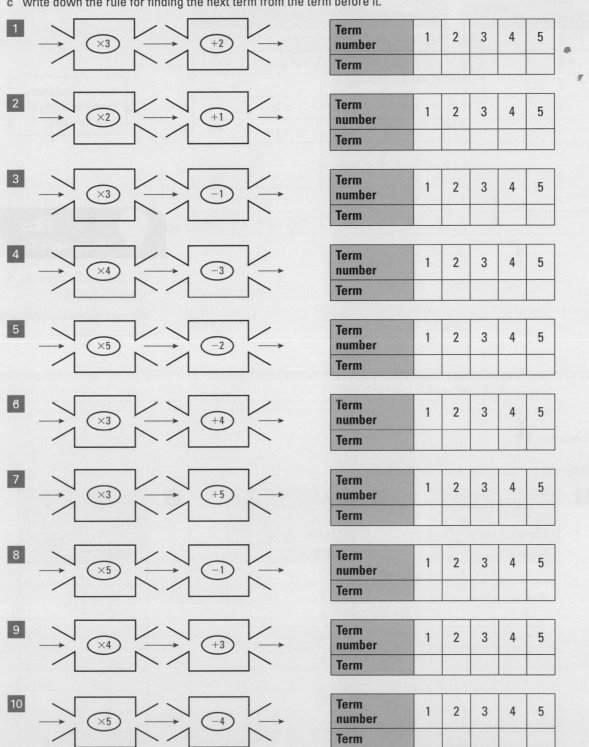

1 →(×3)→(+2)→

Term number	1	2	3	4	5
Term					

2 →(×2)→(+1)→

Term number	1	2	3	4	5
Term					

3 →(×3)→(−1)→

Term number	1	2	3	4	5
Term					

4 →(×4)→(−3)→

Term number	1	2	3	4	5
Term					

5 →(×5)→(−2)→

Term number	1	2	3	4	5
Term					

6 →(×3)→(+4)→

Term number	1	2	3	4	5
Term					

7 →(×3)→(+5)→

Term number	1	2	3	4	5
Term					

8 →(×5)→(−1)→

Term number	1	2	3	4	5
Term					

9 →(×4)→(+3)→

Term number	1	2	3	4	5
Term					

10 →(×5)→(−4)→

Term number	1	2	3	4	5
Term					

Example 8

Complete this table of values for the number pattern with term number to term rule 'Multiply by 4 and subtract 2'.

×4 → −2	
Term number	Term
1	2
2	6
3	
4	
5	
↓	↓
8	
↓	↓
	38

Term number	Term	Working
1	2	1 × 4 − 2 = 2
2	6	2 × 4 − 2 = 6
3	10	3 × 4 − 2 = 10
4	14	4 × 4 − 2 = 14
5	18	5 × 4 − 2 = 18
↓		↓
8	30	8 × 4 − 2 = 30
↓		↓
10	38	10 × 4 − 2 = 38

You can find these terms by using the rule × 4 then − 2.

Results **Plus**
Examiner's Tip

Don't forget Bidmas: you do the × before the −.
You met Bidmas in Chapter 8.

Exercise 9H

Copy and complete these tables of values.

1

×3 → +1	
Term number	Term
1	4
2	
3	
4	
5	
↓	↓
10	
↓	↓
	34

2

×2 → −1	
Term number	Term
1	1
2	
3	
4	
5	
↓	↓
10	
↓	↓
	25

3

×5 → +3	
Term number	Term
1	8
2	
3	
4	
5	
↓	↓
10	
↓	↓
	78

E

4

×4 → −3	
Term number	Term
1	1
2	
3	
4	
5	
↓	↓
10	
↓	↓
	45

5

×10 → +1	
Term number	Term
1	11
2	
3	
4	
5	
↓	↓
10	
↓	↓
	151

6

×5 → −3	
Term number	Term
1	2
2	
3	
4	
5	
↓	↓
10	
↓	↓
	67

7 a Find the 10th number in this number pattern.   3, 7, 11, 15, …
   b What is the term number for the term that is 47?

8 a Find the 10th number in this number pattern.   4, 9, 14, 19, …
   b What is the term number for the term that is 69?

9 a Find the 10th number in this number pattern.   8, 11, 14, 17, …
   b What is the term number for the term that is 50?

9.3 Finding the *n*th term of a number pattern

◉ Objective

○ You can use the first difference to find the *n*th term of a number pattern and use the *n*th term to find any number in a number pattern or sequence.

⟐ Why do this?

This may be useful when your teacher is dividing the class into groups, so that you can work out which group you are going to be in, or make sure you will be in a group with your friends.

⟐ Get Ready

1. Write down the difference between each term in these number patterns.
   a 5, 10, 15, 20, 25, 30, …
   b 40, 35, 30, 25, 20, …
   c 4, 7, 10, 13, 16, …
   d 7, 11, 15, 19, 21, …
   e 50, 47, 44, 41, 37, …
2. Find the 10th term in each of the number patterns in question **1**.

🔍 Key Point

◉ The first difference can be used to find the *n*th term of a number pattern and then the *n*th term can be used to find any number in a sequence.

Example 9

Here is a number pattern   4, 7, 10, 13, 16, …

   a   Find the *n*th term in this pattern.

   b   Find the 20th term in this number pattern.

a

Term number	Term	Difference
1	4	+3
2	7	+3
3	10	+3
4	13	+3
5	16	
n	3*n* + 1	

Step 1
Put the number pattern into a table of values.

Step 2
Find the difference between the terms in the number pattern. In this case it is +3.

Step 3
Multiply each term number by the difference to get a new pattern.
3, 6, 9, 12, 15 …

b   The 20th term is 61.

Step 4
Compare your new pattern with the original one and see what number you need to add or subtract to/from each term to get the original number pattern. In this case it is +1.
The *n*th term is 3*n* + 1.
You replace the *n* by 20 in the *n*th term to find the 20th term. It is 3 × 20 + 1 = 61

Exercise 9I

C

1   For questions 1, 2 and 3 in Exercise 9H, find the *n*th term of each of the number patterns.

2   Write each pattern in a table and use the table to find the *n*th term of these number patterns. Use your *n*th term to find the 20th term in each of these number patterns.

   a   1, 3, 5, 7, 9, 11, …   **b**   3, 5, 7, 9, 11, 13, …

   c   2, 5, 8, 11, 14, 17, …   **d**   5, 8, 11, 14, 17, 20, …

   e   1, 5, 9, 13, 17, 21, …   **f**   2, 6, 10, 14, 18, 22, …

   g   2, 7, 12, 17, 22, 27, …   **h**   4, 9, 14, 19, 24, 29, …

   i   8, 13, 18, 23, 28, …   **j**   5, 7, 9, 11, 13, …

   k   40, 35, 30, 25, 20, …   **l**   38, 36, 34, 32, 30, …

   m   35, 32, 29, 26, 23, …   **n**   20, 18, 16, 14, 12, …

   o   19, 17, 15, 13, 11, …   **p**   190, 180, 160, 150, …

ResultsPlus
Examiner's Tip

To find the *n*th term of a sequence that gets smaller you subtract a multiple of *n* from a fixed number.
e.g. 15 − 2*n* is the *n*th term of 13, 11, 9, 7, …

3 Here is a pattern made from sticks.

 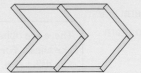 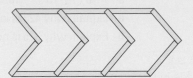

Pattern number 1   Pattern number 2   Pattern number 3

a Draw pattern number 4.

b Copy and complete this table of values for the number of sticks used to make the patterns.

Pattern number	1	2	3	4	5	6
Number of sticks	6	10				

c Write, in terms of n, the number of sticks needed for pattern number n.

d How many sticks would be needed for pattern number 20?

9.4 Deciding whether or not a number is in a number pattern

⊙ Objective

○ You can use number patterns or use the nth term to identify whether a number is in the pattern.

⊘ Why do this?

This is useful when you want to work out what will happen in the future, for example, you could work out whether next year will be a leap year as this happens every four years.

◈ Get Ready

1. Write each of these patterns in a table and use the table to find the nth term.
Use your nth term to find the 20th term in each pattern.

a 4, 7, 10, 13, ...   b 3, 8, 13, 18, ...   c 13, 15, 17, 19, ...

⊛ Key Points

⊙ Number patterns or the nth term can be used to identify whether a number is in the pattern.

⊙ Sometimes you will be asked how you know if a number is part of a sequence. You would then have to explain why the number is in the sequence or, even, why it is not in the sequence.

Example 10 Here is a number pattern.

3   8   13   18   23

a Explain why 423 is in the pattern.

b Explain why 325 is not in the pattern.

a 423 is in the number pattern.

Every odd term ends in 3 and goes up 3, 13, 23, etc,
so 423 will be a member as it ends in a 3.

> There are other ways of answering questions like these. For example, you could identify the nth term.
> The nth term is $5n - 2$ if
> $$5n - 2 = 423$$
> $$5n = 425 \text{ so } n = 85$$
> so 423 is the 85th term.

b 325 is not in the number pattern.

325 ends in a 5 and every member
of the pattern ends in either a 3 or
an 8 so 325 cannot be in the pattern.

> The nth term is $5n - 2$ so if 325 is in the pattern.
> $$5n - 2 = 325$$
> $$5n = 327 \quad \text{so} \quad n = 65.4$$
> If 325 is in the pattern n must be a whole number.
> 65.4 is not a whole number so 325 is not in the pattern.

Exercise 9J

For each of these number patterns, explain whether each of the numbers in brackets are members of the number pattern or not.

1 1, 3, 5, 7, 9, 11, …   (21, 34)

2 3, 5, 7, 9, 11, 13, …   (63, 86)

3 2, 5, 8, 11, 14, 17, …   (50, 66)

4 5, 8, 11, 14, 17, 20, …   (50, 62)

5 1, 5, 9, 13, 17, 21, …   (101, 150)

6 2, 6, 10, 14, 18, 22, …   (101, 98)

7 2, 7, 12, 17, 22, 27, …   (97, 120)

8 4, 9, 14, 19, 24, 29, …   (168, 169)

9 40, 35, 30, 25, 20, …   (85, 4)

10 38, 36, 34, 32, 30, …   (71, 82)

11 3, 7, 11, 15, 19, 21, …   (46, 79)

12 5, 11, 17, 23, 29, …   (119, 72)

Chapter review

- A **sequence** is a number or shape pattern which follows a rule.
- Number patterns can be continued by adding, subtracting, multiplying and dividing.
- Patterns using pictures can be continued by finding the rule for continuing the pattern.
- The numbers in a number pattern are called terms.
- The **term to term rules** for continuing number patterns can be given.
- The term number tells you the position of each term in the pattern.
- You can use number machines to produce a number pattern and write down the rule (term number to term rule)
- You can complete a table of values using a number machine and write down the rule (term number to term).
- The first difference can be used to find the nth term of a number pattern and then the nth term can be used to find any number in a sequence.
- Number patterns or the nth term can be used to identify whether a number is in the pattern.

Review exercise

1 Here are some patterns made of squares.

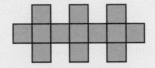

Pattern number 1   Pattern number 2   Pattern number 3

The diagram below shows part of Pattern number 4.

a Copy and complete Pattern number 4.

Pattern number 4

b Find the number of squares used for Pattern number 10.

Nov 2008 adapted

A02

2 Here are the first 4 terms in a number sequence.

    124   122   120   118

a Write down the next term in this number sequence.

b Write down the 7th term in this number sequence.

c Can 9 be a term in this number sequence? You must give a reason for your answer.

May 2009

D

A03

3 The diagram shows a mathematical rule.

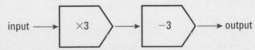

input ⟶ ×3 ⟶ −3 ⟶ output

It multiplies a number by 3 and then subtracts 3
Copy and complete the diagram in each case.

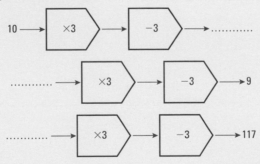

10 ⟶ ×3 ⟶ −3 ⟶ ....

.... ⟶ ×3 ⟶ −3 ⟶ 9

.... ⟶ ×3 ⟶ −3 ⟶ 117

Nov 2008

4 Here is a table for a 2-stage number machine.
It subtracts 5 and then multiplies by 2

a Copy and complete the table.

b The input is n.
Write down an expression in terms of n for the output.

− 5 then × 2	
Input	**Output**
4	−2
2	....
−3	....

Nov 2006

F

5 The nth term of a sequence is $n^2 + 4$.

Alex says 'The nth term of the sequence is always a prime number when n is an odd number'.

Is Alex correct? You must give a reason for your answer. *Nov 2008 adapted*

6 Here are the first 5 terms of a sequence.

     1   1   2   3   5

The rule for the sequence is 'The first two terms are 1 and 1. To get the next term add the two previous terms'.

a Find the 6th term and the 7th term.

b Find the 10th term.

7 Dylan and Evie are studying a number pattern.

The first three numbers in the number pattern are 1, 2, 4.

Dylan says that the next number is 8.

Evie says that the next number is 7.

Explain how both Dylan and Evie could be right.

8 Here are the first four terms of an arithmetic sequence.

     5   8   11   14

Is 140 a term in the sequence? You must give a reason for your answer.

9 The first term of a sequence is x. To get the next term, multiply the previous term by 2 and add 1.

The third term of the sequence is 21. Find the value of x.

10 Here are the first five terms of a number sequence.

     3   7   11   15   19

a Work out the 8th term of the number sequence.

b Write down an expression, in terms of n, for the nth term of the number sequence. *Nov 2006*

10 GRAPHS 1

'Battleships' is a guessing game involving coordinates. It dates back to before World War I when it was played with pen and paper. It has since been developed into a board game as shown in the picture, and more recently has become available on the Playstation 2, the Xbox 360, the Wii, many mobile phones and some social networking sites.

◉ Objectives

In this chapter you will:
- ◉ write down and plot the coordinates of a point in any of the four quadrants
- ◉ find the coordinates of the midpoint of a line.

◈ Before you start

You need to know:
- ◉ how to find points along a number line
- ◉ how to use maps and plans to find a position.

10.1 Coordinates of points in the first quadrant

⊙ Objectives

○ You can write down the coordinates of a point in the first quadrant.

○ You can plot the coordinates of a point in the first quadrant.

⟐ Why do this?

GPS systems use coordinates to find your position, for example so that the emergency services can find you.

⬆ Get Ready

Tall Tree Island

On this map if you start at *O* and go 1 square to the right and 4 squares up you get to the Lookout.

1. Describe how you get from
   a the Lookout to the Tall Tree
   b the Tall Tree to the Beach
   c the Water Hole to the Beach.

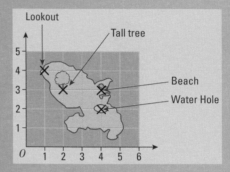

⬤ Key Points

⊙ The position of a point on a coordinate grid is described as two numbers.

⊙ The number of units across (the **x-coordinate**) is written first, and the number of units up (the **y-coordinate**) is written second (*x, y* is in alphabetical order).

⊙ Given **coordinates** can be plotted on a grid.

⬤ Example 1

Here is the plan of part of a zoo.
It is drawn on a coordinate grid.
Write down the names of the animals at the following coordinates.

a (1, 2)   b (4, 3)
c (6, 7)   d (0, 5)

a lions   b elephants
c penguins   d gorillas

Example 2 ▶ On squared paper, draw a coordinate grid and number it to 10 across the page and up the page. Join these points up in the order given.

(2, 4)   (4, 6)   (7, 4)   (4, 2)   What shape have you drawn?

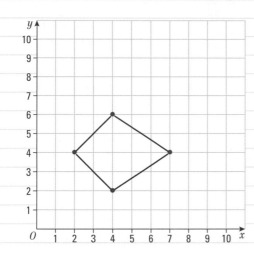

The shape is a kite.

Exercise 10A

Questions in this chapter are targeted at the grades indicated.

1   Here is the plan of part of a theme park. It is drawn on a coordinate grid.

   a Write down the names of the attractions at the following coordinates.

     i (1, 2)   **ii** (4, 3)   **iii** (6, 7)

     iv (0, 6)   **v** (3, 6)

   b Write down the coordinates where you will find the:

     i entrance   **ii** big dipper   **iii** exit

     iv ghost ride   **v** swinging boats.

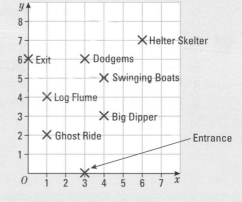

G

2   Here is a coordinate grid.

   a Write down the letter of the point with the following coordinates.

     i (1, 2)   **ii** (3, 4)   **iii** (5, 7)

     iv (0, 4)   **v** (0, 0)

   b Write down the coordinates of the following points.

     i O   **ii** C   **iii** D

     iv G   **v** E

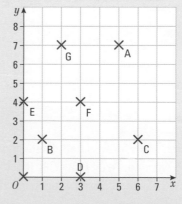

G

3   Here is a coordinate grid.

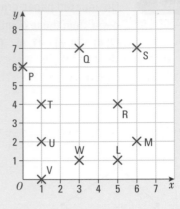

a   Write down the letter of the point with coordinates
   i (1, 2)   ii (1, 4)   iii (6, 7)
   iv (0, 6)   v (1, 0)

b   Write down the coordinates of the points
   i W   ii M   iii R
   iv L   v Q

4   On squared paper, draw a coordinate grid and number it from 0 to 10 across the page and 0 to 10 up the page. Join these points up in the order given.
   (5, 3)   (1, 3)   (2, 1)   (7, 1)   (9, 3)   (5, 3)   (5, 9)   (9, 4)   (0, 4)   (5, 9)

5   On squared paper, draw a coordinate grid and number it from 0 to 10 across the page and 0 to 10 up the page. Join each of these sets of points in the order given.
   a   (1, 1)   (4, 1)   (4, 4)   (1, 4)   (1, 1)   b   (6, 1)   (9, 1)   (9, 6)   (6, 6)   (6, 1)
   c   (0, 6)   (5, 6)   (0, 10)   (0, 6)   d   (5, 7)   (5, 10)   (8, 10)   (5, 7)

6   On squared paper, draw a coordinate grid and number it from 0 to 8 across the page and 0 to 8 up the page.
   a   Plot the points P at (1, 2), Q at (7, 2) and R at (7, 5).
   b   Mark the position of point S so that PQRS is a rectangle.
   c   Write down the coordinates of point S.

7   On squared paper, draw a coordinate grid and number it from 0 to 8 across the page and 0 to 8 up the page.
   a   Plot the points A at (1, 2), B at (6, 2) and C at (8, 5).
   b   Mark the position of point D so that ABCD is a parallelogram.
   c   Write down the coordinates of point D.

10.2 Coordinates of points in all four quadrants

◎ Objectives

○ You can write down the coordinate of a point in any of the four quadrants.
○ You can plot the coordinate of a point in any of the four quadrants.

◈ Get Ready

On squared paper, draw a coordinate grid and number it from 0 to 4 across the page and 0 to 4 up the page.
1. Plot the points A at (1, 3), B at (3, 3), C at (3, 1) and D at (1, 1).
2. Write down the coordinates of the point at the centre of the shape you have drawn.

Key Points

- The horizontal **axis** is called the *x*-**axis**.
- The vertical axis is called the *y*-**axis**.
- A coordinate grid is divided into four regions (**quadrants**) by the *x*- and *y*-axes.
- The point O is called the **origin** and has coordinates (0, 0).

Example 3   Write down the coordinates of the points P, Q, R and S.

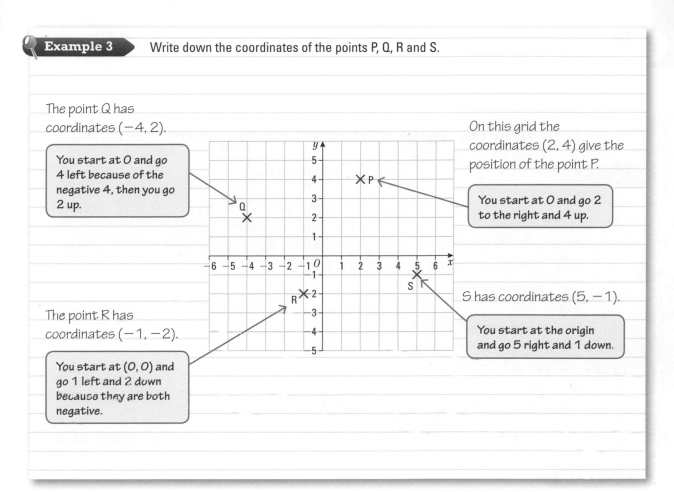

The point Q has coordinates $(-4, 2)$.

You start at O and go 4 left because of the negative 4, then you go 2 up.

On this grid the coordinates (2, 4) give the position of the point P.

You start at O and go 2 to the right and 4 up.

The point R has coordinates $(-1, -2)$.

You start at $(0, 0)$ and go 1 left and 2 down because they are both negative.

S has coordinates $(5, -1)$.

You start at the origin and go 5 right and 1 down.

Exercise 10B

1 Write down the coordinates of all the points A to L marked on the coordinate grid.

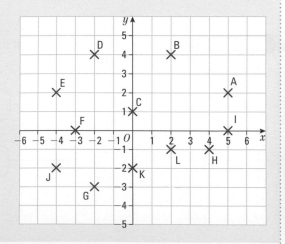

F

F

2 Draw a coordinate grid similar to the one above with the x-axis marked from -6 to $+6$ and the y-axis marked from -4 to $+5$.

Plot the following points and join them up in the order given.

$(-6, 1)$	$(-4, -2)$	$(4, -2)$	$(5, 1)$	$(-6, 1)$	$(-2, 1)$	$(0, 4)$
$(4, 4)$	$(4, 1)$	$(5, 1)$	$(5, 4)$	$(6, 4)$	$(6, 3)$	$(5, 3)$

A02

3 On a coordinate grid make your own picture and write down the coordinates of all the points in your picture.

10.3 Finding the midpoint of a line segment

◎ Objectives

○ You can find the coordinates of the midpoint of a line in the first quadrant.

○ You can find the coordinates of the midpoint of a line in all four quadrants.

◈ Why do this?

You might need to find the midpoint if you and a friend have agreed to meet halfway between your house and theirs.

◈ Get Ready

The number 5 is halfway between 0 and 10.
The number 15 is halfway between 10 and 20.
To find the halfway point add the two numbers and divide by 2.

1. Find the number halfway between:

a 20 and 30   **b** 30 and 40   **c** 4 and 6   **d** 5 and 7

e -2 and 4   **f** -2 and 6   **g** 4 and -2 **h** -2 and 5.

◈ Key Points

◉ The **midpoint** of a line is halfway along the line.

◉ To find the midpoint you add the x-coordinates and divide by 2 and add the y-coordinates and divide by 2.

Example 4   Work out the coordinates of the midpoint of the line segment PQ where P is (2, 3) and Q is (7, 11).

x-coordinate $2 + 7 = 9$ ←   Add the x-coordinates and divide by 2.
$9 \div 2 = 4\frac{1}{2}$

y-coordinate $3 + 11 = 14$ ←   Add the y-coordinates and divide by 2.
$14 \div 2 = 7$

Midpoint is $(4\frac{1}{2}, 7)$.

Example 5 Work out the midpoint of the line AB.

M is the midpoint of the line AB.

You can find this by measuring or by looking at the coordinates.
A has coordinates (1, 1)
B has coordinates (4, 5)

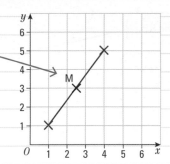

x-coordinate $1 + 4 = 5$ $5 \div 2 = 2.5$ ← Add the x-coordinates and divide by 2.

y-coordinate $1 + 5 = 6$ $6 \div 2 = 3$ ← Add the y-coordinates and divide by 2.

M has coordinates $(2.5, 3)$ or $(2\frac{1}{2}, 3)$.

Exercise 10C

1 Work out the coordinates of the midpoint of each of the line segments shown on the grid.

a UA   b BC   c DE
d FG   e HJ

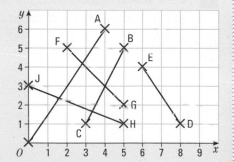

2 Work out the coordinates of the midpoint of each of these line segments.

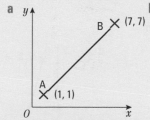 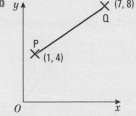 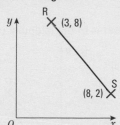 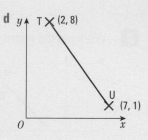

3 Work out the coordinates of the midpoint of each of these line segments.

a AB when A is (1, 1) and B is (9, 9)

b PQ when P is (2, 4) and Q is (6, 9)

c ST when S is (5, 8) and T is (2, 1)

d CD when C is (1, 7) and D is (7, 2)

e UV when U is (2, 3) and V is (6, 8)

f GH when G is (2, 6) and H is (7, 3)

D

C

Example 6 Find the midpoint of RS.

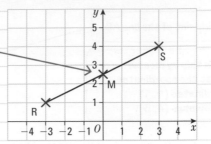

M is the midpoint of the line RS.

You can find this by measuring or by looking at the coordinates.
R has coordinates $(-3, 1)$
S has coordinates $(3, 4)$

x-coordinate $-3 + 3 = 0$ $0 \div 2 = 0$ ← Add the x-coordinates and divide by 2.

y-coordinate $1 + 4 = 5$ $5 \div 2 = 2.5$ ← Add the y-coordinates and divide by 2.

M has coordinates $(0, 2.5)$ or $(0, 2\frac{1}{2})$.

Exercise 10D

1 Work out the coordinates of the midpoint of each of the line segments shown on the grid.

a OA   b BC   c DE
d FG   e HJ   f KL
g MN   h PQ   i ST
j UV

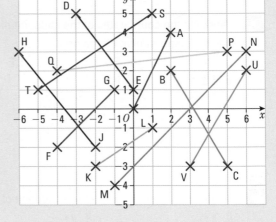

2 Work out the coordinates of the midpoint of each of these line segments.

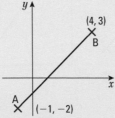 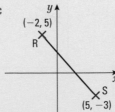 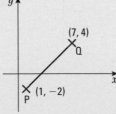 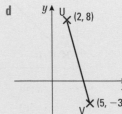

3 Work out the coordinates of the midpoint of each of these line segments.

a AB when A is $(-1, -1)$ and B is $(9, 9)$
b PQ when P is $(2, -4)$ and Q is $(-6, 9)$
c ST when S is $(5, -8)$ and T is $(-2, 1)$
d CD when C is $(1, 7)$ and D is $(-7, 2)$
e UV when U is $(-2, 3)$ and V is $(6, -8)$
f GH when G is $(-2, -6)$ and H is $(7, 3)$

Chapter review

- The position of a point on a coordinate grid is described as two numbers.
- The number of units across (the x-**coordinate**) is written first, and the number of units up (the y-**coordinate**) is written second.
- Given **coordinates** can be plotted on a grid.
- The horizontal **axis** is called the x-**axis**.
- The vertical axis is called the y-**axis**.
- A coordinate grid is divided into four regions (**quadrants**) by the x- and y-axes.
- The point O is called the **origin** and has coordinates (0, 0).
- The **midpoint** of a line is halfway along the line.
- To find the midpoint you add the x-coordinates and divide by 2 and add the y-coordinates and divide by 2.

Review exercise

1. **a** Write down the coordinates of the point A.
    b Write down the coordinates of the point B.

    N is the point $(-3, 2)$.
    c On a copy of the grid, mark the point N with a cross ($\times$).
    Label it N.

    M is another point.
    The x-coordinate of M is the same as the x-coordinate of N.
    The y-coordinate of M is the same as the y-coordinate of B.
    d Write down the coordinates of the point M.

    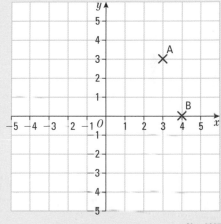

    Nov 2007

    F

2. **a** Write down the coordinates of the points
    i P   **ii** Q   **iii** R
    b **i** Copy the graph and join the points
    to form triangle PQR.
    ii Write down the mathematical name of
    triangle PQR.
    c Write down the coordinates of the
    midpoint of
    i the side QR
    ii the side PQ
    iii the side PR.
    d PQRS is a rectangle. Write down the
    coordinates of the point S.

    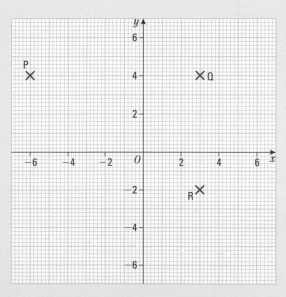

    D

C

3 Here is a grid of centimetre squares.
   a Write down the coordinates of the points
      i B   **ii** C.
ABCD is a quadrilateral.
The area of ABCD is 24 cm².
   b Write down the coordinates of D.
E is another point on the grid.
The coordinates of the midpoint of AE are (−1, 2).
   c Find the coordinates of E.

A03

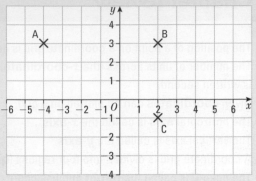

Specimen Paper 2009

11 GRAPHS 2

Marathon runners often use a graph which shows how their pace varied throughout a race. Looking at a graph is the easiest way to see their running pattern quickly and easily and help them plan the next one.

◉ Objectives

In this chapter you will:
- write down the equations of vertical and horizontal lines and draw them
- draw straight-line graphs of the form $y = mx + c$ and $x + y = k$, with or without using a table of values
- give the equation of any straight line drawn on a coordinate grid.

◑ Before you start

You need to know:
- how to read and plot the coordinates of a point in any of the four quadrants.

11.1 Drawing and naming horizontal and vertical lines

⊙ Objectives

⊙ You can write down the equations of vertical and horizontal lines.

⊙ You can draw the vertical and horizontal lines with equations of the form $x = n$ and $y = m$.

⚡ Why do this?

This would allow you to plot a route as you would be able to describe exactly where you were, where you were going and which route you would take to get there.

⬦ Get Ready

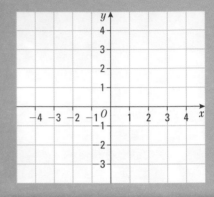

1. Make a copy of the coordinate grid and plot the following sets of points: (2, −3), (2, −2), (2, −1), (2, 0), (2, 1), (2, 2), (2, 3), (2, 4). Join the points up with a straight line.

2. On the same grid, plot these points: (−3, 3), (−2, 3), (−1, 3), (0, 3), (1, 3), (2, 3), (3, 3), (4, 3). Join the points up with a straight line.

3. Write down the coordinates of the point where the two lines meet.

🔑 Key Points

⊙ A vertical line on the grid has the form $x = n$. For example, on this graph all the x-coordinates are 2 so the line is called $x = 2$. The equation of the line is $x = 2$.

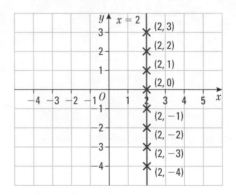

⊙ A horizontal line on the grid has the form $y = m$. In this graph all the y-coordinates are −2 so the line is called $y = -2$. The equation of the line is $y = -2$.

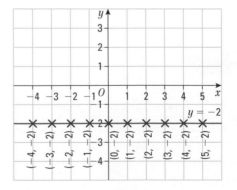

Exercise 11A

Questions in this chapter are targeted at the grades indicated.

E

1 Write down the equations of the lines marked **a** to **d** in this diagram.

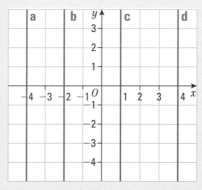

> **ResultsPlus**
> **Examiner's Tip**
>
> The x-axis has equation $y = 0$.
> The y-axis has equation $x = 0$.

2 Write down the equations of the lines labelled **a** to **d** in this diagram.

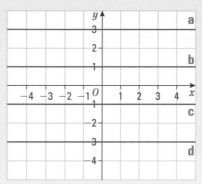

3 Draw a coordinate grid with x- and y-axes labelled from -5 to 5. On the grid draw and label the graphs of
   a $x = 4$ b $x = -2$ c $x = -4$ d $x = 1$

4 Draw a coordinate grid with axes labelled from -5 to 5. On the grid draw and label the graphs of
   a $y = 4$ b $y = -2$ c $y = -4$ d $y = 1$

5 Draw a coordinate grid with axes labelled from -5 to 5. On the grid draw and label the graphs of
   a $y = 3$ b $x = -1$
   c Write down the coordinates of the point where the two lines cross.

11.2 Drawing slanting lines

◉ Objectives

- ◉ You can use a table of values to draw straight-line graphs with positive gradients.
- ◉ You can use a table of values to draw straight-line graphs with negative gradients.

◈ Why do this?

In science, you might be asked to plot the results of your experiments on a graph. The graph might be a straight line.

◈ Get Ready

Draw these lines.

1. $y = 2$ 2. $x = -4$ 3. $y = \frac{1}{2}$ 4. $y = -3$
                                                                              $x = 3$

Key Points

● Lines that slant upwards ⟋ have a positive **gradient**.

● Lines that slant downwards ⟍ have a negative gradient.

● To draw a straight-line graph with a given equation:
  ● make a table of values, selecting some values for x
  ● substitute the values of x into the equation
  ● plot the points from the table of values on the grid
  ● draw in the line.

Example 1 ▶ Draw the graph of $y = 2x - 1$.

1. Make a table of values, selecting some values for x.

x	-1	0	1	2	3
$y = 2x - 1$					

2. Substitute the values of x into $y = 2x - 1$.

x	-1	0	1	2	3
$y = 2x - 1$	-3	-1	1	3	5

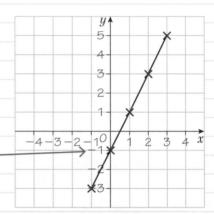

3. Plot the points on the grid.

4. Draw in the line.

ResultsPlus
Watch Out!

Don't forget to draw in the line.

Exercise 11B

D

1   **a**   Copy and complete the tables of values for the straight-line graphs below.

    b   On a coordinate grid with the x-axis drawn from -3 to $+3$ and y-axis drawn from -10 to $+10$, draw the graphs of $y = x - 1$, $y = 2x - 4$, $y = 3x + 1$, $y = x + 4$ and $y = 4x + 1$.

i

x	-3	-2	-1	0	1	2	3
$y = x - 1$		-3		-1		1	

ii

x	-3	-2	-1	0	1	2	3
$y = 2x - 4$		-8		-4		0	

iii

x	−3	−2	−1	0	1	2	3
$y = 3x + 1$	−8			1			10

iv

x	−3	−2	−1	0	1	2	3
$y = x + 4$		2		4			7

v

x	−2	−1	0	1	2
$y = 4x + 1$	−7				9

2 Draw the graphs of these straight lines on a coordinate grid with axes drawn from −3 to +3 on the x-axis and −10 to +10 on the y-axis.

  a $y = 2x + 1$ **b** $y = 2x + 3$ **c** $y = 2x - 3$ **d** $y = 2x + 2$

3 Draw the graphs of these straight lines on a coordinate grid with axes drawn from −3 to +3 on the x-axis and −6 to +6 on the y-axis.

  a $y = x + 1$ **b** $y = x + 3$ **c** $y = x - 3$ **d** $y = x + 2$

4 Draw the graphs of these straight lines on a coordinate grid with axes drawn from −4 to +4 on the x-axis and −10 to +10 on the y-axis.

  a $y = 2x - 1$ **b** $y = 3x + 2$ **c** $y = 2x - 2$ **d** $y = 3x - 2$

5 Draw the graphs of these straight lines on a coordinate grid with axes drawn from −4 to +4 on the x-axis and −6 to +6 on the y-axis.

  a $y = \frac{1}{2}x + 1$ **b** $y = \frac{1}{2}x + 3$ **c** $y = \frac{1}{2}x - 3$ **d** $y = \frac{1}{2}x - 2$

Example 2   Draw the graph of $y = -2x - 1$.

1. Make a table of values, selecting some values for x.

x	−3	−2	−1	0	1
$y = -2x - 1$					

2. Substitute the values of x into $y = -2x - 1$.

x	−3	−2	−1	0	1
$y = -2x - 1$	5	3	1	−1	−3

3. Plot the points on the grid.

4. Draw in the line.

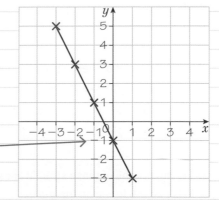

Exercise 11C

D

1 **a** Copy and complete the tables of values for the straight-line graphs below.

b Draw the graphs of these straight lines on a coordinate grid with the x-axis drawn from -3 to $+3$ and y-axis drawn from -10 to $+10$.

i

x	-3	-2	-1	0	1	2	3
$y = -x - 1$		1		-1		-3	

ii

x	-3	-2	-1	0	1	2	3
$y = -2x - 4$		0		-4		-8	

iii

x	-3	-2	-1	0	1	2	3
$y = -3x + 1$	10			1			-8

iv

x	-3	-2	-1	0	1	2	3
$y = -x + 4$		6		4			1

v

x	-2	-1	0	1	2
$y = -4x + 1$	9				-7

2 Draw the graphs of these straight lines on a coordinate grid with axes drawn from -3 to $+3$ on the x-axis and -10 to $+10$ on the y-axis.

  a $y = -2x + 1$ **b** $y = -2x + 3$

  c $y = -2x - 3$ **d** $y = -2x + 2$

3 Draw the graphs of these straight lines on a coordinate grid with axes drawn from -3 to $+3$ on the x-axis and -6 to $+6$ on the y-axis.

  a $y = -x + 1$ **b** $y = -x + 3$

  c $y = -x - 3$ **d** $y = -x + 2$

4 Draw the graphs of these straight lines on a coordinate grid with axes drawn from -4 to $+4$ on the x-axis and -10 to $+10$ on the y-axis.

  a $y = -2x - 1$ **b** $y = -3x + 1$

  c $y = -2x - 2$ **d** $y = -3x - 2$

5 Draw the graphs of these straight lines on a coordinate grid with axes drawn from -4 to $+4$ on the x-axis and -6 to $+6$ on the y-axis.

  a $y = -\frac{1}{2}x + 1$ **b** $y = -\frac{1}{2}x + 3$

  c $y = -\frac{1}{2}x - 3$ **d** $y = -\frac{1}{2}x - 2$

11.3 Drawing straight-line graphs without a table of values

◎ Objectives

- ⦿ You can draw straight-line graphs of the form $y = mx + c$ using the intercept and gradient when the gradient is positive.
- ⦿ You can draw straight-line graphs of the form $y = mx + c$ using the intercept and gradient when the gradient is negative.
- ⦿ You can draw straight-line graphs by using intercepts on the x- and y-axes for graphs of the type $x + y = k$.

◈ Why do this?

This could be a good way of predicting the results of an experiment before you complete it.

◈ Get Ready

Use a table of values to draw these straight lines.
1. $y = x + 4$ **2.** $y = x + 5$ **3.** $y = 2x + 4$
What do you notice about the lines in questions **1** and **3**?

◈ Key Points

- ⦿ The equation of a straight-line graph can be written in the form $y = mx + c$. The number on its own (c) tells you where the straight line crosses the y-axis.

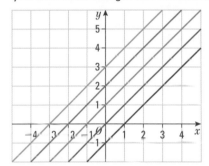

Here are five straight-line graphs.
$y = x + 3$
$y = x + 2$
$y = x + 1$
$y = x + 0$
$y = x - 1$

- ⦿ The number (m) in front of the x tells you the gradient (steepness) of the line. If the number is positive, for each square you move to the right you move up by the number in front of the x. If the number is negative, for each square you move to the right you move down by the number in front of the x.

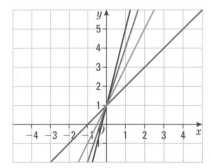

Here are four straight-line graphs.
$y = x + 1$
$y = 2x + 1$
$y = 3x + 1$
$y = 4x + 1$

- ⦿ To draw a straight line from the equation $y = mx + c$:
  - ⦿ mark the point (c) where the line will cross the y-axis
  - ⦿ find out the gradient – how many squares you go up (or down) each square you move to the right – from the number (m) in front of the x
  - ⦿ join up the points with a straight line.
- ⦿ When the line is in the form $x + y = c$ the c tells you where the graph cuts the x-axis and the y-axis.

Example 3   On the grid, draw the graph of $y = 2x + 1$.

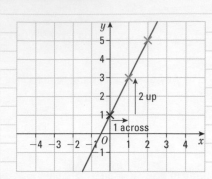

Step 1   Mark the point **X** where the line will cross the y-axis.

Step 2   For each square you move to the right you then go up 2 because there is a 2 in front of the x.

Step 3   Join up the points with a straight line.

Exercise 11D

For each question, draw the graphs of all the straight lines on the same coordinate grid with the x-axis drawn from -3 to $+3$ and y-axis drawn from -10 to $+10$. What do you notice about the set of graphs for each question?

1   **a** $y = x + 1$ **b** $y = x + 2$ **c** $y = x + 3$
    d $y = x - 1$ **e** $y = x - 2$

2   **a** $y = 2x + 1$ **b** $y = 2x + 2$ **c** $y = 2x + 3$
    d $y = 2x - 1$ **e** $y = 2x - 2$

3   **a** $y = 3x + 1$ **b** $y = 3x + 2$ **c** $y = 3x + 3$
    d $y = 3x - 1$ **e** $y = 3x - 2$

4   **a** $y = 4x + 1$ **b** $y = 4x + 2$ **c** $y = 4x + 3$
    d $y = 4x - 1$ **e** $y = 4x - 2$

5   **a** $y = \frac{1}{2}x + 1$ **b** $y = \frac{1}{2}x + 2$ **c** $y = \frac{1}{2}x + 3$
    d $y = \frac{1}{2}x - 1$ **e** $y = \frac{1}{2}x - 2$

Example 4   On the grid, draw the graph of $y = -2x + 1$.

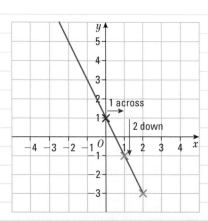

Step 1   Mark the point **X** where the line will cross the y-axis.

Step 2   For each square you move to the right you go down 2 because there is a -2 in front of the x.

Step 3   Join up the points with a straight line.

Exercise 11E

For each question, draw the graphs of all the straight lines on the same coordinate grid with the x-axis drawn from -3 to $+3$ and y-axis drawn from -10 to $+10$. What do you notice about the set of graphs for each question?

1 **a** $y = -x + 1$ **b** $y = -x + 2$ **c** $y = -x + 3$
   d $y = -x - 1$ **e** $y = -x - 2$

2 **a** $y = -2x + 1$ **b** $y = -2x + 2$ **c** $y = -2x + 3$
   d $y = -2x - 1$ **e** $y = -2x - 2$

3 **a** $y = -3x + 1$ **b** $y = -3x + 2$ **c** $y = -3x + 3$
   d $y = -3x - 1$ **e** $y = -3x - 2$

4 **a** $y = -4x + 1$ **b** $y = -4x + 2$ **c** $y = -4x + 3$
   d $y = -4x - 1$ **e** $y = -4x - 2$

5 **a** $y = -\frac{1}{2}x + 1$ **b** $y = -\frac{1}{2}x + 2$ **c** $y = -\frac{1}{2}x + 3$
   d $y = -\frac{1}{2}x - 1$ **e** $y = -\frac{1}{2}x - 2$

D

Example 5   On the grid, draw the graph of $x + y = 4$.

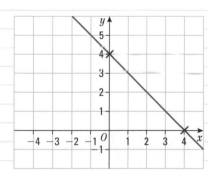

Step 1 Mark the value of the number on the x-axis and the y-axis.

Step 2 Join these points up with a straight line.

 Exercise 11F

For each question, draw the graphs of all the straight lines on the same coordinate grid with the x-axis drawn from -3 to $+8$ and y-axis drawn from -3 to $+8$.

1 **a** $x + y = 3$ **b** $x + y = 5$ **c** $x + y = 2$

2 **a** $x + y = -2$ **b** $x + y = -3$ **c** $x + y = 1$

3 **a** $x + y = -1$ **b** $x + y = 2.5$ **c** $x + y = 0$

D

11.4 Naming straight-line graphs

Objective

- You can give the equation of any straight line drawn on a coordinate grid.

Why do this?

If you have plotted a graph from values you have found in an experiment, you might need to find the relationship between them. This is what you do when you give the equation of a straight line.

Get Ready

Draw these graphs.

1. $y = -4x + 2$ **2.** $y = 2x + 5$ **3.** $x + y = 6$

Key Points

- When you have a straight line for which you need to find the equation, you need to work out the gradient (the m in the equation) and look to see what the **intercept** on the y-axis is (the c in the equation). You then put these values in the equation $y = mx + c$ to find the equation of the line.

- The equation of a line can be used to describe a line of reflection or the line of symmetry of an object on a coordinate grid. See Unit 2 Section 15.6 Line symmetry and Unit 3 Section 11.4 Reflections.

Example 6   Write down the equations of these straight lines.

Line **a** cuts the y-axis at 1.
Gradient (slope) is 1 (it goes up 1 for every 1 across).
Equation is $y = x + 1$.

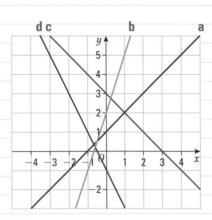

Line **b** cuts the y-axis at 2.
Gradient (slope) is 3 (it goes up 3 for every 1 across).
Equation is $y = 3x + 2$.

Line **c** cuts the x-axis at 3.
Cuts the y-axis at 3.
Equation is $x + y = 3$.

Line **d** cuts the y-axis at -1.
Gradient (slope) is -2 (it goes down 2 for every 1 across).
Equation is $y = -2x - 1$.

Exercise 11G

1 Write down the equations of these straight lines.

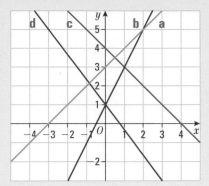

2 Write down the equations of these straight lines.

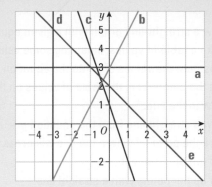

3 Write down the equations of these straight lines.

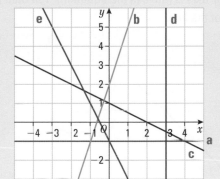

4 Write down the equations of these straight lines.

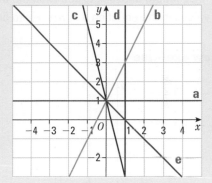

Chapter review

- Horizontal lines are $y =$ lines.
- Vertical lines are $x =$ lines.

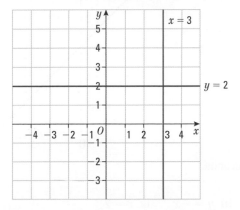

⦿ Lines that slant upwards ╱ have a positive **gradient**.

⦿ Lines that slant downwards ╲ have a negative gradient.

⦿ To draw a straight-line graph with a given equation:
  ⦿ make a table of values, selecting some values for x
  ⦿ substitute the values of x into the equation
  ⦿ plot the points from the table of values on the grid
  ⦿ draw in the line.
⦿ To draw a straight line from the equation $y = mx + c$:
  ⦿ mark the point (c) where the line will cross the y-axis
  ⦿ find out the gradient
  ⦿ join up the points with a straight line.
⦿ When the line is in the form $x + y = c$ the c tells you where the graph cuts the x-axis and the y-axis.
⦿ When you have a straight line for which you need to find the equation, you need to work out the gradient (the m in the equation) and look to see what the **intercept** on the y-axis is (the c in the equation). You then put these values in the equation $y = mx + c$ to find the equation of the line.

Review exercise

1

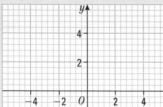

A

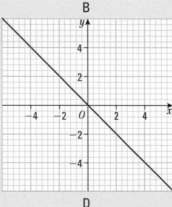

B

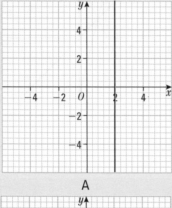

C

D

The diagrams show four graphs.
Here are four equations.

i $y = x$ **ii** $y = -2$ **iii** $y = -x$ **iv** $x = 2$

Match the letter of each graph with its equation.

E

2 **a** Copy and complete the table of values for $y + 2x = 6$

x	−1	0	1	2	3
y		6	4		

  b Draw a graph of $y + 2x = 6$ for values of x from −1 to 3.
  c Use your graph to find an estimate for the value of y when $x = 2.4$.
  d Use your graph to find an estimate for the value of x when $y = 6.6$.

3 On the coordinate grid draw the graph of $y = 2x - 3$.
Use values of x from -2 to $+2$.

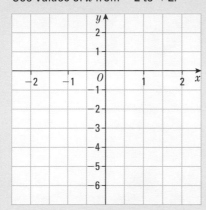

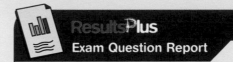

89% of students scored poorly on this question. The most common incorrect answer was to plot just one point, generally (2, −3), sometimes joining this point to the axes.

4 Draw the graph of $y = 2x - 3$ for values of x from −1 to 3.   *May 2009*

5 **a** Draw a graph of $2y + x = 8$ for values of x from −2 to 4.
  b On the same axes draw the graph of $y = x$ for values of x from −2 to 4.
  c P is the point on the graph of $2y + x = 8$ for which the value of x is the same as the value of y.
    Estimate the value of x.

6 Write down the equation of each of the lines shown in the grid.

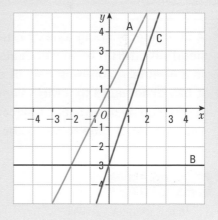

A swimmer is doing a heart rate training session which involves working at her maximum heart rate (MHR) minus 10 beats. MHR is the highest number of times your heart can beat in a minute. A common way of estimating MHR is to subtract your age from 220. However, it has been found that MHR varies widely from person to person, so this estimation is dismissed by scientists as inaccurate. She could use a graph to show her heart rate over a set amount of time to see if she is effectively training her heart.

Objectives

In this chapter you will:
- draw and interpret a range of linear and non-linear graphs
- draw and use conversion graphs
- draw and interpret distance–time graphs, including those that include sections which are curved.

Before you start

You need to know:
- how to plot a coordinate
- how to substitute a number for x in a linear equation.

12.1 Interpreting and drawing the graphs you meet in everyday life

Objectives

⦿ You can interpret and draw straight-line graphs through the origin, e.g. to find the cost of items.

⦿ You can interpret and draw straight-line graphs not through the origin, e.g. to find the total bill given a basic cost plus cost per unit in gas bills or mobile phone bills.

Why do this?

You will often see graphs in newspapers and you need to be able to interpret the information on them.

Get Ready

1. Cereal bars cost 20p each.
   Copy and complete this table for the cost of buying cereal bars.

Number of cereal bars	1	2	3	4	5	6	7	8	9	10
Cost in pence	20									200

Key Points

⦿ A straight-line graph that goes through the origin means that the more items you buy, the more it will cost you. The cost is related to the number of items you buy. This is the type of graph you get with a Pay as you go mobile phone. If you don't use the phone there is no cost.

Pay as you go

The more minutes you use the more it costs.

If you don't use the phone there is no cost.

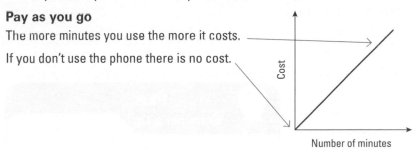

⦿ A straight-line graph that does not cross the vertical axis at the origin means there is a basic charge and then the cost is related to the number of items you buy. This is the type of graph you get with a mobile phone on a contract. If you don't use the phone you still have to pay the monthly cost.

Contract

The more minutes you use the more it costs.

If you don't use the phone you still have to pay the monthly cost.

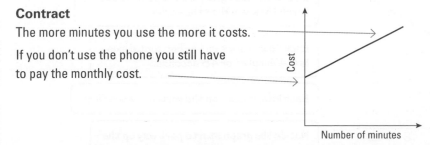

Example 1

Sam sells packets of crisps at 30p each.

He makes a table of values to help him remember what to charge people when they buy different numbers of packets of crisps.

Plot a graph to show this information.

This pattern goes up in 30s.

Number of packets	1	2	3	4	5	6	7	8
Cost in pence	30	60	90	120	150	180	210	240

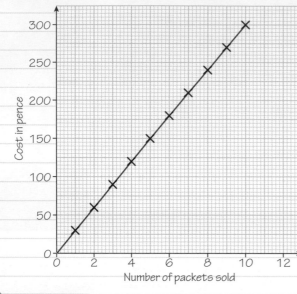

You plot the number of packets against the cost in pence.

When the points are plotted and joined you get a straight line. This is why it is a linear relationship.

You can extend the plotted points to 9 and 10 packets and so read off the cost of these numbers of packets.

Example 2

Sharon is a taxi driver. She charges £2 before the journey starts and then £1.50 for every 2 minutes of the journey.

Use the information to make up a table of values for taxi rides up to 20 minutes long.

Plot a graph to show this information.

Journey time in minutes	0	2	4	6	8	10	12	14	16	18	20
Cost in pounds	2	3.50	5	6.50	8	9.50	11	12.50	14	15.50	17

ResultsPlus
Examiner's Tip

Make sure you understand the scale of your graph before you draw it or read values from it.

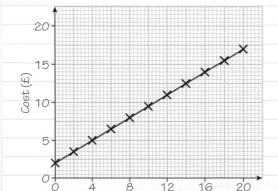

On a coordinate grid plot the points but be careful how you read the scales.

Each little square on the horizontal axis is 0.4 minutes or 48 seconds.

Each little square on the vertical axis is 50p.

Notice the graph starts part way up the vertical axis and not at the origin.

Exercise 12A

F

1 The table shows the cost of potatoes per kg.

Weight in kg	1	2	3	4	5
Cost in pence	30	60	90	120	150

a Draw a graph for this table.
b Work out how much 2.5 kg of potatoes would cost.
c Extend the graph to work out the cost of 6 kg of potatoes.

2 The table shows the cost of ice lollies.

Number of ice lollies	1	2	3	4	5
Cost in pence	25	50	75	100	125

a Draw a graph for the cost of ice lollies from the table.
b Extend the graph and then use it to work out the cost of
  i 8 ice lollies   ii 6 ice lollies.

3 The table shows the number of litres of petrol left in a car's petrol tank on a journey.

Travelling time in hours	1	2	3	4	5	6	7	8
Number of litres left	55	50	45	40	35	30	25	20

a Draw a graph from the information given in the table.
b How many litres were in the tank at the start of the journey (after 0 hours)?
c How many litres were in the tank after $5\frac{1}{2}$ hours?

4 A car uses 2 litres of petrol for every 5 km it travels.
a Copy and complete the table showing how much petrol the car uses.

Distance travelled in km	0	5	10	15	20	25
Petrol used in litres	0	2	4			

b Draw a graph from the information in your table.
c Work out how much petrol is used to travel 4 km.
d Work out how many kilometres have been travelled by the time 15 litres of petrol have been used.

5 The water in a reservoir is 144 m deep. During a dry period the water level falls by 4 m each week.
a Copy and complete this table showing the expected depth of water in the reservoir.

Week	0	1	2	3	4	5	6	7	8
Expected depth of water in m	144	140							

b Draw a graph from the information in your table.
c How deep would you expect the reservoir to be after 10 weeks?
If the water level falls to 96 m, the water company will divert water from another reservoir.
d After how long will the water company divert water?

F

6   Sally has a Pay as you go mobile phone. She pays 40p for each minute she uses her phone.

   a   Copy and complete this table of values for the cost of using Sally's phone.

Minutes used	0	5	10	15	20	25	30	35	40	45	50
Cost in pounds	0		4								20

   b   Plot the points in the table on a coordinate grid and draw a graph to show the cost of using Sally's phone.

   c   Use your graph to find the cost of using her phone for 32 minutes.

   d   One month Sally paid £8.40 to use her phone. For how many minutes did Sally use her phone that month?

7   Bob has a contract phone. He pays £15 each month and then 10p for each minute he uses his phone.

   a   Copy and complete this table of values for the cost of using Bob's phone.

Minutes used	0	5	10	15	20	25	30	35	40	45	50
Cost in pounds	15		16								20

   b   Plot the points in the table on a coordinate grid and draw a graph to show the cost of using Bob's phone.

   c   Use your graph to find the cost of Bob using his phone for 32 minutes.

   d   One month Bob paid £17 to use his phone. For how many minutes did Bob use his phone that month?

8   Kieran buys his gas from a company that charges 50p for each unit of gas he uses.

   a   Copy and complete this table of values for the cost of gas used by Kieran.

Units used	0	10	20	30	40	50	60	70	80	90	100
Cost in pounds	0		10								50

   b   Plot the points in the table on a coordinate grid and draw a graph to show the cost of using gas.

   c   Use your graph to find the cost of using 32 units of gas.

   d   One month Kieran paid £45 for gas. How many units of gas did Kieran use that month?

E

9   Jamie buys his electricity from a company that charges £20 each month and then 25p for each unit of electricity he uses.

   a   Copy and complete this table of values for the cost of using electricity for Jamie.

Units used	0	10	20	30	40	50	60	70	80	90	100
Cost in pounds	20		25								45

   b   Plot the points in the table on a coordinate grid and draw a graph to show the cost of using electricity.

   c   Use your graph to find the cost of using 32 units of electricity.

   d   One month Jamie paid £38 for electricity. How many units of electricity did Jamie use that month?

10 The graph shows the cost of using a mobile phone for one month on three different tariffs.

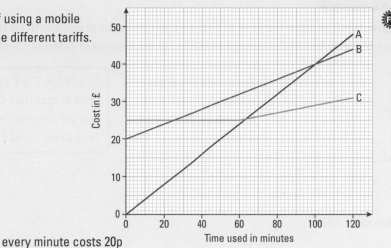

The three tariffs are

Tariff 1   Rental £20   every minute costs 20p
Tariff 2   Pay as you go   every minute costs 40p
Tariff 3   Rental £25   first 60 minutes free then each minute costs 10p

a Match each tariff with the letter of its graph.

Fiona uses her mobile phone for about 60 minutes each month.

b Explain which tariff would be the cheapest for her to use.
You must give the reasons for your answer.

11 Jodie wants to buy a new phone. She has a choice of one of these tariffs.

Tariff	Monthly payment	Cost of calls per minute
A	£0	35p
B	£10	20p
C	£20	10p
D	£25	5p

Jodie uses her phone for about 50 minutes each month.
Which tariff should Jodie choose?

12.2 Drawing and interpreting conversion graphs

Objectives

- You can use a conversion graph to change one unit to another.
- You can draw a conversion graph to change one unit to another.

Why do this?

A conversion graph can help you convert money from one currency to another when you go on holiday.

Get Ready

One euro is worth 80 pence. This can be written as €1 = 80p or £0.80.

1. Write in pence or pounds   **a** €5   **b** €10   **c** €20   **d** €100
2. Write in euros   **a** £1.60   **b** £8   **c** £24   **d** £16

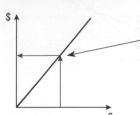

Key Point

● **Conversion graphs** are used to change measurements in one unit to measurements in a different unit. They can also be used to change between money systems in different countries.

To change £ to $ you read up from the £ to the line and then read across to the $.

To change $ to £ you read across from the $ to the line and then read down to the £.

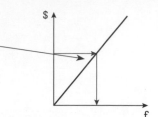

Exercise 12B

1 This graph can be used to change between pounds (£) and Hong Kong dollars.

   a Use the graph to change these amounts to Hong Kong dollars (HK$).
      i £10 ii £5 iii £8 iv £100 v £200

   b Use the graph to change these amounts to pounds.
      i HK$60 ii HK$30 iii HK$90
      iv HK$600 v HK$1200

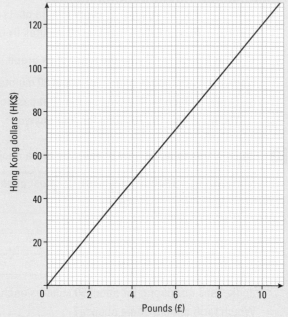

2 Copy the table and use the temperature conversion graph to complete it.

°C	5	20		28			35	80		40
°F			80		50	100			200	

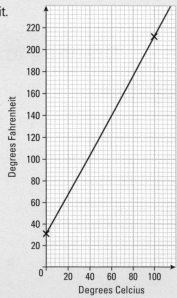

3   **a** Draw a conversion graph from pounds to kilograms. Use the fact that 0 pounds is 0 kilograms and 50 kg is approximately 110 pounds.

On your graph draw axes for kilograms and pounds using scales of 1 cm = 10 pounds and 1 cm = 10 kg. Plot the points (0, 0) and (50, 110) and join them with a straight line.

b Copy and complete this table using your conversion chart to help you.

Kilograms	0			45	30	15			35	50
Pounds	0	10	20				50	14		110

4   Copy this table and then use the information in the table to draw a conversion graph from inches into centimetres. Use your graph to help you fill in the missing values.

Inches	0	1	2				9	8		12
Centimetres	0			10	15	20			25	30

5   Copy this table and then use the information in the table to draw a conversion graph from miles into kilometres. Use your graph to help you fill in the missing values.

Miles	0	5		40		30			24	50
Kilometres	0		16		36		72	20		80

6   Copy this table and then use the information in the table to draw a conversion graph from acres into hectares. Use your graph to help you fill in the missing values.

Hectares	0			12	15	17			3	20
Acres	0	20	30				24	45		50

12.3 Drawing and interpreting distance–time graphs

Objectives

- You can interpret straight-line distance–time graphs.
- You can draw straight-line distance–time graphs.
- You can interpret distance–time graphs where the line is curved.

Why do this?

Graphs help us to understand information more easily.

Get Ready

1. Lauren travelled 90 miles in 3 hours. What was her average speed?
2. Anna travelled at 50 miles per hour for 2 hours. How far did she travel?
3. Idris travelled 100 miles at 50 miles per hour. How long did it take him?

Key Points

- On **distance–time graphs**:
  - time is always on the horizontal axis
  - distance is always on the vertical axis
  - a slanting line means movement is taking place
  - a horizontal line means no movement is taking place, the object is stationary.
- To work out speed you divide the distance travelled by the time taken. (Speed is covered in Section 17.5).

Example 3

Mary travels to work by bus.

She walks the first 750 metres in 10 minutes, waits at the bus stop for 5 minutes, then travels the remaining 3000 metres by bus. She arrives at the work bus stop 21 minutes after she set off from home.

  a Draw a distance–time graph of her journey.

  b Work out the average speed of the bus in kilometres per hour.

a

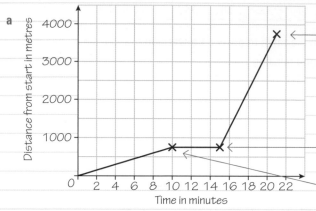

Plot the final point at (21 minutes, 750 + 3000 metres).

Plot the point (15 minutes, 750 metres).

Plot the point (10 minutes, 750 metres).

b In 6 minutes, 3000 metres travelled

In 60 minutes, 3000 × 10 = 30 000 m

30 000 ÷ 1000 = 30 km per hour

Join the points.
Waiting at the bus stop means that the line is horizontal for 5 minutes.

Example 4

This is a graph showing the journey made by an ambulance.

On the graph from 0 to A the ambulance travels 10 km in 10 minutes. From A to B the ambulance travels 20 km in 10 minutes. From B to C the ambulance does not go anywhere for 5 minutes. The 30 km journey back to base takes 15 minutes.

Work out the speed of the ambulance for each part of the journey.

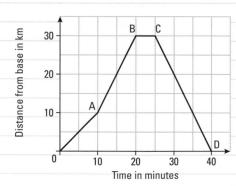

0 to A

10 km in 10 minutes

In 60 minutes 60 km could be travelled

Speed = 60 km per hour

A to B

20 km in 10 minutes

In 60 minutes 120 km could be travelled

Speed = 120 km per hour

B to C

0 km in 5 minutes

Speed = 0 km per hour

C to D

30 km in 15 minutes

In 60 minutes 120 km could be travelled

Speed = 120 km per hour

Example 5

The distance fallen by a stone when it is dropped from a cliff is shown on this graph.

a What distance did the stone fall in 2 seconds?

b How long did the stone take to fall 32 metres?

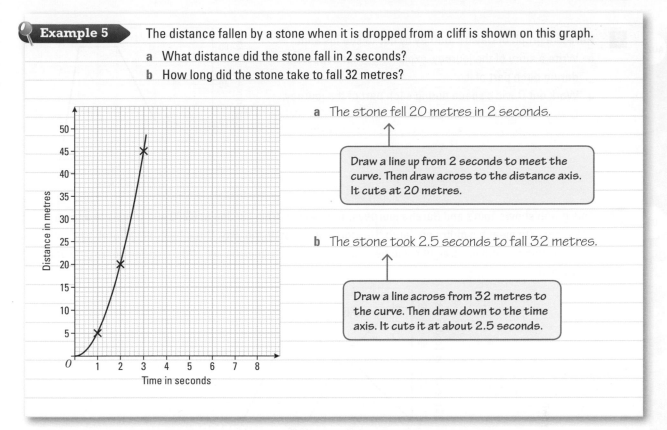

a The stone fell 20 metres in 2 seconds.

Draw a line up from 2 seconds to meet the curve. Then draw across to the distance axis. It cuts at 20 metres.

b The stone took 2.5 seconds to fall 32 metres.

Draw a line across from 32 metres to the curve. Then draw down to the time axis. It cuts it at about 2.5 seconds.

Exercise 12C

1 Jane walked to the shops, did some shopping, then walked home again.

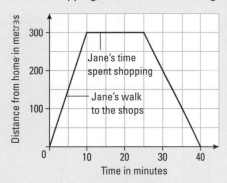

a How many minutes did it take Jane to walk to the shops?

b How far away were the shops?

c How many minutes did Jane spend shopping?

d How many minutes did it take Jane to walk home?

e Work out the speed at which Jane walked to the shops.
First give your answer in metres per minute, then change it to km per hour.

f Work out the speed at which Jane walked back from the shops.
First give your answer in metres per minute, then change it to km per hour.

C

A02

2   Here is a graph of David's journey by car to see his aunt.

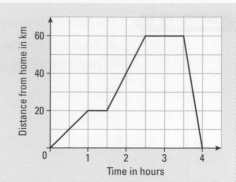

   a   Write a story of the journey, explaining what happened
       during each part of it.
   b   Work out David's speed during each part of the journey.
       Give your answers in km per hour.

A02

3   This graph shows Tom's and Sarah's journeys. Tom sets off from London at 08:00 and travels to a town
    90 km away to meet his girlfriend Sarah. He stops for a rest on the way. Once he gets to Sarah's he
    turns around and drives straight home because he discovers that she set off for London some time ago
    to see him.

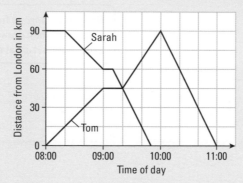

   a   Describe Tom's journey in detail, explaining after what distance he stopped on the way and for
       how long.
   b   Describe Sarah's journey in detail, explaining after what distance she stopped on the way and for
       how long.
   c   At what time did Sarah and Tom pass each other and what distance were they from London when
       they passed each other?

A02

* 4   Imran has a bath. The graph shows the depth of the bath water.

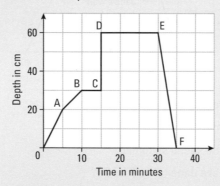

He starts at 0 by turning the hot and cold water taps on.
Between 0 and point A on the graph the depth of water goes up 20 cm in 5 minutes.
Explain what happens between points A and B, B and C, C and D, D and E, and E and F on the graph and
how long each part of the process lasts.

* 5 Daniel walked to the post box near his house to post a letter. It took him 4 minutes to walk to the post box, which was 400 m away. Daniel chatted to a friend for 2 minutes and he walked home in 3 minutes. Use graph paper to draw a distance–time graph for this journey.

A02

* 6 Kirsty took a trip in a hot air balloon. The balloon rose 400 metres in the air in one hour and stayed at this height for two and a half hours. The balloon then came back to the ground in half an hour. The hot air balloon company wants to display the journey as a distance–time graph. Draw a distance–time graph for this balloon flight.

A02

7 Annabel travels to school. She walks 250 metres to the bus stop in 4 minutes, waits at the bus stop for about 5 minutes and then travels the remaining 1000 metres by bus. She arrives at the bus stop outside the school 15 minutes after she sets off from home.
  a Draw a distance–time graph of the journey.
  b Work out the speed of the bus, first in metres per minute, then in km per hour.

A02

* 8 Mae went shopping by car. She drove the 10 miles to the shops in 30 minutes. She stayed at the shops for 30 minutes and then started to drive home. The car then broke down after 5 minutes when she had travelled 4 miles from the shops. It took 10 minutes to repair the car and another 5 minutes to get home. Draw a distance–time graph for Mae's journey.

A02
A03

9 Use the graph in Example 6 to find
  a the distance fallen by the stone in
    i 1.5 seconds   **ii** 3 seconds
  b the time taken for the stone to fall
    i 40 metres   **ii** 25 metres.

A03

10 Karen skis down a mountain. The graph shows her run.

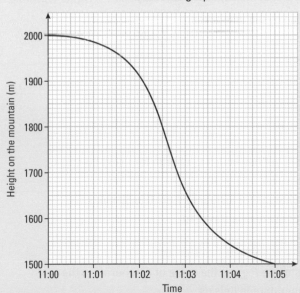

  a From the graph, write down the height Karen was at after
    i 1 minute   **ii** 2 minutes 30 seconds   **iii** 4 minutes 15 seconds.
  b Use the graph to write down the time at which Karen was at the following heights.
    i 1900 m   **ii** 1750 m   **iii** 1625 m

C

11 The speed of a ball when it is dropped is shown in the following table of values.

Distance in metres	0	5	10	15	20	25
Speed in metres per second	0	10	14	17	20	22

a Draw a graph using the information given in the table.
b Use the graph to work out the speed when the distance fallen is 12 metres.
c Use the graph to work out the distance fallen when the speed is 18 metres per second.

Chapter review

⦿ A straight-line graph that goes through the origin means that the more items you buy, the more it will cost you. The cost is related to the number of items you buy.

⦿ A straight-line graph that does not cross the vertical axis at the origin means there is a basic charge and then the cost is related to the number of items you buy.

⦿ **Conversion graphs** are used to change measurements in one unit to measurements in a different unit. They can also be used to change between money systems in different countries.

⦿ On **distance–time graphs**:
  ◉ time is always on the horizontal axis
  ◉ distance is always on the vertical axis
  ◉ a slanting line means movement is taking place
  ◉ a horizontal line means no movement is taking place, the object is stationary.

⦿ To work out speed you divide the distance travelled by the time taken.

 Review exercise

D

1 Dave drives a truck. He uses this rule to work out how much to charge for using his truck.
    Total charge (£) = number of miles travelled × 2 + 10
    a Draw a graph to show how much Dave charges for distances from 0 to 50 miles.
Nick also owns a truck. He charges £2.50 for every mile travelled.
    b When is it cheaper to use Nick's truck?

 2 The formula $F = 2C + 30$ can be used to estimate F given the value of C, where F is the temperature in Fahrenheit and C is the temperature in Celsius.
Copy and complete the table and use it to draw the graph of F against C for values of C from 0 to 100.

C	0	20	40	60	80	100
F			110			

3 The graph shows the cost, C, of a hiring a sander for d days from Hire It.

Find:

a a formula linking C with d

b the cost of hiring the sander for 10 days.

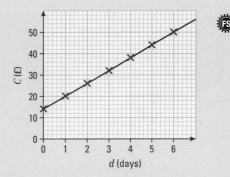

4 Josh is presented with a graph showing him a choice of two different mobile phone tariffs. The graph shows the cost, C, against the number of minutes, m, spent on his mobile in a particular month.

a Find the formulae of C against m for both tariffs.

b Explain in words how the tariffs are calculated.

c Advise Josh which scheme he should choose.

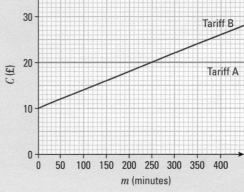

* **5** Abbie has the option of joining two health clubs:

Hermes has a joining fee of £100 plus a fee of £5 per session.

Atlantis has a joining fee of £200 with a fee of £3 per session.

Which health club should she choose?

You must show all calculations and fully explain your solution.

13 FORMULAE

You could use a formula to work out the distance you could drive before running out of fuel. It might be good idea if you were driving somewhere remote like Death Valley in the USA. Temperatures have been known to reach 57°C (134°F) in July and there is no shade. It is still known for people to die if they break down or run out of fuel in one of the most remote areas.

⊙ Objectives

In this chapter you will:
- use formulae
- write word formulae
- substitute numbers into expressions
- use and write algebraic formulae.

◁▷ Before you start

You need to know:
- the order of operations
- the difference between expressions, variables and terms.

13.1 Using word formulae

Objectives

- You can understand word formulae.
- You can substitute values into word formulae.
- You can evaluate word formulae.
- You can write a formula in words to represent a problem.

Why do this?

Word formulae can help you work out a value when you know all the other variables. For example, you can work out total pay when you know the hours worked and rate of pay.

Get Ready

Work out

1. $15 \times 25p$ **2.** $36 \times £5.90$ **3.** 3.14×12

Key Points

- A **word formula** uses words to represent a relationship between two quantities.
- When using formulae you need to remember the order of operations, BIDMAS (see Unit 2 Chapter 8).
- Formulae are used when calculating:
  - area of triangles (see Unit 2 Section 18.3)
  - angles in polygons (see Unit 3 Section 7.1)
  - area of circles (see Unit 3 Chapter 8).
- Make sure the units of measure are consistent when calculating using formulae.

Example 1

This word formula can be used to work out the perimeter of an equilateral triangle.

Perimeter = 3 × length of side

Work out the perimeter of an equilateral triangle with sides of 9 cm.

9 cm   9 cm

9 cm

Perimeter = 3 × 9 = 27 cm   ← Put 9 into the word formula for length of side.

Example 2

Suki is paid £6.50 per hour.

a Write a word formula for her weekly pay.

b Work out her pay for a week when she works 24 hours.

a Weekly pay = hourly rate × hours worked per week   ← A word formula must be in the form something = something else.

b Pay = £6.50 × 24   ← Put 24 and £6.50 into the formula. Do not forget the units.

= £156

Exercise 13A

Questions in this chapter are targeted at the grades indicated.

G

1 This word formula can be used to work out the perimeter of a regular pentagon.

Perimeter = 5 × length of side

Work out the perimeter of a regular pentagon with sides of 7 cm.

2 Gwen uses this word formula to work out her wages.

Wages = rate per hour × number of hours worked

Gwen's rate per hour is £5 and she works for 37 hours.

Work out her wages.

3 This word formula can be used to work out the area of a triangle.

Area = $\frac{1}{2}$ base × vertical height

A triangle has base 8 cm and vertical height 10 cm.

Work out the area of the triangle.

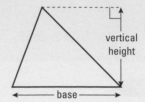

vertical height

base

F

4 This word formula can be used to work out the perimeter of a rectangle.

Perimeter = 2 × length + 2 × width

Work out the perimeter of a rectangle with a length of 8 cm and width of 3 cm.

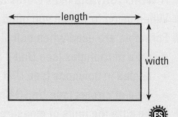

length

width

5 Owen uses this word formula to work out his phone bill.

Total bill = cost per minute × number of minutes + monthly charge

The cost per minute is 6p. Owen made 70 minutes of calls.

The monthly charge is £15. Work out his total bill.

6 Kirsty buys 12 stamps at 42p each.

a Write a word formula for the total cost of the stamps.

b Work out the cost of the stamps.

7 **a** Write a word formula to work out the number of packets of crisps left in a machine after a number have been sold.

b Use your formula to work out the number of packets left in a machine that holds 56 packets, when 27 have been sold.

8 Georgina uses this word formula to work out her take-home pay.

Take-home pay = rate per hour × number of hours worked − deductions

Georgina's rate per hour is £7. She worked for 40 hours and her deductions were £96.

Work out her take-home pay.

9 For each part write a word formula, then use it to calculate the answers.

a Alex shared a bag of sweets equally between herself and her three brothers. There were 84 sweets in the bag. How many did each person have?

b At a buffet lunch there were 36 slices of pizza. Every person at the lunch had 3 slices. All the slices were eaten. How many people were at the lunch?

10 This word formula can be used to work out the average speed for a journey.

Average speed = $\dfrac{\text{total distance travelled}}{\text{time taken}}$

Raj travels 215 miles in 5 hours. Work out his average speed in miles per hour (mph).

11 This word formula can be used to work out the angle sum, in degrees, of a polygon.
Angle sum = (number of sides − 2) × 180
Work out the angle sum of a polygon with 7 sides.

12 This word formula can be used to work out the size, in degrees, of each exterior angle of a regular polygon.

Exterior angle = $\dfrac{360}{\text{number of sides}}$

Work out the size of each exterior angle of a regular polygon with 8 sides.

13 This word formula can be used to work out the area inside a circle.
Area = π × radius × radius
Work out the area of a circle with a radius of 4 cm.
Give your answer in terms of π.

13.2 Substituting numbers into expressions

Objectives

- You can work out the value of an algebraic expression by substituting in the values of each letter or letters.
- You can substitute negative numbers into expressions.

Why do this?

You could work out pocket or birthday money if you get a certain amount of money per year depending on your age.

Get Ready

1. Use the word formula perimeter = 2 × length + 2 × width to work out the perimeter of a rectangle with:

    a length 4 cm, width 2 cm   b length 5 cm, width 3 cm   c length 7 cm, width 5 cm.

Key Point

- You can **substitute** values into an algebraic expression in the same way as you substitute values into a word formula.

Example 3 $a = 4$ and $b = 5$

Work out the value of

    a $3a - b$ **b** $3ab$ **c** $a^2 + 2b$

a $3a - b = 3 \times 4 - 5$ ←   The numbers replace the letters so $3a = 3 \times 4$ and $b = 5$.

       $= 12 - 5$ ←   Remember: multiplication before subtraction.

       $= 7$

b $3ab = 3 \times 4 \times 5 = 60$ ← $3ab = 3 \times a \times b$

c $a^2 + 2b = 4 \times 4 + 2 \times 5$ ← $a^2 = a \times a = 4 \times 4$

          $= 16 + 10$

          $= 26$

Exercise 13B

1 $p = 3$, $q = 2$, $r = 5$ and $s = 0$

Work out the value of the following expressions.

  a $p + r$ **b** rs **c** $5p - 2r$

  d pqr **e** $pr - pq$ **f** $6pq + 3qr$

  g $4(p + 7)$ **h** $p(q + 4)$ **i** $p(r - q)$

  j $r^2 + 1$ **k** $(q + 1)^2$ **l** $2p^3$

  m $(p + r)^2$ **n** $(p - q)^3$

2 $p = \frac{3}{4}$, $q = \frac{1}{4}$, $r = 2$ and $s = 1$

Work out the value of the following expressions.

  a $4q$ **b** $7qr$ **c** $4p - 6q$

  d qrs **e** $qr - pq$ **f** $4qr - 5pq$

  g $6(r - 2)$ **h** $r(r - 2)$ **i** p^2

  j $3r^3 - 2$ **k** $(r - 4)^2$ **l** $6q^3$

  m $(p + q)^3$

3 $p = 0.5$, $q = 2$, $r = 3$ and $s = 1.25$

Work out the value of the following expressions.

  a pq **b** $6p + 4q$ **c** $7p + 8s$

  d $pr + 4q$ **e** $qr + rs$ **f** $6pr - 7qs$

  g $5(p + q)$ **h** $q(p + r)$ **i** $5p^2$

  j $5q^2 + 7$ **k** r^3 **l** $p^2 + r^2$

  m $p^3 - q^3$

Example 4 $p = 2$, $q = -3$ and $r = -5$

Work out the value of

| **a** $p + r$ | **b** $q - r$ | **c** pq | **d** qr |
| **e** $p(q + r)$ | **f** $r^2 + 6r$ | **g** $(p + q)^2$ | **h** $4r^3$ |

a $p + r = 2 + (-5)$
$\quad = 2 - 5$
$\quad = -3$

> Adding a negative number is the same as subtracting a positive number.

b $q - r = -3 - (-5)$
$\quad = -3 + 5$
$\quad = 2$

> Subtracting a negative number is the same as adding a positive number.

c $pq = 2 \times -3$
$\quad = -6$

> When you multiply two numbers which have different signs the answer is negative. Remember: 2 means $+2$.

d $qr = -3 \times -5$
$\quad = 15$

> When you multiply two numbers which have the same sign the answer is positive.

e $p(q + r) = 2 \times (-3 + -5)$
$\quad = 2(-8)$
$\quad = -16$

> A bracket means times.
> Work out the value of the bracket first.

f $r^2 + 6r = (-5)^2 + 6 \times -5$
$\quad = (5 \times -5) - 30$
$\quad = 25 - 30$
$\quad = -5$

> Square and multiply before adding.

g $(p + q)^2 = (2 + (-3))^2$
$\quad = (2 - 3)^2$
$\quad = (-1)^2$
$\quad = -1 \times -1$
$\quad = 1$

> Work out the value of the bracket first.

> Square.

h $4r^3 = 4 \times (-5)^3$
$\quad = 4 \times (-5 \times -5 \times -5)$
$\quad = 4 \times -125$
$\quad = -500$

> $(-5)^3$ means cube of -5.
> Cube before you multiply.

Exercise 13C

In this exercise $a = -5$, $b = 6$, $c = -2$, $d = \frac{1}{2}$ and $e = 1$.

Work out the value of the following expressions.

1	$a + b$	2	$a - b$	3	$b - a$	4	$a - c$
5	$b - c$	6	$a + b + c$	7	$3a + 7$	8	$4a + 3b$
9	$2b + 5c$	10	$2a - 5c$	11	$3b - 2a$	12	ab
13	acd	14	$3bc$	15	$bd - 1$	16	$ab - bc$
17	$2ab + 3ac$	18	$3ac - 2bc$	19	$abcd$	20	$3d(a + 1)$
21	$b(c - a)$	22	$c(a + b)$				
23	$5(c - 1)$	24	a^2	25	$3c^2d$	26	$4a^2 - 3$
27	$(a + 1)^2$	28	$(a + b)^2$	29	$(c - a)^2$	30	$2b^3$
31	$6c^3$	32	$2(b + c)^2$	33	$a^2 - b^2$	34	$(e - d)^2$

13.3 Using algebraic formulae

◎ Objective

○ You can substitute number values for the letters in a formula to work out a quantity.

⬦ Why do this?

If you are given an algebraic formula to work out your pay, you need to know how to use it!

⬦ Get Ready

$a = -5$, $b = 6$, $c = \frac{1}{2}$

Work out

1. $a + (b \times c)$

2. abc

3. $\dfrac{(a + b)}{c}$

🌐 Key Point

● An **algebraic formula** uses letters to show a relationship between quantities, e.g. $A = lw$.

The letter that appears on its own on one side of the $=$ sign and does not appear on the other side is called the **subject** of the formula.

In the formula above, A is the subject of the formula.

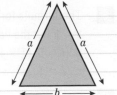

Example 5 The formula for the perimeter of this isosceles triangle is $P = 2a + b$.
Work out the value of P when $a = 8$ and $b = 5$.

$P = 2 \times 8 + 5 = 16 + 5 = 21$ ← Put $a = 8$ and $b = 5$ into the formula to find P.

Example 6 Harry's pay is worked out using the formula $P = hr + b$
where P = pay
h = hours worked
r = rate of pay
b = bonus.
Work out Harry's pay when $h = 35$, $r =$ £10 and $b =$ £45.

A03

$P = hr + b$
$P = 35 \times 10 + 45$
$P = 350 + 45$ ← Substitute $h = 35$, $r = 10$ and $b = 45$.
$P =$ £395

Exercise 13D

1. $y = 2x + 3$ is the equation of a straight line.
Work out the value of y when
   a $x = 4$
   b $x = 6$
   c $x = 10$
   d $x = 7.5$

F

2. The formula for the area of a parallelogram is $A = bh$.
Work out the value of A when
   a $b = 7$ and $h = 3$
   b $b = 9$ and $h = 7$
   c $b = 6$ and $h = 3.7$
   d $b = 8.4$ and $h = 4.5$

E

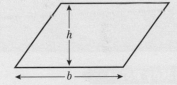

3. The formula for the circumference of a circle is $C = \pi d$.
Work out, to the nearest whole number, the value of C when
   a $d = 2$
   b $d = 5$
   c $d = 3.7$
   d $d = 9.3$

4. Euler's formula for the number of edges of a solid is $E = F + V - 2$, where E is the number of edges,
F is the number of faces and V is the number of vertices.
Work out the value of E when
   a $F = 6$ and $V = 8$
   b $F = 16$ and $V = 19$

5. The formula $F = 1.8C + 32$ can be used to convert a temperature from degrees Celsius to degrees
Fahrenheit. Work out the value of F when
   a $C = 10$
   b $C = 100$
   c $C = -30$
   d $C = 0$

D

6 The formula for the volume of a cuboid is $V = lwh$.
Work out the value of V when
a $l = 5, w = 4$ and $h = 2$ b $l = 8, w = 5$ and $h = 3.5$
c $l = 10, w = 6$ and $h = 4$ d $l = 9.3, w = 4.2$ and $h = 5.1$

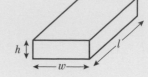

7 The formula $v = u + at$ can be used to work out velocity.
Velocity means speed in a particular direction.
Work out the value of v when
a $u = 8, a = 4$ and $t = 3$ b $u = 0, a = 10$ and $t = 2$
c $u = 7, a = 2.6$ and $t = 5$ d $u = 12, a = 10$ and $t = 4.7$

8 The formula $T = 15(W + 1)$ can be used to work out the time needed to cook a turkey.
Work out the value of T when
a $W = 5$ b $W = 8$ c $W = 12$ d $W = 18$

9 The formula for the volume of a cuboid is $V = l^2h$. Work out the value of V when
a $l = 2$ and $h = 5$ b $l = 7$ and $h = 9.8$ c $l = 2.5$ and $h = 4.3$

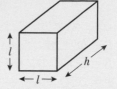

13.4 Writing an algebraic formula to represent a problem

◎ Objective

○ You can form a simple formula including squares, cubes and roots.

⑦ Why do this?

You could write a formula to work out the total cost of your holiday including air fares, hotel costs and entertainment costs, then you could alter any one of these to find how it affects the total cost.

⬆ Get Ready

Use the formula $C = \pi d$ to work out C when
1. $d = 3$ 2. $d = 8$ 3. $d = 10$

🔑 Key Point

◉ You can use information given in words to write an algebraic formula to solve a problem.

🔍 Example 1

Florence's pay is worked out using the formula
pay = number of hours work × rate per hour + commission.

a Write this as an algebraic formula.
b Work out Florence's pay when she works 30 hours at £7.50 per hour and gets a commission of £20.

a $P = nr + c$ ⟵ | Use letter for pay $= P$, number of hours worked $= n$, rate per hour $= r$ and commission $= c$.

b $P = nr + c$ ⟵ | Substitute $n = 30, r = £7.50$ and $c = 20$.
$P = 30 \times 7.50 + 20$
$P = 225 + 20$
$P = £245$

Exercise 13E

1 Write a formula for the perimeter P of this regular hexagon, with side l.
Work out the value of P when
   a $l = 3$ **b** $l = 7$
   c $l = 29$ **d** $l = 8.6$

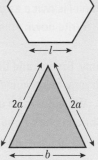

2 Write a formula for the perimeter of this isosceles triangle.
Work out the perimeter when
   a $a = 6$ and $b = 4$ **b** $a = 12$ and $b = 7$
   c $a = 5.3$ and $b = 3.4$ **d** $a = 4.7$ and $b = 8.5$

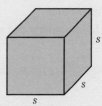

3 **a** Write an algebraic formula for the price of a number of pens that cost 70p each.
   b Use your formula to work out the cost of:
      i 4 pens   **ii** 6 pens   **iii** 12 pens.

4 Write a formula for the volume of this cube.
Work out the volume when
   a $s = 2\,\text{cm}$ **b** $s = 4.5\,\text{cm}$

5 Write a formula for the surface area of the cube in question 4.
Work out the surface area when
   a $s = 2\,\text{cm}$ **b** $s = 4.5\,\text{cm}$

6 Write a formula for the length of the side of a square given the area.
Work out the length of the side when the area is
   a $4\,\text{cm}^2$ **b** $1.44\,\text{cm}^2$

Chapter review

- A **word formula** uses words to represent a relationship between two quantities.
- When using formulae you need to remember the order of operations, BIDMAS.
- Formulae are used when calculating:
  - area of triangles
  - angles in polygons
  - area of circles.
- Make sure the units of measure are consistent when calculating using formulae.
- You can **substitute** values into an algebraic expression in the same way as you substitute values into a word formula.
- An **algebraic formula** uses letters to show a relationship between quantities, e.g. $A = lw$.
  The letter that appears on its own on one side of the $=$ sign and does not appear on the other side is called the **subject** of the formula. In the formula above, A is the subject.
- You can use information given in words to write an algebraic formula to solve a problem.

⚙ **Review exercise**

F

A03

1 Nathan is three years younger than Ben.
   a Write down an expression for Nathan's age.
   Daniel is twice as old as Ben.
   b Write down an expression for Daniel's age.

Nov 2008

A03

2 Amy, Bryony and Christina each collect bracelet charms.
   Bryony has twice as many charms as Amy.
   a Write down an expression for the number of charms that Bryony has.
   Christina has 7 charms less than Amy.
   b Write down an expression for the number of charms that Christina has.

A03

3 Adam, Brandon and Charlie each buy some stamps.
   Brandon buys three times as many stamps as Adam.
   a Write down an expression for the number of stamps Brandon buys.
   Charlie buys 5 more stamps than Adam.
   b Write down an expression for the number of stamps Charlie buys.

E

4 To calculate the cost of printing leaflets for a school fair, the printer uses the formula:
   $C = 40 + 0.05n$
   where C is the cost in pounds and n is the number of leaflets printed.
   a How much would it cost to print 200 leaflets?
   b Can you suggest what 40 and 0.05 represent?

5
   a Write a formula for the area of a square of side l.
   b Work out the area when $l = 9$.

6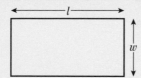
   a Write a formula for the perimeter of a rectangle.
   b Work out the perimeter when
      i $l = 9$ and $w = 4$
      ii $l = 6.7$ and $w = 3.4$

D

7 If $a = 1$, $b = 2$ and $c = -3$, find the value of
   a $\dfrac{c - ab}{c + ab}$ b $3(a + b)^2 - 2(b - c)^2$

8 $a = 4$, $b = \frac{1}{4}$, $c = -3$

Work out

a $\dfrac{5a}{b} + 7$

b $6a + \dfrac{2c}{3}$

c $3a - 6b + c$

d $a(a - 8b)$

9 $s = ut + \frac{1}{2}at^2$ is a formula for working out the distance, s, moved by an object.

Work out s when

a $u = 4.2$, $a = 10$ and $t = 3$

b $u = 5$, $a = -10$ and $t = 5.7$

c $u = -3$, $a = -32$ and $t = 6$

10 $A = \frac{1}{2}bh$ is the formula for working out the area of a triangle. Work out the area of a triangle when

a $b = 30\,\text{cm}$ and $h = 20\,\text{cm}$

b $b = 15\,\text{cm}$ and $h = 26\,\text{cm}$

c $b = 7.3\,\text{cm}$ and $h = 2.9\,\text{cm}$

d $b = 2.3\,\text{cm}$ and $h = 1.3\,\text{cm}$

11 $v = u + at$ is a formula for finding the speed of an object.

Find v when

a $u = 6$, $a = 10$ and $t = 5$

b $u = 8$, $a = -10$ and $t = 6$

c $u = 20$, $a = -32$ and $t = 4\frac{1}{2}$

12 $p = 2$

Work out the value of $5p^2$.

13 $v = u + 10t$

Work out the value of v when

a $u = 10$ and $t = 7$

b $u = -2.5$ and $t = 3.2$

14   a   Work out the value of $2a + ay$ when $a = 5$ and $y = -3$.

      b   Work out the value of $5t^2 - 7$ when $t = 4$.

15   There are many factors that determine how long it would take to climb a mountain.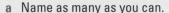

a   Name as many as you can.

In 1892 a Scottish mountaineer named William Naismith devised a rule which stated that you must allow 1 hour for every 5 km you walk forward, plus $\frac{1}{2}$ hour for every 300 metres of ascent.

Estimate, using Naismith's rule, how long it would take to

b   walk 10 km with 900 m of ascent   c   walk 20 km with 300 m of ascent

d   walk 12 km with 1000 m of ascent   e   walk 18 km with 450 m of ascent.

D

16 The cost of hiring a car can be worked out using this rule.

$$\text{Cost} = \text{£90} + \text{50p per mile}$$

Bill hires a car and drives 80 miles.

a Work out the cost.

The cost of hiring a car is C pounds.

b Write the formula for C.

Nov 2007

In this exercise $a = -5$, $b = 6$, $c = -2$, $d = \frac{1}{2}$ and $e = 1$.

Work out the value of the following expressions.

C

17 $a(b + c)$ **18** $5c^2 + 3c$ **19** $2a^2 - 3a$ **20** $(c + 3)^2$

21 $(a + c)^2$ **22** $3a^3$ **23** $(a - c)^3$ **24** $(c - 5d)^2$

14 ANGLES 1

Makers of fairground rides often make them more thrilling by sending the riders through a series of sharp bends and around loops at high speed. The turns in the track are angles and they can be measured in degrees.

⊙ Objectives

In this chapter you will:
- learn how to measure angles
- name different types of angles
- use angle facts to solve problems.

⟠ Before you start

You need to know:
- how to add and subtract numbers to 360
- how to draw lines, using a ruler, accurate to the nearest 2 mm.

14.1 Fractions of a turn and degrees

◎ Objective

○ You know that turns are measured in degrees.

◈ Why do this?

Understanding how turns are measured is a useful skill. You use fractions of a turn when map reading or orienteering.

◈ Get Ready

1. a 360 ÷ 4   **b** 60 × 3   **c** 180 ÷ 4   **d** 360 ÷ 2

● Key Points

◉ There are 360 **degrees** in a full turn. This is written as 360°.

◉ A half turn is 180°.

180°

◉ A quarter turn is 90°. A quarter turn is called a right angle.

90°

◉ A right angle is shown on diagrams with a small square.

⚙ Exercise 14A

Questions in this chapter are targeted at the grades indicated.

G

1 Write down the size of the following angles in degrees.

a

b

c

F

2 Write down how much a compass turns between:
a N and E
b E and SW
c SE and SW.

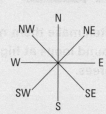

3 On a clock, how many degrees does the hour hand turn between:
a 2 pm and 8 pm   **b** 7 am and 10 am   **c** midnight and midday?

C

4 Karen goes for a walk. She walks 2 km due west and then 2 km due south.
She then walks back to the start by the shortest distance.
What compass bearing does she use to walk back to the start?

5   At 10 am Jean drives to a shop. It takes 5 minutes to drive to the shop.
    Jean is at the shop for 40 minutes. She then takes 5 minutes to drive home.
    On a clock, how many degrees has the minute hand turned from the time Jean drives to the shop to the
    time she gets home?

14.2 **What is an angle?**

⊙ Objectives

⊙ You know what an angle is.
⊙ You can identify acute, obtuse and reflex angles.

⊘ Why do this?

Many people use and measure angles, particularly
architects, designers and artists.

⬆ Get Ready

1. Write down how many degrees a compass point makes between:
   a S and W   b N and SE   c SW and NE.

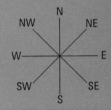

🔍 Key Points

⊙ An **angle** is a measure of turn.
An angle is formed when two lines meet.
These three angles are all the same
size. The length of the line and position of
the angle do not change the size.

⊙ There are different types of angle.

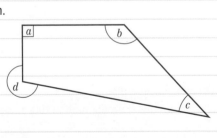

Acute angle   **Obtuse angle**   **Reflex angle**   **Right angle**
Less than a $\frac{1}{4}$ turn.   More than a $\frac{1}{4}$ turn.   More than a $\frac{1}{2}$ turn. $\frac{1}{4}$ turn.

🔍 Example 1   Name the different types of angle in this diagram.

a = right angle   ←   It is a $\frac{1}{4}$ turn.

b = obtuse angle   ←   It is more than a $\frac{1}{4}$ turn.

c = acute angle   ←   It is less than a $\frac{1}{4}$ turn.

d = reflex angle   ←   It is more than a $\frac{1}{2}$ turn.

Exercise 14B

1   Which angle is the odd one out?

a   b   c

2   Write down the special name for each of the following angles.

a   b   c

d   e   f

g   h   i

14.3 Naming sides and angles

Objective

- You can name sides and angles.

Why do this?

You can use letters to represent sides of shapes on a map or scale drawing.

Get Ready

1. Which angle is the odd one out?

a   b   c

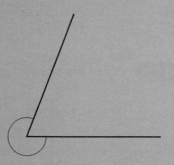

2. Draw an acute angle.
3. Write down the special name of this angle.

⊙ **Key Point**

◉ You can use letters to name the sides and angles of shapes.

  ◉ This shape is named ABCD using the letters for the corners and going round clockwise.

  ◉ Lines are named using the letters they start and finish with.

  ◉ Angles are named using the three letters of the lines that make the angle. The angle is always at the middle letter.

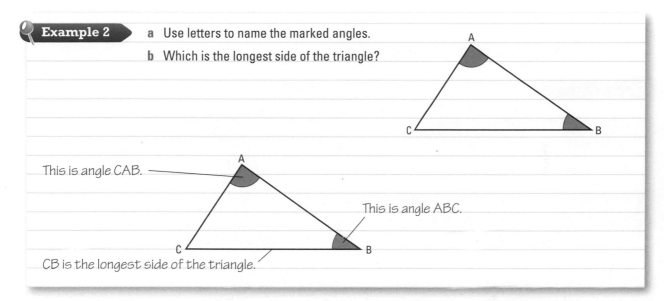

This is angle ABC.

This is line AD.

This is angle BCD.

This is line DC.

⊙ **Example 2**   **a** Use letters to name the marked angles.

b Which is the longest side of the triangle?

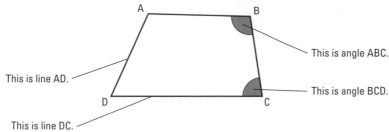

This is angle CAB.

This is angle ABC.

CB is the longest side of the triangle.

⚙ **Exercise 14C**

☐**1**   Write down the names of the marked angles in each of these diagrams.

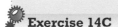

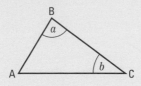

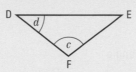

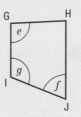

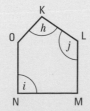

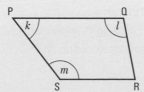

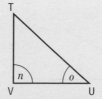

G

2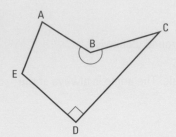

Write down the names of the

a acute angles

b right angle

c obtuse angle

d reflex angle

e longest side

f shortest side.

F

A02

3 Draw a sketch of triangle ABC, where AB is the longest side and angle BCA is an obtuse angle.

14.4 Estimating angles

Objective

○ You can estimate the size of angles.

Why do this?

Sports players need to estimate angles in order to pass the ball and score goals.

Get Ready

1. Write down the size of the following angles in degrees.

a

b

c

Key Point

● You should always estimate the size of angles before measuring them. This enables you to check that your answer is sensible.

Example 3

Estimate the size of this angle.

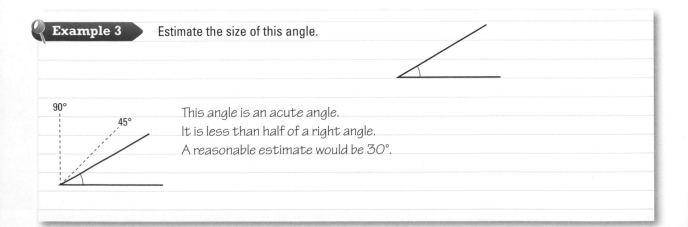

This angle is an acute angle.

It is less than half of a right angle.

A reasonable estimate would be 30°.

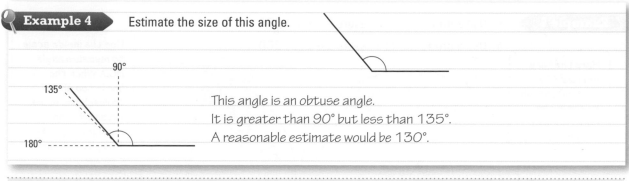

Example 4   Estimate the size of this angle.

This angle is an obtuse angle.
It is greater than 90° but less than 135°.
A reasonable estimate would be 130°.

Exercise 14D

Estimate the size of the following six angles.

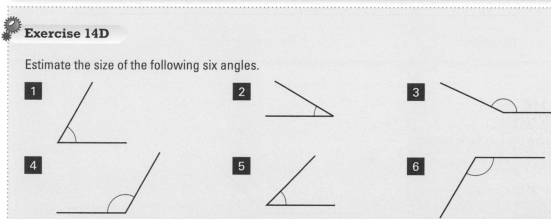

F

14.5 **Measuring angles**

Objective

● You can use a protractor to measure angles.

Why do this?

Landscape gardeners need to measure angles to produce accurate drawings of a garden design.

Get Ready

1. Estimate the size of the following angles.

a   b   c

Key Point

● A protractor is an instrument used to measure angles.

Use the inside scale to measure anticlockwise turns ↺

Use the outside scale to measure clockwise turns ↻

Place the cross at the point of the angle you are measuring.

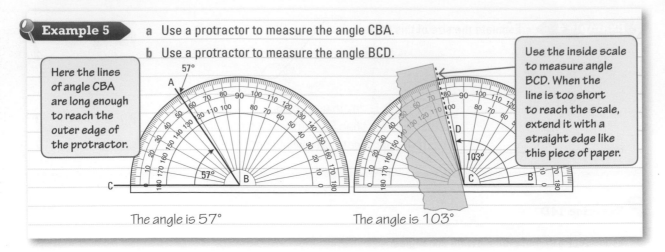

Example 5   **a** Use a protractor to measure the angle CBA.

           b Use a protractor to measure the angle BCD.

Here the lines of angle CBA are long enough to reach the outer edge of the protractor.

Use the inside scale to measure angle BCD. When the line is too short to reach the scale, extend it with a straight edge like this piece of paper.

The angle is 57°

The angle is 103°

Exercise 14E

1 Measure the following angles.

    a

    b

2 Measure the angles:

    a ABC
    b CDA
    c DAB.

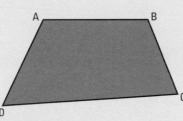

3 Measure the angles:

    a STU
    b UVR
    c RST.

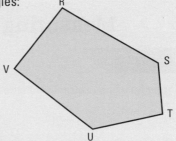

4 The table shows some information about the angles used in cutting different materials with a chisel.

Material	Aluminium	Medium steel	Mild steel	Brass	Copper	Cast iron
Cutting angle	30°	65°	55°	50°	45°	60°
Angle of inclination	22°	39.5°	34.5°	32°	29.5°	37°

The diagram shows the angle of inclination for the chisel.

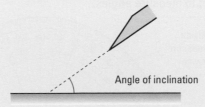

Angle of inclination

By measuring the angle shown in the diagram, write down the material being cut by the chisel.

14.6 Drawing angles

Objective

- You can use a protractor to draw angles.

Why do this?

A manufacturer of toys would have to be able to draw accurate angles and shapes when designing the toys.

Get Ready

1. Draw a line 6 cm long.   **2.** Draw a line 2.5 cm long.   **3.** Draw a line 8.3 cm long.

Key Point

- You will be expected to draw angles accurately. Angles must be accurate to within 2°.

Example 6   Draw an angle of 125°.

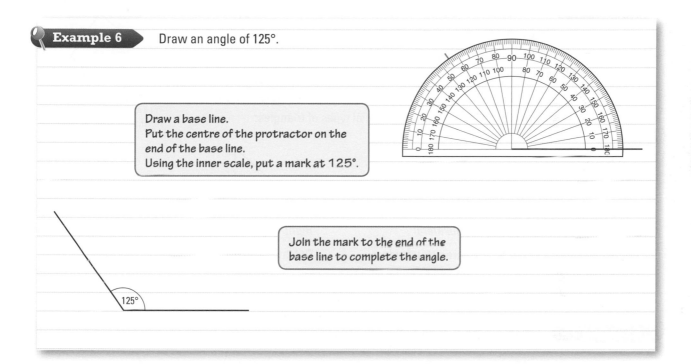

Draw a base line.
Put the centre of the protractor on the end of the base line.
Using the inner scale, put a mark at 125°.

Join the mark to the end of the base line to complete the angle.

125°

Exercise 14F

1   Use a protractor to draw the following angles.

  a 50°   **b** 160°   **c** 55°   **d** 115°   **e** 43°

  f 67°   **g** 117°   **h** 163°   **i** 17°   **j** 84°

2   Draw and label the following angles.

  a ABC = 30°   **b** DEF = 105°   **c** GHK = 65°

  d LMN = 48°   **e** PQR = 162°   **f** STU = 97°

F

14.7 Special triangles

Objectives

- You can identify right-angled, equilateral and isosceles triangles.
- You can find missing angles in triangles.

Why do this?

You can see examples of special triangles in fashion, construction and art.

Get Ready

1. Make an accurate drawing of this triangle.
2. Measure the two unmarked angles and the unmarked side on your diagram.
3. What do you notice about the sides and angles?

Key Points

- The **interior angles** of a triangle add up to 180°. You can see this if you cut out a triangle and tear the corners off as in the diagram.

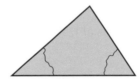

Tear these corners off.

Put all three corners together. They make a straight line which is an angle of 180°.

- You need to be able to recognise the following special types of triangles:

Isosceles

Equilateral

Right-angled

Two equal sides
Two equal angles

Three equal sides
All angles equal 60°

One angle 90°

Example 7 Work out the missing angle in this triangle.

The two given angles add up to 135°.
The missing angle is 180° − 135° = 45°.

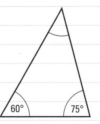

Example 8 Work out the missing angles in this triangle.

Triangle ABC has two equal sides.
ABC is an isosceles triangle.
ACB = ABC
ABC = 50°
CAB = 180° − (50° + 50°)
CAB = 80°

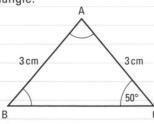

Exercise 14G

1 Work out the missing angles in the triangles below.

a
30°
80°

b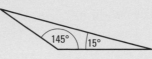
145° 15°

c
50°
87°

d
20°

e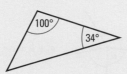
100° 34°

f
96°
13°

2 Work out the missing angles in the following triangles.

a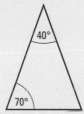
40°
70°

b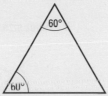
60°
60°

c
30°
80°

3 Use your answers to question 2 to write down the special name for each triangle.

4 Find angle ABC.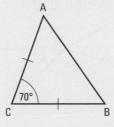

A
70°
C B

5 Find angles DEF and FDE.

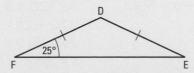
D
25°
F E

14.8 Angle facts

◉ Objective

○ You can use angle facts to work out missing angles.

? Why do this?

Triangles are used in the support structures of fairground rides.

⬆ Get Ready

Without using a calculator, write down the answers to the following questions.

1. a $180 - 40$ **b** $180 - 67$ **c** $180 - 132$
2. a $360 - 47$ **b** $360 - 108$ **c** $360 - 247$

🔍 Key Points

◉ The **angles on a straight line** add up to 180°.

 $a + b = 180°$

◉ The **angles around a point** add up to 360°.
◉ Where two straight lines cross, the **opposite angles** are equal. They are called **vertically opposite angles**.

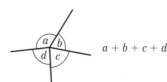 $a + b + c + d = 360°$

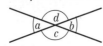

Angles a and b are the same.
Angles c and d are the same.

Example 9

a What size is angle a?

The angles make a straight line so
$58 + a = 180$
$a = 180 - 58 = 122°$

b What size is angle b?

The three angles make a straight line so
$45 + b + 67 = 180$
$b = 180 - 45 - 67 = 68°$

Example 10

a What size is angle a?

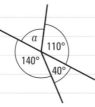

All the angles add up to 360°.
So, $140 + 40 + 110 + a = 360°$
$290° + a = 360°$
$a = 360 - 290 = 70°$

b What size is angle b?

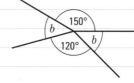

All the angles add up to 360°.
So, $120 + b + 150 + b = 360$
$270 + 2b = 360$
$2b = 360 - 270 = 90$
$b = 90 \div 2 = 45°$

angles on a straight line   angles around a point   opposite angles   vertically opposite angles

Example 11 Find all the missing angles in the diagram below. Give reasons for your answers.

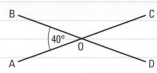

Results**Plus**
Examiner's Tip

Always remember to give reasons for your answers when you can.

Angle AOB = 40°
So, angle COD = 40° (vertically opposite angle AOB)
Angle AOD = 180 − 40 = 140° (the angles make a straight line)
So, angle BOC = 140° (vertically opposite angle AOD)

Exercise 14H

1 In each diagram, find the value of the letter.

2 Find the value of the letter in each of the following diagrams.

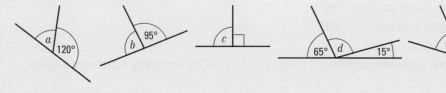

3 Find the value of the letters in the diagrams below.
Give reasons for your answers.

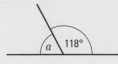

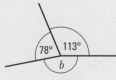

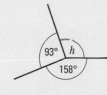

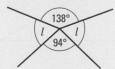

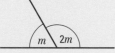

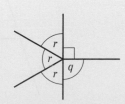

Chapter review

- There are 360° in a full turn.
- A half turn is 180°.
- A quarter turn is 90° and is called a **right angle**, shown with a small square.
- An **angle** is a measure of turn. An angle is formed when two lines meet.
- An angle that is less than a right angle is called an **acute angle**.
- An angle that is more than a quarter turn is called an **obtuse angle**.
- An angle that is more than a half turn is called a **reflex angle**.
- You can use letters to name the sides and angles of shapes.
- You should always estimate the size of angles before measuring them to check that your answer is sensible.
- A protractor is an instrument used to measure angles.
- You will be expected to draw angles. They must be accurate to within 2°.
- The **interior angles** of a triangle add up to 180°.
- An isosceles triangle has 2 equal angles and 2 equal sides.
- An equilateral triangle has 3 equal sides and 3 equal angles of 60°.
- A right-angled triangle contains an angle of 90°.
- The **angles on a straight line** add up to 180°.
- The **angles at a point** add up to 360°.
- When two lines cross, the **opposite angles** are equal. They are called **vertically opposite angles**.

Review exercise

G

1   Measure the length of the line AB.
    Give your answer in centimetres.

    A ————————————————————— B

    Nov 2008 adapted

2   What type of angle is this?

    June 2008

3   Here is a shape drawn on a square grid.
    On the shape
    a   Mark with the letter A an acute angle.
    b   Mark with the letter B an obtuse angle.

    March 2008 adapted

G

4 Here is a diagram drawn on a square grid.
  a Mark, with the letter A, an acute angle.
  b Mark, with the letter O, an obtuse angle.

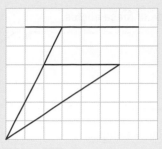

Nov 2007 adapted

5 **a** Draw a line 6 cm long.
    Start the line from a point labelled A.
  b Mark with a cross (×) the point on your line which is 2 cm from the point A. *June 2007*

F

6 **a** **i** Write down the value of x.
      ii Give a reason for your answer.

Diagram **NOT** accurately drawn

A03

  This diagram is **wrong**.
  b Explain why. *June 2008*

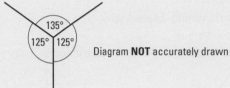

Diagram **NOT** accurately drawn

7 Work out the size of angle y.

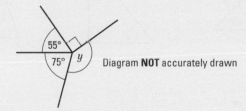

Diagram **NOT** accurately drawn

March 2008

8 **a** Measure the length of PQ.
    State the units with your answer.

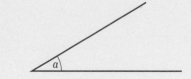

  b Measure the size of angle a.

June 2007

9

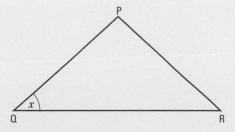

  a Measure the length, in centimetres, of QR.
  b Measure the size of angle x. *March 2007*

F

10 Estimate the size of these angles.

a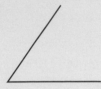

b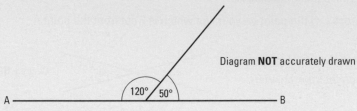

11 Draw and label these angles.
   a ABC = 50°
   b DEF = 170°
   c GHI = 37°
   d JKL = 90°
   e MNO = 143°
   f PQR = 77°

E

12

Diagram **NOT** accurately drawn

A ——————————— 120° 50° ——————————— B

AB is a straight line.
This diagram is wrong. Explain why.

Nov 2008

13 Work out the missing angles in these triangles.

a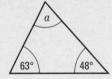

b

c

15 TWO-DIMENSIONAL SHAPES

BBC ACTIVE

Fashion designers may get the fame and fortune, but it is their pattern makers' painstaking work that is responsible for how well the clothes fit the models. They take the designer's sketches and make a pattern out of card. This pattern is then used to cut the right shapes out of the fabric and it is these shapes that are put together to make the creations that we see on the catwalk.

◎ Objectives

In this chapter you will:
- look at the properties of triangles
- learn about similar shapes
- look at the symmetry of various shapes
- look at the properties of quadrilaterals
- identify and name parts of circles
- construct circles.

◑ Before you start

You need to know:
- one-dimensional (1D) means it only has length
- two-dimensional (2D) shapes have area
- three-dimensional (3D) shapes have volume
- two lines are parallel if they can never meet
- two lines are perpendicular if they are at right angles to each other
- how to use a pair of compasses
- how to draw and measure angles and lines.

15.1 Triangles

○ You can identify different types of triangles.

Why do this?

People sometimes need to identify or describe the shape of different objects. For example, pool balls are racked in a triangle and some road signs are triangular.

Get Ready

1. Write down if each of these angles is acute, right-angled or obtuse.

 a   b   c

Key Points

◉ A **triangle** is any three-sided shape. Some triangles have special names.
Any triangle will have two special names: one that describes its sides and another that describes its angles.

Equilateral triangle

All three sides are the same length.
All three angles are 60°.

Isosceles triangle

Two of the sides are the same length.
Two of the angles are equal.

Scalene triangle

None of the sides or angles are equal.

Right-angled triangle

One of the angles is 90°.

Obtuse-angled triangle

One of the angles is obtuse (more than 90°).

Acute-angled triangle

All of the angles are acute (less than 90°).

Exercise 15A

Questions in this chapter are targeted at the grades indicated.

G

1 Match each triangle (A to F) with its mathematical names (1 to 6).
Each shape can be matched with two names.

A   B   C   D   E   F

(1) Scalene   (2) Isosceles   (3) Right-angled   (4) Equilateral   (5) Obtuse-angled   (6) Acute-angled

2

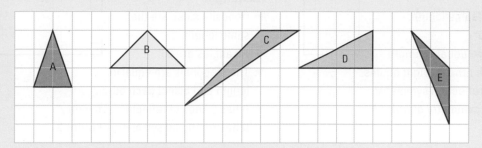

- **a** Two of these triangles are right-angled triangles. Which two?
- **b** Which of these triangles are isosceles triangles?
- **c** Two of these triangles are obtuse-angled triangles. Which two?
- **d** Which triangles are scalene triangles?

3 Draw a sketch of an obtuse-angled triangle that is also isosceles.

4 Jemma says this triangle is an isosceles triangle.
Jemma is wrong. Explain why.

A03

5 Clare draws an equilateral triangle.
Gerry says that the triangle she has drawn is an acute-angled triangle.
Gerry is correct. Explain why.

A03

15.2 **Quadrilaterals**

⊙ Objective

⊚ You know the properties of quadrilaterals.

⊘ Why do this?

Regular quadrilaterals are useful shapes. We see them in sports fields, windows and furniture.

⬥ Get Ready

Sketch the following triangles, marking on any special triangles.
1. Isosceles triangle   2. Right-angled triangle   3. Equilateral triangle

🔵 Key Points

⊚ A **quadrilateral** is any four-sided shape. There are some examples of common quadrilaterals on the following page.
⊚ The two **diagonals** of a quadrilateral go from one corner to the opposite corner.

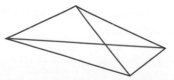

Square
- All sides are the same length.
- All angles are 90°.
- The diagonals are equal in length and **bisect** each other at right angles.
- It has 4 lines of reflection symmetry.
- It has rotational symmetry of order 4.

Rectangle
- Opposite sides are the same length.
- All angles are 90°.
- The diagonals are equal in length and bisect each other.
- It has 2 lines of reflection symmetry.
- It has rotational symmetry of order 2.

Parallelogram
- Opposite sides are parallel and are the same length.
- The diagonals bisect each other.
- It has no lines of reflection symmetry.
- It has rotational symmetry of order 2.

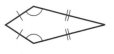

Trapezium
- One pair of opposite sides are parallel.

Kite
- Two pairs of **adjacent** sides are equal.
- One pair of opposite angles are equal.
- Diagonals cross each other at right angles.
- It has 1 line of reflection symmetry.

Rhombus
- All sides are the same length.
- Opposite angles are equal.
- Diagonals bisect each other at right angles.
- It has 2 lines of reflection symmetry.
- It has rotational symmetry of order 2.

Exercise 15B

G

1 Match each shape (A to H) with its mathematical name (1 to 6). Some shapes will have the same name.

A   B   C   D   E   F   G   H

1 Rhombus   2 Rectangle   3 Kite
4 Trapezium   5 Square   6 Parallelogram

F

2
a Write down the mathematical name for each type of quadrilateral.
b Which five of these shapes have two pairs of parallel sides?
c Which shape has just one pair of opposite angles the same size?

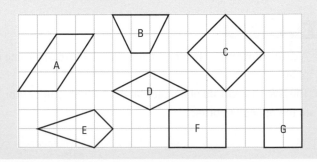

F

3 Name each of the following shapes.

   a This shape has four equal sides and its diagonals are the same length.

   b This shape has twice the number of sides as a quadrilateral.

   c This shape has one pair of parallel sides and its diagonals are not the same length.

   d This shape has half the number of sides as a hexagon and has one angle of 90°.

4 ABCD is a rhombus.

Which of the following statements are true and which are false?

   a AB = DC   **b** angle BAD = angle ADC

   c BC ∥ DA   **d** AC is perpendicular to BD

5 EFGH is a kite.

FH and EG meet at J.

   a Name a side equal in length to HG.

   b Is angle EJH the same size as angle GJH?

   c Name an angle equal to angle EFG.

   d Is triangle EFG a scalene triangle?

   e Write down the mathematical name for triangle FGH.

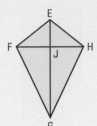

6 Which one of the following pairs of lines could be the diagonals of a parallelogram?

A03

   a   **b**    **c**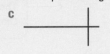

7 Pietro designs a company badge.

He is told that the badge must have 5 sides and must be made up of two different quadrilaterals and the longest side must be the same length as the total length of two different sides.

The diagram shows his design.

Copy the diagram and show the two different quadrilaterals.

Write down the mathematical names for the two different types of quadrilateral Pietro has used.

A02
A03

E

15.3 Similar shapes

Objective

- You can identify similar shapes.

Why do this?

You need to be able to identify similar shapes to play puzzle games, such as Tetris, or to complete jigsaw puzzles.

Get Ready

1. What features in common do these three squares have?

Key Points

- When one shape is an enlargement of another, the shapes are called **similar** shapes.
- The angles in two similar shapes are the same, but the lengths of the sides are not.

Example 1

Which of these shapes are similar to shape A?

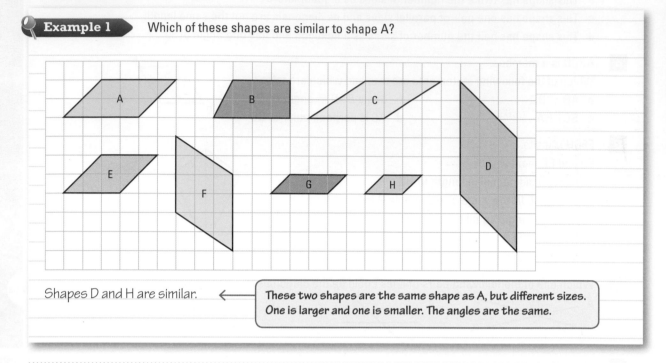

Shapes D and H are similar. ← These two shapes are the same shape as A, but different sizes. One is larger and one is smaller. The angles are the same.

Exercise 15C

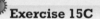

1 Write down the letters of the two pairs of shapes that are similar.

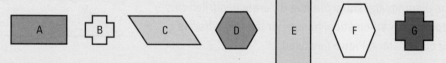

2 Which pairs of shapes are similar?

a

b

c

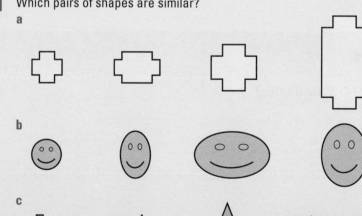

E

3   Copy the grid, and draw a shape that is similar to the shaded shape, but has been turned so that it is not the same way up.

15.4 Circles

◉ Objective

⊙ You can identify and name parts of a circle.

⑦ Why do this?

The circle is a particularly important shape. What wheels have you seen today?

🕐 Key Point

⊙ There are several key words associated with **circles**.

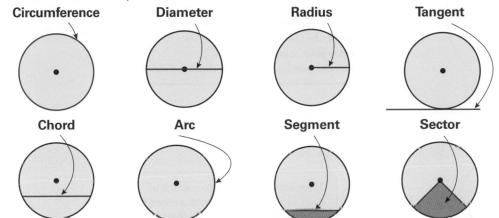

Circumference   **Diameter**   **Radius**   **Tangent**

Chord   **Arc**   **Segment**   **Sector**

⚙ Exercise 15D

Copy and complete the sentences below. Use the correct word chosen from the following list:

circumference   diameter   radius   tangent
chord   arc   segment   sector

G

1   A line through the centre of a circle that touches the circumference at each end is the

    ....

2   A line outside a circle that touches the circle at only one point is called a ....

3   A line that does not pass through the centre of a circle but touches the circumference at each end is

    called a ....

4   A line from the centre of a circle that is half the length of the diameter is the ....

5   A part of a circle that has a chord and an arc as its boundary is called a ....

6   A part of a circle that has two radii and an arc as its boundary is called a ....

15.5 **Drawing circles**

◎ **Objectives**

○ You can draw accurately a circle with a given radius.
○ You can draw accurately an arc of a given radius and angle.

⑦ Why do this?

Landscape gardeners use circles to represent plants and sculptures in their design plans.

⊕ Get Ready

1. Sketch 3 circles. On each of these circles draw

   a a radius   **b** a chord   **c** a sector.

🔍 **Example 2**   Draw a circle with a diameter of 8 cm.

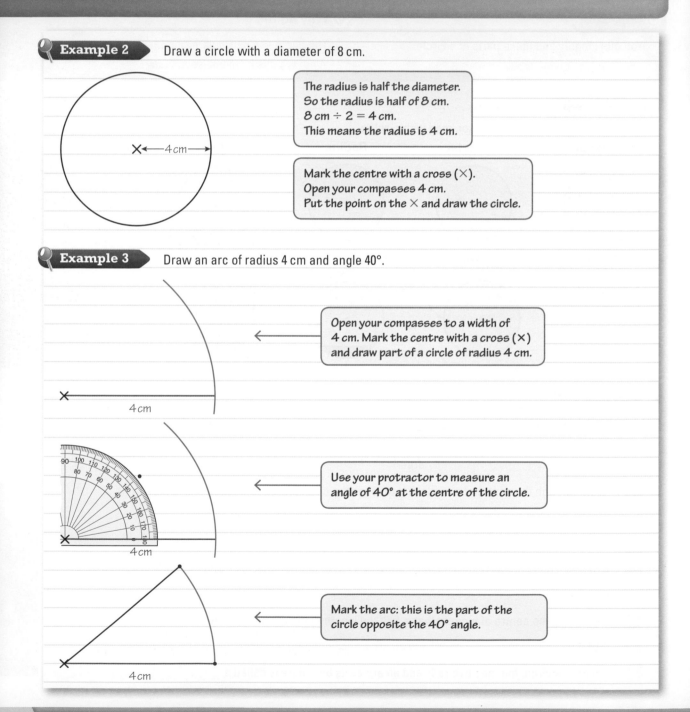

The radius is half the diameter.
So the radius is half of 8 cm.
8 cm ÷ 2 = 4 cm.
This means the radius is 4 cm.

Mark the centre with a cross (✕).
Open your compasses 4 cm.
Put the point on the ✕ and draw the circle.

🔍 **Example 3**   Draw an arc of radius 4 cm and angle 40°.

Open your compasses to a width of
4 cm. Mark the centre with a cross (✕)
and draw part of a circle of radius 4 cm.

Use your protractor to measure an
angle of 40° at the centre of the circle.

Mark the arc: this is the part of the
circle opposite the 40° angle.

Exercise 15E

1 Draw a circle of radius 5.3 cm.

2 Draw a circle with a diameter of 12 cm.

3 **a** Draw a circle of radius 6.7 cm.   **b** Shade a segment of your circle.

4 **a** Draw a circle of diameter 15.4 cm.   **b** Shade a sector of your circle.

5 **a** Draw a circle of diameter 11.6 cm.
    b On your circle, draw and label   **i** a radius   **ii** a chord   **iii** a tangent.

6 Draw arcs with the following measurements.
    a radius 3 cm, angle 30°   **b** radius 4 cm, angle 80°   **c** radius 5 cm, angle 60°
    d radius 4 cm, angle 75°   **e** radius 3 cm, angle 40°   **f** radius 4.5 cm, angle 65°

15.6 Line symmetry

◉ Objective

● You can understand line symmetry and identify and draw lines of symmetry on a 2D shape.

❓ Why do this?

Symmetry and balance tend to be closely related – this accounts for the symmetry of the human body.

◈ Get Ready

1. In each of these questions, does the red line in the middle act as a mirror?

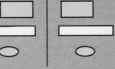 **a**   **b**   **c**

🔍 Key Points

● A shape is **symmetrical** if you can fold it in half and one half is the **mirror image** of the other half. The dividing line is called a **line of symmetry** or a **mirror line**.

● You can use tracing paper to help you. Trace the diagram and then fold it in half on the mirror line. You can then check if each half folds exactly onto the other half.

🔎 Example 4   Draw all the lines of symmetry for: **a** a kite;   **b** a rectangle.

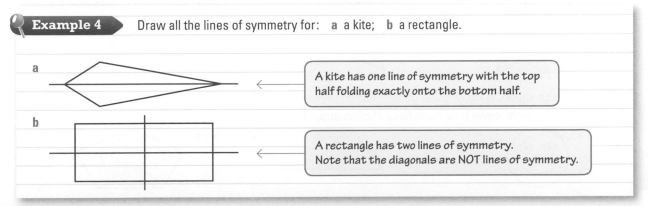

a

A kite has one line of symmetry with the top half folding exactly onto the bottom half.

b

A rectangle has two lines of symmetry.
Note that the diagonals are NOT lines of symmetry.

Example 5   Reflect the shaded shape in the mirror line.

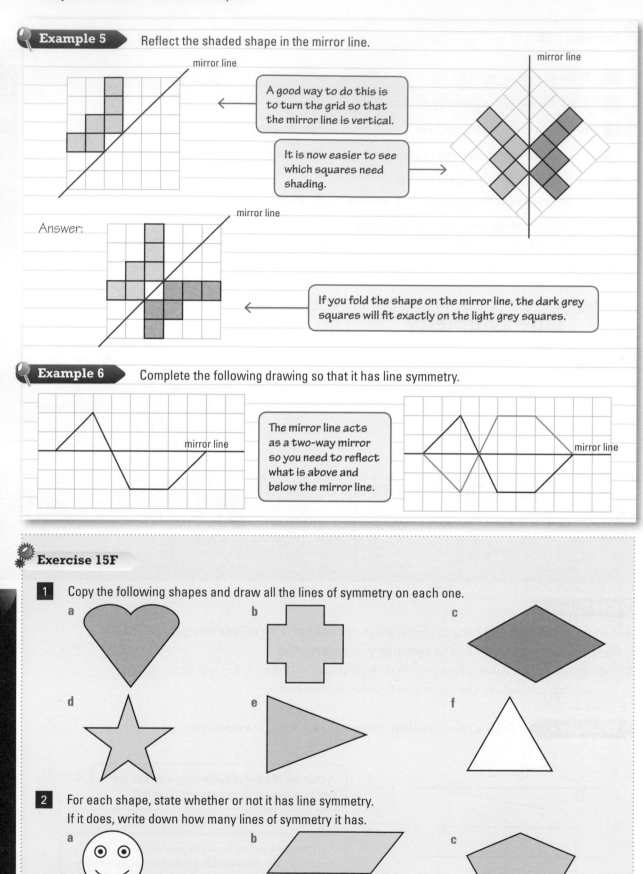

A good way to do this is to turn the grid so that the mirror line is vertical.

It is now easier to see which squares need shading.

Answer:

If you fold the shape on the mirror line, the dark grey squares will fit exactly on the light grey squares.

Example 6   Complete the following drawing so that it has line symmetry.

The mirror line acts as a two-way mirror so you need to reflect what is above and below the mirror line.

Exercise 15F

1 Copy the following shapes and draw all the lines of symmetry on each one.

a   b   c

d   e   f

2 For each shape, state whether or not it has line symmetry.
   If it does, write down how many lines of symmetry it has.

a   b   c

F

d   e 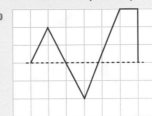 f

3 Copy and complete each drawing so that it has line symmetry.

a 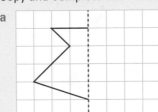 b 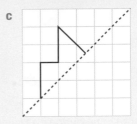 c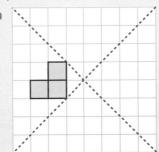

E

4 Copy and complete each drawing so that the final pattern is symmetrical about both lines.

a 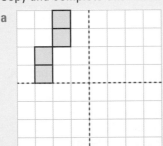 b

15.7 Rotational symmetry

◉ Objective

● You can understand rotational symmetry and can identify the order of rotational symmetry of a 2D shape.

❓ Why do this?

Rotational symmetry can be found all around us. For example, it occurs in the sails of a windmill or a hubcap on a car.

◈ Get Ready

1. How many degrees are there in

a a full turn

b a quarter turn

c a half turn?

🔍 Key Points

● To see if a shape has **rotational symmetry**, rotate it one full turn and see how many times along the rotation the shape still looks the same.

● When a rectangle is turned through 360° (one full turn) around its centre you can see that it looks the same twice:

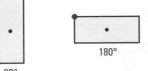

90°   180°   270°   360°

Therefore a rectangle has rotational symmetry of **order** 2.

Example 7 ▶ Write down the order of rotational symmetry of this shape.

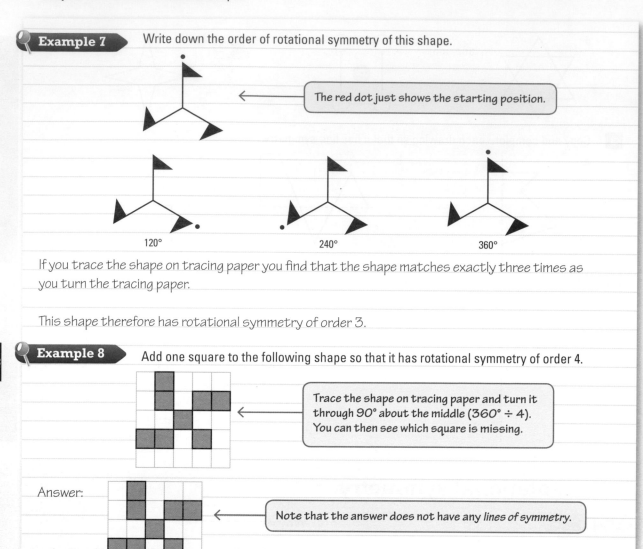

The red dot just shows the starting position.

120° 240° 360°

If you trace the shape on tracing paper you find that the shape matches exactly three times as you turn the tracing paper.

This shape therefore has rotational symmetry of order 3.

A03

Example 8 ▶ Add one square to the following shape so that it has rotational symmetry of order 4.

Trace the shape on tracing paper and turn it through 90° about the middle (360° ÷ 4). You can then see which square is missing.

Answer:

Note that the answer does not have any *lines of symmetry.*

Exercise 15G

F

1 For each letter, write down if it has rotational symmetry or not.
If it does, write down the order of rotational symmetry.

a **H** b **T** c **X** d **Z**

2 Write down the order of rotational symmetry for each of the following shapes.

a b c  d

3 On a copy of this grid, add one square to the shape so that it has rotational symmetry of order 2.

4 On a copy of this grid, add three squares to the shape so that it has rotational symmetry of order 4.

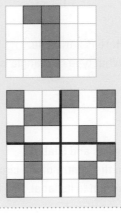

Exercise 15H

1 Copy and complete the following table.

Shape	Name of shape	Number of lines of symmetry	Order of rotational symmetry
		2	
	Equilateral triangle		
			2
	Hexagon		

Chapter review

- A **triangle** is any three-sided shape. Some triangles have special names: **equilateral**, **isosceles**, **scalene**, **right-angled**, acute-angled and obtuse-angled.
- A **quadrilateral** is any four-sided shape. Some examples of quadrilaterals are: **square**, **rectangle**, **parallelogram**, **trapezium**, **kite**, **rhombus**.
- The two **diagonals** of a quadrilateral go from one corner to the opposite corner.
- When one shape is an enlargement of another the shapes are called **similar** shapes.
- The angles in two similar shapes are the same, but the lengths of the sides are not.
- There are several key words associated with **circles**: **circumference**, **diameter**, **radius**, **tangent**, **chord**, **arc**, **segment** and **sector**.
- A shape is **symmetrical** if you can fold it in half and one half is the **mirror image** of the other half. The dividing line is called a **line of symmetry** or a **mirror line**.
- You can use tracing paper to help you. Trace the diagram and then fold it in half on the mirror line. You can then check if each half folds exactly onto the other half.
- To see if a shape has **rotational symmetry**, rotate it one full turn and see how many times along the rotation the shape still looks the same.

G

1 Here is a triangle.
What type of triangle is it?

Nov 2007

2 Here are 5 diagrams and 5 labels.
In each diagram the centre of the circle is marked with a cross (×).
Match each diagram to its label. One has been done for you.

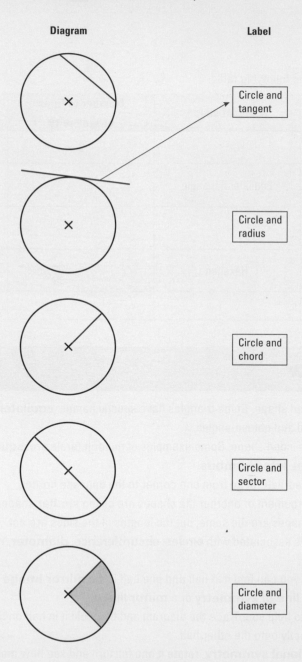

Diagram   Label

Circle and
tangent

Circle and
radius

Circle and
chord

Circle and
sector

Circle and
diameter

June 2008

3 **a** Here is a quadrilateral.
What type of quadrilateral is it?

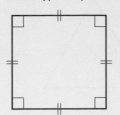

b On a copy of the grid, draw a trapezium.

Nov 2007

4 Here are some quadrilaterals.
Draw an arrow from each quadrilateral to its mathematical name.
The square has been done for you.

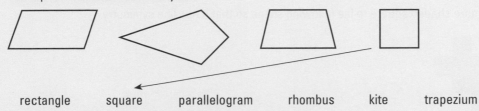

rectangle   square   parallelogram   rhombus   kite   trapezium   *June 2007*

5 Write down which of the triangles below are isosceles triangles.

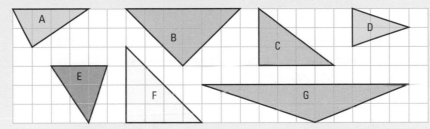

6 **a** On the diagram below, shade **one** square so that the shape has exactly **one** line of symmetry.

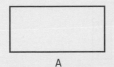

b On the diagram below, shade **one** square so that the shape has rotational symmetry of order 2.

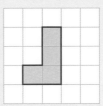

Nov 2008

7 Here are four shapes.

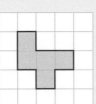

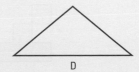

Write down the letter of the shape which has
i exactly **one** line of symmetry   **ii** **no** lines of symmetry   **iii** exactly **two** lines of symmetry.

Nov 2008

8   **a**   Reflect the shaded shape in the mirror line.

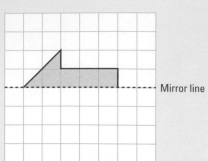

Mirror line

b   Draw the lines of symmetry on this triangle.

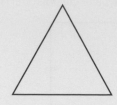

May 2008 adapted

9   Add one more shaded square to the following shape so that it has line symmetry.

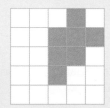

10   The shape below has one line of symmetry.
  a   On the grid, draw this line of symmetry.

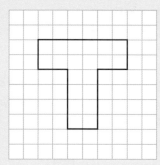

The shape below has rotational symmetry.
b   Write down the order of rotational symmetry.

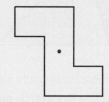

Nov 2007

11   **a**   On a copy of the diagram, shade **one** more square to make a pattern with 1 line of symmetry.

b   On a copy of the diagram, shade **one** more square to make a pattern with rotational symmetry of order 2.

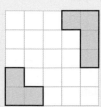

June 2007

12 The diagram shows part of a shape.
The shape has rotational symmetry of order 3 about the point P.
On a copy of the grid, complete the shape.

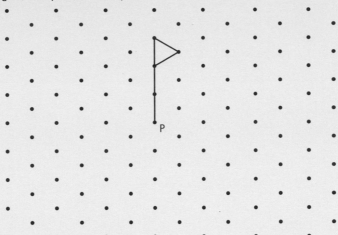

Nov 2006

13 a On a copy of this diagram, shade **one** more square so that the shape has exactly **one** line of symmetry.

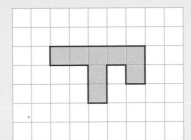

b On a copy of this diagram, shade **one** more square so that the shape has rotational symmetry of order **2**.

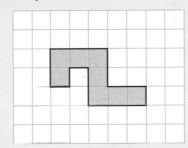

Nov 2006

14 Here are five shapes.

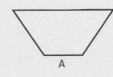
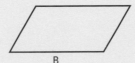
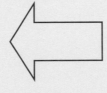
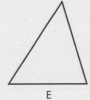

Two of these shapes have only **one** line of symmetry.
a Write down the letter of each of these two shapes.

Two of these shapes have rotational symmetry of order 2.
b Write down the letter of each of these **two** shapes.

June 2007

E

15 Add one more shaded square to the following shape so that it has rotational symmetry of order 2.

16 ANGLES 2

Rugby players need to think about angles when they are converting a try or taking a penalty; 45° is the optimum angle at which the player should kick the ball. Getting the ball between the posts is a trade-off between energy and distance travelled, and an angle of 45° gives enough height and force for the ball to clear the horizontal bar.

⊙ Objectives

In this chapter you will:
- learn how to use angle facts to solve problems
- learn how to demonstrate proofs about simple angle facts.

◈ Before you start

You need to:
- know how to find missing angles on a straight line and in triangles.

16.1 Angles in quadrilaterals

Objective

○ You can use angle facts to find missing angles in quadrilaterals.

Why do this?

Carpenters need to be able to work out angles in quadrilaterals so that they can work out the angles in joints.

Get Ready

1. Find the missing angles in the diagrams below.

a

b

c

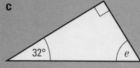

Key Point

⊙ The interior angles of a quadrilateral (a four-sided shape) always add up to 360°.

You can see this by measuring the angles…

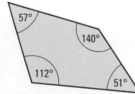

… or by dividing the quadrilateral into two triangles…

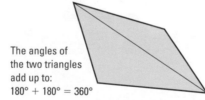

The angles of the two triangles add up to:
180° + 180° = 360°

… or by tearing off the four corners of a quadrilateral.

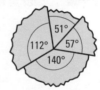

Put the angles together. They make a full turn of 360°.

Example 1

Find the missing angle in this quadrilateral.

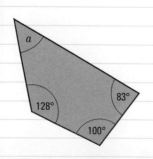

The total of the three angles marked = 311°
So a = 360° − 311°
a = 49°

Exercise 16A

Questions in this chapter are targeted at the grades indicated.

Find the missing angles in the following quadrilaterals.

1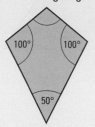

2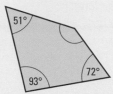

3

E

4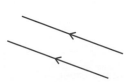

5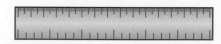

16.2 **Perpendicular and parallel lines**

◎ Objective

⦿ You know what perpendicular and parallel lines are and can mark them in diagrams.

⧉ Why do this?

Parallel lines are used on running tracks to make sure that runners stay in their lanes.

🕓 Key Points

⦿ Two lines are **perpendicular** to each other if they meet at a 90° angle.

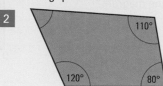

Lines AB and CD are perpendicular to each other.

Line FO is perpendicular to EG.

⦿ Lines that remain the same distance apart are called **parallel** lines. On diagrams this is shown by marking the parallel sides with arrows. If there is a second pair of parallel lines in one diagram these are marked with double arrows.

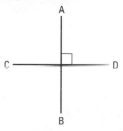

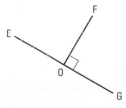

⦿ The distance between the two edges of a ruler is the same all the way along it. Similarly, the distance between the two rails of a train track is the same wherever it is measured.

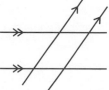

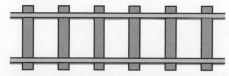

Example 2

a Look at the diagram.
Which sides are perpendicular?

b Which sides are parallel in the diagram?

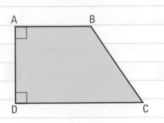

a Sides AB and AD meet at 90° so they are perpendicular to each other.
Sides AD and DC are perpendicular to each other as well.

b Sides AB and DC are parallel. They are the same distance apart all along their length.

Exercise 16B

1 a Find and name as many pairs of parallel lines as you can in the diagram.

b Now find and name as many pairs of perpendicular lines as you can in the diagram.

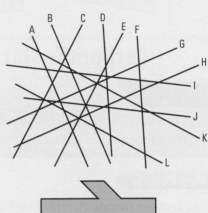

2 a Copy the diagram and mark three pairs of perpendicular lines.

b Now mark three pairs of parallel lines on your diagram.

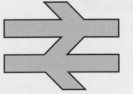

16.3 Corresponding and alternate angles

Objectives

○ You can find corresponding and alternate angles in diagrams.

○ You can use the angle facts of corresponding and alternate angles to find missing angles in diagrams.

Why do this?

It could be useful to be able to calculate angles accurately in Resistant Materials.

Get Ready

In each of the following questions find the value of a. Give a reason for your answer.

1.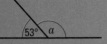

2.

3.

Key Points

○ The marked angles below are equal. They are called **corresponding angles**.

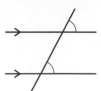 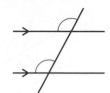 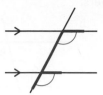

○ The marked angles below are equal. They are called **alternate angles**.

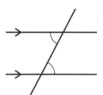 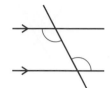 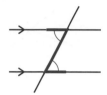

Example 3   **a** Find the size of angle a. Give a reason for your answer.

b Now find the size of angle b, giving a reason for your answer.

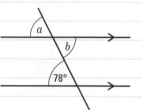

a Angle $a = 78°$ corresponding angles are equal.

b Angle $b = 78°$ alternate angles are equal.

Example 4   Find the size of the lettered angles, giving reasons for your answers.

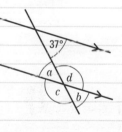

$a = 37°$ alternate angles are equal.
$b = 37°$ corresponding angles are equal.
$c = 143°$ angles on a straight line add up to 180°.
$d = 143°$ vertically opposite angles are equal.

 Exercise 16C

1   **a** In the diagram, which pairs of angles are corresponding angles?

b Which pairs of angles in the diagram are alternate angles?

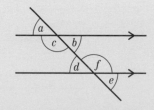

D

D

2 In the diagrams below, find the size of each angle marked with a letter.
Give reasons for your answers.

a

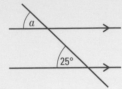

b

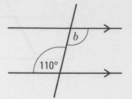

c

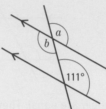

d

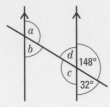

e

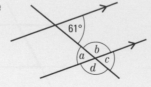

f

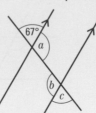

A02 **3** Find the size of the angles marked with a letter in the diagrams below.

a

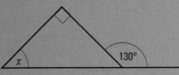

b

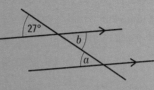

16.4 Proofs

◎ Objective

○ You can demonstrate proofs about angle facts of exterior angles, angle sums of a triangle and opposite angles of a parallelogram.

⓸ Why do this?

Proofs are used in science to prove that a law is true and that a conclusion is correct.

⓸ Get Ready

1. Find the value of angle x.

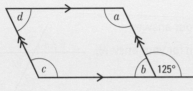

2. Write down the value of a and b. Give a reason for your answer.

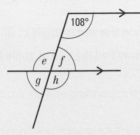

3. ABCD is a parallelogram.
Angle ABC = 81°
Write down the size of the other 3 angles.

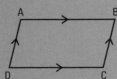

Key Points

⦿ **Proof 1**

Proving the exterior angle of a triangle is equal to the sum of the two interior opposite angles.

Through the point of angle e draw a line parallel to the opposite side of the triangle.

Angle e is now divided into two angles, c and d.

$a = c$ alternate angles

$b = d$ corresponding angles

So $a + b = c + d = e$.

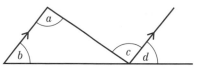

⦿ **Proof 2**

Proving the angle sum of a triangle is 180°.

The angles of the triangle are a, b and x.

$c + d + x = 180°$ angles on a straight line

But $c + d = a + b$ exterior angle = sum of interior opposite angles

Therefore $a + b + x = 180°$.

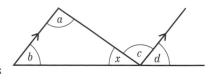

⦿ **Proof 3**

Proving the opposite angles of a parallelogram are equal.

$a = b$ corresponding angles

$b = c$ alternate angles

So $a = c$.

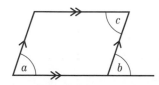

Chapter review

⦿ The interior angles of a quadrilateral (a four-sided shape) always add up to 360°.

⦿ **Perpendicular** lines meet at an angle of 90°.

⦿ Lines which remain the same distance apart along their length are **parallel** lines.

⦿ **Corresponding angles** are equal.

⦿ **Alternate angles** are equal.

⦿ The exterior angle of a triangle is equal to the sum of the two interior opposite angles.

⦿ The angle sum of a triangle is 180°.

⦿ The opposite angles of a parallelogram are equal.

Review exercise

1   **a**   On a copy of the grid, draw a line that is parallel to the line **L**.

b   On a copy of the grid, draw a line perpendicular to the line **P**.

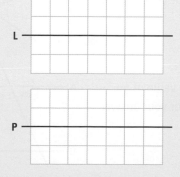

Nov 2007

G

G   **2**   Copy the following flag.
Mark two pairs of parallel lines
and a pair of perpendicular lines.

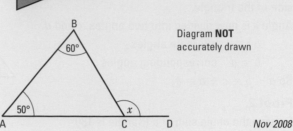

E   **3**   In the diagram, ABC is a triangle.
ACD is a straight line.
Angle CAB = 50°.
Angle ABC = 60°.
Work out the size of the angle marked x.

Diagram **NOT**
accurately drawn

Nov 2008

4   Work out the size of the angle marked x.

Diagram **NOT**
accurately drawn

Nov 2008

5   PQR is a straight line.
PQ = QS = QR.
Angle SPQ = 25°.
   a   **i** Write down the size of angle w.
      ii Work out the size of angle x.
   b   Work out the size of angle y.

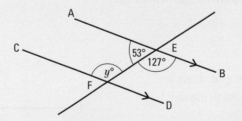

Diagram **NOT**
accurately drawn

Nov 2008

6   In the diagram, ABC is a straight line and BD = CD.
   a   Work out the size of angle x.
   b   Work out the size of angle y.

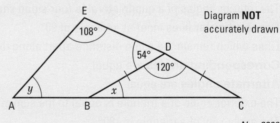

Diagram **NOT**
accurately drawn

Nov 2006

D   **7**   AB is parallel to CD.
Angle BEF = 127°.
   i   Write down the value of y.
   ii   Give a reason for your answer.

Diagram **NOT**
accurately drawn

Nov 2008

8   AB is parallel to CD.
   i   Write down the value of y.
   ii   Give a reason for your answer.

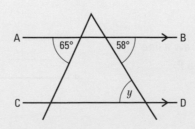

Diagram **NOT**
accurately drawn

June 2008

9 AB is parallel to CD. EF is a straight line.

   i Write down the value of y.

   ii Give a reason for your answer.

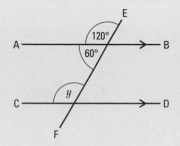

Diagram **NOT** accurately drawn

June 2008

* 10 ABCD is a straight line.

PQ is parallel to RS.

Write down the size of the angles marked x and y and give a reasons for your answers.

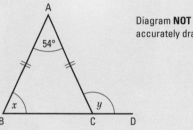

Diagram **NOT** accurately dawn

March 2008 adapted

11 ABC is an isosceles triangle.

BCD is a straight line.

AB = AC.

Angle A = 54°.

   a **i** Work out the size of the angle marked x.

      ii Give a reason for your answer.

   b Work out the size of the angle marked y.

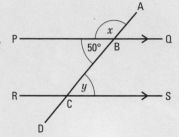

Diagram **NOT** accurately drawn

June 2007

12 Find the size of the marked angles. Give reasons for your answers.

   a

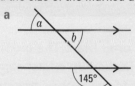

   b

* 13 Prove that the angle sum of a triangle is 180°.

D

A02

C

237

In 1999, NASA spent $125 million on a space probe designed to orbit Mars. The mission ended in disaster after the probe steered too close to Mars and burned up whilst skimming the planet's thin atmosphere. Apparently navigation commands to the probe's engines were provided in imperial units rather then the metric ones that NASA had been using since at least 1990.

◎ Objectives

In this chapter you will:
- learn to take and estimate readings from dials and scales
- learn how to choose the most appropriate unit for taking a measurement and convert between metric units of measurement
- work with time and solve problems relating to speed
- learn how to convert between imperial units and between metric and imperial units.

◈ Before you start

You need to be able to:
- estimate simple measurements and larger measurements
- carry out simple measurements using rulers and weighing scales
- tell the time using clock faces or digital clocks.

17.1 Reading scales

◎ Objectives

- ◉ You can take readings from dials and scales.
- ◉ You can estimate readings from dials and scales.

⑦ Why do this?

Diabetics need to accurately monitor their blood sugar levels and asthmatics use a peak flow meter to monitor their asthma. In both cases this will help doctors treat them effectively.

◈ Get Ready

1. Measure the length and width of this book, using suitable units of measure.

🔍 Key Points

- ◉ Lines can be measured using a ruler. The following ruler is marked off in centimetres (cm) but the smaller marks show you millimetres (mm) (10 mm = 1 cm).

  This line can be measured as 8.5 cm or 85 mm.
- ◉ Some dials have a scale. You need to interpret the scale to take an accurate reading from the dial.

  On this scale there are 5 spaces between 10 and 20, so each mark shows 2 units.

  The reading is 18 kg.

- ◉ Sometimes you will need to estimate a reading.

  The arrow on this scale is more than halfway between 40 and 50, so it is more than 45 but less than 50. It is also some distance from 50, so it is not 49.

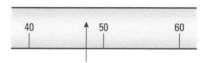

  A good estimate is 47 or 48.

🔍 Example 1

Measure and write down the reading from this ruler. Give your answer in cm and mm.

The reading from this ruler is 3.5, which is 3.5 cm.
In millimetres it is 35 mm.

Results Plus
Examiner's Tip

Always remember to show the units with your answer.

Exercise 17A

G

1 Measure and write down the lengths of these lines in cm and mm.

a ——————————

b ——————————————

c ————————————————————

d ——————————————————————

e —————————————————————————————

f ——————————————————

g ——————————————————————————

h ————————————————————————

i ——————————————————————

j ————————————

k ——————————————————

l ————————————————————————

2 Draw and label lines with the following lengths.

a	6 cm	b	3 cm	c	4 cm	d	6.3 cm
e	5.1 cm	f	6.8 cm	g	3.4 cm	h	7.9 cm
i	3.5 cm	j	4.2 cm	k	5.7 cm	l	7.6 cm

3 Draw and label lines with the following lengths.

a	40 mm	b	70 mm	c	14 mm	d	35 mm
e	12 mm	f	31 mm	g	48 mm	h	26 mm
i	27 mm	j	59 mm	k	73 mm	l	66 mm

F

4 Write down the readings shown on the following scales.

a    b 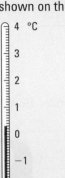   c    d 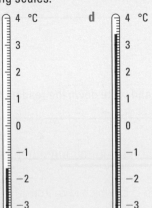   e

5 Write down the readings shown on the following scales.

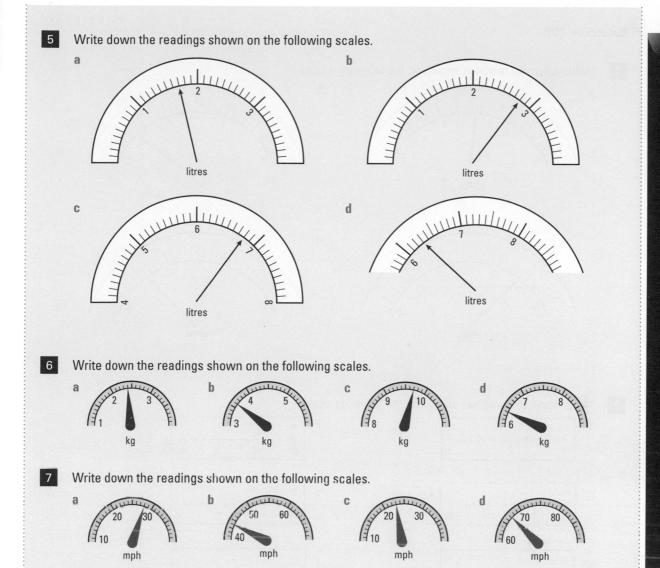

a

litres

b

litres

c

litres

d

litres

6 Write down the readings shown on the following scales.

a
kg

b
kg

c
kg

d
kg

7 Write down the readings shown on the following scales.

a
mph

b
mph

c
mph

d
mph

Example 2 Write down the reading shown on this dial.

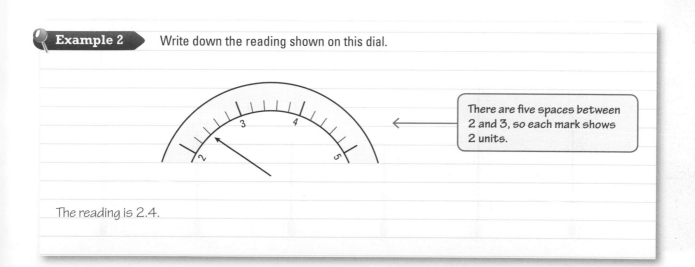

There are five spaces between 2 and 3, so each mark shows 2 units.

The reading is 2.4.

Exercise 17B

G

1 Write down the readings shown on the following scales.

a

amps

b

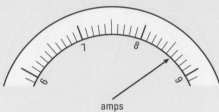

amps

c

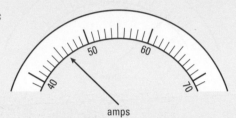

amps

d

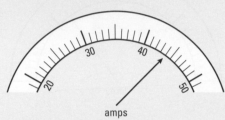

amps

F

2 Write down the readings shown on the following scales.

a

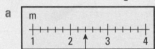

b

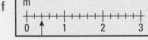

c

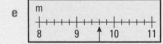

d

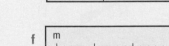

e

f

3 Write down the readings shown on the following scales.

a

b

c
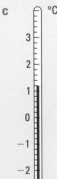

d

e

f

4 Write down the readings shown on the following scales.

a
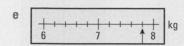 kg

b
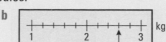 kg

c
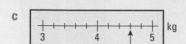 kg

d
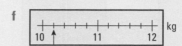 kg

e
 kg

f
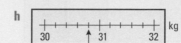 kg

g
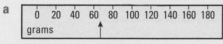 kg

h
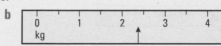 kg

---

Example 3   Estimate the reading shown on this dial.

mph

ResultsPlus
Examiner's Tip

Use an estimate when you are not able to take an accurate reading.

The speed is more than 25 and less than 30.
It is nearer to 25 than 30.
A good estimate is 26 or 27 mph.

Exercise 17C

1 Estimate the reading shown on the following scales.

a
| 0 | 20 | 40 | 60 | 80 | 100 | 120 | 140 | 160 | 180 |
grams

b
| 0 | 1 | 2 | 3 | 4 |
kg

c
| 0 | 5 | 10 | 15 | 20 | 25 | 30 | 35 | 40 |
grams

d
| 0 | 20 | 40 | 60 | 80 | 100 | 120 | 140 | 160 | 180 |
grams

2 Estimate the reading shown on the following scales.

a
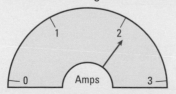

b
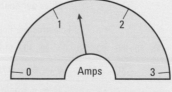

c
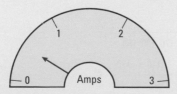

d
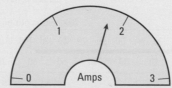

F

3 Estimate the readings shown on the following scales. Give the readings in both °C and °F.

a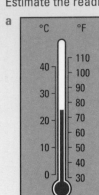

b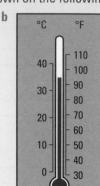

c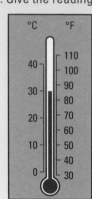

d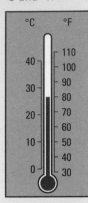

e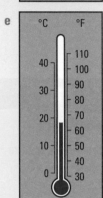

f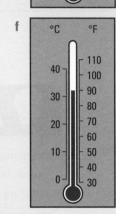

g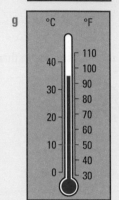

h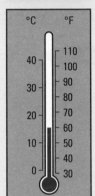

4 Estimate the readings shown on the following scales.

a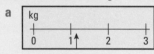

b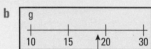

c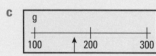

d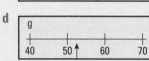

5 Estimate the readings shown on the following scales. Give the readings in both kilometres per hour and miles per hour.

a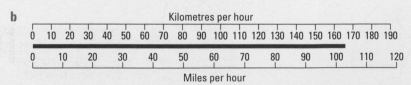

b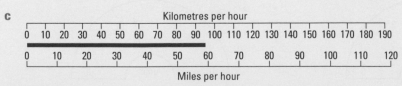

c

17.2 Time

⊙ Objectives

- You can read time from a variety of clocks.
- You can convert between 12-hour time and 24-hour time.
- You can work out durations of time.
- You can solve problems involving time and timetables.

❔ Why do this?

You need to use time in order to read a bus or train timetable or to plan how long to spend on each question in an exam.

⬆ Get Ready

1. Write down the times at which you have your meals during the day, and find the duration of time between each one.

🔑 Key Points

- A 12-hour clock goes from 0 to 12, using am for times before noon, and pm for times after noon.
- A 24-hour clock goes from 0 to 24, where 24 is midnight.

- You will need to know the following **conversions** involving time:

  60 seconds = 1 minute   366 days = 1 leap year
  60 minutes = 1 hour   3 months = 1 quarter
  24 hours = 1 day   12 months = 1 year
  7 days = 1 week   52 weeks = 1 year
  365 days = 1 year

🔍 Example 4   The time shown on this clock is in the afternoon.
Write the time using both 12-hour time and 24-hour time.

ResultsPlus
Examiner's Tip

Make sure you write time using the correct notation.

The time is quarter past five, or 5.15 pm, in 12-hour time.
In 24-hour time, this time is 17:15 hours.

G

Exercise 17D

1 These clock faces show times in the morning.

a  b  c

d 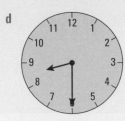 e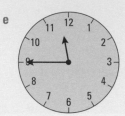

Write each of the times in both 12-hour time and in 24-hour time.

2 These clock faces show times in the afternoon.

a 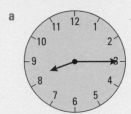 b 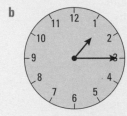 c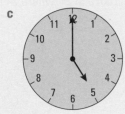

d 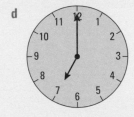 e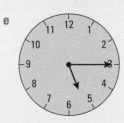

Write each of the times in both 12-hour time and 24-hour time.

3 Change the following 12-hour times to 24-hour times.

a 8 am	b 11.15 am	c 3.40 pm	d 8.20 am
e 8.55 pm	f 3.25 pm	g 2.30 am	h 5.25 pm
i 10.15 pm	j 7.20 am	k 9.45 am	l 1.15 pm
m 11.25 pm	n 2.50 am	o 1.50 pm	p 12.20 pm

4 Change the following 24-hour times to 12-hour times.

a 11:10 h	b 08:20 h	c 07:40 h	d 23:35 h
e 14:17 h	f 09:35 h	g 18:16 h	h 17:25 h
i 13:20 h	j 13:10 h	k 08:30 h	l 13:35 h
m 03:42 h	n 22:16 h	o 09:17 h	p 13:37 h

5   Write down the correct time in the following questions.

a   4 hours before 3.15 pm

b $4\frac{1}{2}$ hours before 16:40 h

c $3\frac{1}{4}$ hours after 11.15 am

d $2\frac{1}{4}$ hours before 1 pm

e $1\frac{1}{2}$ hours before 13:00 h

f   3 hours after 2.40 am

g $2\frac{1}{2}$ hours after 11.20 am

h $4\frac{1}{4}$ hours before 14:45 h

i $5\frac{1}{2}$ hours after 10:55 h

j $6\frac{1}{2}$ hours before 15:10 h

k $5\frac{1}{4}$ hours before 3 am

l $2\frac{3}{4}$ hours after 22:00 h

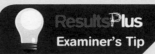
Examiner's Tip

It may be useful to use a clock face to help you answer this type of question. There will usually be a clock in the exam room that you can use.

6   Change the units of time in the following questions.

a   3 years into weeks

b $3\frac{1}{2}$ hours into minutes

c   5 minutes into seconds

d   5 years into months

e   4 days into hours

f   540 minutes into hours

g   312 weeks into years

h $2\frac{1}{2}$ minutes into seconds

i   90 seconds into minutes

j   4 years into weeks

k   8 hours into minutes

l $2\frac{1}{2}$ days into hours

m $3\frac{1}{2}$ years into weeks

n   2 years into days

7   a   A man buys three packets of crisps each week. How many does he buy in a year?

b   It takes 3 minutes for Bill to make a toy. How many can he make in 1 hour?

c   Sally pays her gas bill every month. How many gas bills does she pay in a year?

d   A clock chimes every hour. How many times does the clock chime in a week?

A02

Example 5

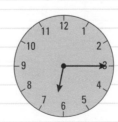

Watch Out!

Do not use a calculator to solve time problems.

These two clocks both show times in the morning.
What is the time difference between the clocks?

3.30 am to 4.00 am is 30 minutes   ← | Start with the earliest time and count on to the next hour. |

4.00 am to 6.00 am is 2 hours   ← | Count the number of full hours. |

6.00 am to 6.15 am is 15 minutes   ← | Count the minutes to the end time. |

Total time: 30 minutes + 2 hours + 15 minutes = 2 hours 45 minutes.

Exercise 17E

1 Work out the time difference between each of the following times:

a 9.50 am to 10.15 am   b 07:30 h to 08:20 h

c 12.15 pm to 2.25 pm   d 13:45 h to 00:15 h

e 10 pm Monday to 7 am Tuesday   f 09:17 h to 13:27 h

g 9.37 am to 1.15 pm   h 03:42 h to 22:14 h

i 8.30 am to 11.15 pm   j 9.15 am to 1.05 pm

k 09:15 Tue to 08:05 Wed   l 18:35 h to 21:50 h

2 Saima leaves her house at 6.15 am. She travels by aeroplane to her parents' home, arriving at 10.35 pm. How long does her journey take?

3 A man arrives at work at 08:55 h and leaves at 05:05 h. How long is he at work?

4 A ferry sets sail from Portsmouth at 08:50 h and arrives in France at 15:40 h. How long does the crossing take?

5 The following table shows information about flight times to Aberdeen. Ella wants to fly to Aberdeen. What time does her flight depart and arrive if she wants to take the fastest flight?

Flight number	Departure time	Arrival time	Flight time
BA52	22:20	04:45	
XA160	05:42	09:14	
FC492	14:15		4 h 40 min
TC223	10:02		4 h 23 min
AL517		07:59	5 h 37 min
AB614	19:17	05:21	
FX910	02:43		3 h 51 min
BI451		12:17	2 h 32 min
AE105		02:25	5 h 29 min
DA452	15:39		6 h 48 min

Example 6   Part of a bus timetable is shown.

a Ali gets the 09:28 bus from Burton. How long does it take him to get to Didcom?

b Chelsea arrives at her bus stop in Camberley at 09:40. She is going to Earlstown. What time will she get there?

Aldwich	09:16	10:05
Burton	09:28	10:17
Camberley	09:35	10:34
Didcom	09:55	10:54
Earlstown	10:08	11:07

a The 09:28 bus arrives in Didcom at 09:55. 09:28 to 09:55 is 27 minutes.

b Chelsea has just missed a bus! The next bus to Earlstown leaves Camberley at 10:34 and arrives at 11:07.

Exercise 17F

1 Use the bus timetable to answer the questions below.

Bus timetable: Ordsall to Bury

Ordsall, Salford Quays			07:30		08:30		18:30	19:00	20:00	21:00	22:00
Trafford Rd			07:35		08:35		18:35	19:05	20:05	21:05	22:05
Pendleton Precinct arr.			07:41		08:41		18:41				
Pendleton Precinct dep.	06:43	07:13	07:43	08:13	08:43	and	18:43	19:10	20:10	21:10	22:10
Lower Kersal	06:54	07:24	07:54	08:24	08:54	every	18:54	19:19	20:19	21:19	22:19
Agecroft	06:57	07:27	07:57	08:27	08:57	30	18:57	19:22	20:22	21:22	22:22
Butterstile Lane	07:04	07:34	08:04	08:34	09:04	mins	19:04				
Prestwich	07:10	07:40	08:10	08:40	09:10	until	19:10				
Besses o'th' Barn	07:14	07:44	08:14	08:44	09:14		19:14				
Unsworth	07:24	07:54	08:24	08:54	09:24		19:24				
Bury, Interchange	07:40	08:10	08:40	09:10	09:40		19:40				

a How long does it take the 06:43 Pendleton bus to get to Besses o'th' Barn?

b How long do buses wait at Pendleton?

c At what time does the last bus arrive in Bury?

d At what time does the first bus call at Trafford Road?

e How long does it take to travel from Lower Kersal to Unsworth?

f How long does it take to travel from Ordsall to Prestwich?

g How many buses call at Trafford Road before 10:00?

h How many buses call at Butterstile Lane during the day?

i Shaun arrives at his bus stop at Prestwich at 7.30 am.
How long will he have to wait for a bus to Unsworth?

j Jane arrives at her bus stop in Trafford Road at 7.50 am.
How long will she have to wait for a bus to Agecroft?

k Umar lives 5 minutes away from his bus stop in Agecroft.
What is the latest time he can leave his house to get to Bury by 9 am?

l Eko wants to catch a bus from Pendleton to Prestwich, to arrive in Prestwich no later than 12 noon.
What is the departure time of the latest bus he can catch from Pendleton?

A02
A03

2 The train route diagram shows the times it takes to travel from Manchester Victoria to all stations on the line. Use the information in the diagram to answer the questions below.

a How long does it take to travel between:
  i Manchester Victoria and Hindley
  ii Swinton and Daisy Hill
  iii Wigan and Farnworth
  iv Bolton and Salford Crescent?

b James is planning a trip from Swinton to Bolton. He will have to wait 12 minutes at Salford Crescent to change trains. What will be his total journey time from Swinton to Bolton?

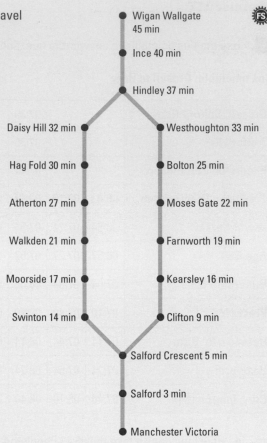

c Copy and complete the following timetables.

Manchester Victoria	09:05	11:35	Wigan Wallgate	08:30	11:55
Salford			Ince		
Salford Crescent			Hindley		
Swinton			Westhoughton		
Moorside			Bolton		
Walkden			Moses Gate		
Atherton			Farnworth		
Hag Fold			Kearsley		
Daisy Hill			Clifton		
Hindley			Salford Crescent		
Ince			Salford		
Wigan Wallgate			Manchester Victoria		

17.3 **Metric units**

⊙ Objectives

- ⊙ You can choose an appropriate unit to use for measurement.
- ⊙ You can convert between metric units.

⊘ Why do this?

It would be useful to understand metric units when you go on holiday as many European countries use kilometres rather than miles on their road signs.

⊕ Get Ready

1. Write down a list of objects you could measure using the following metric units of measure: centimetres, metres, grams, kilograms and litres.

⊗ Key Points

⊙ The following facts show how to change from one **metric unit** to another metric unit.

Length	Weight	Capacity
10 mm = 1 cm	1000 mg = 1 g	100 cl = 1 litre
100 cm = 1 m	1000 g = 1 kg	1000 ml = 1 litre
1000 mm = 1 m	1000 kg = 1 tonne	1000 l = 1 cubic metre
1000 m = 1 km		1000 cm^3 = 1 litre

The pictures below show examples of everyday items along with appropriate units of measurement.

A ruler is about 30 cm long.

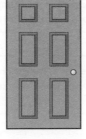

A door has a height of about 2 m.

A can of cola is about 300 ml.

A bag of sugar is about 1 kg.

🔍 Example 7

Write down the metric unit you would choose to measure the following:

A02

a the length of your classroom
b the weight of a pen
c the amount of water in a bucket.

a metres   b grams   c litres

Exercise 17G

Write down the metric unit you would use to take the measurements listed below:

1 an amount of medicine
2 the length of a finger nail
3 the weight of a house brick
4 the weight of a lorry
5 the amount of petrol in a car's petrol tank
6 the length of a bus
7 the length of a ballpoint pen
8 the weight of 30 of these books
9 the weight of a £1 coin
10 the length of an ant
11 the capacity of a kettle
12 the weight of a human hair
13 the distance from home to school
14 the weight of a box of cornflakes
15 the height of a person.

Example 8   a   Change 3 kilometres to metres.
                b   Change 450 mm to cm.

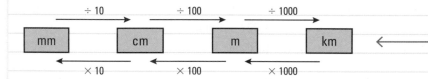

When you divide by 10, 100 or 1000 everything moves to the right by 1, 2 or 3 places.

a   3 km = 3 × 1000 = 3000 m.

Kilometres are longer than metres so you should expect to get more metres than kilometres. 1 km = 1000 m

b   450 mm = 450 ÷ 10 = 45 cm.

Millimetres are smaller than centimetres so you should expect to get fewer centimetres than millimetres. 10 mm = 1 cm

Exercise 17H

1 Change the following lengths to centimetres.
   a   4 m
   b   50 mm
   c   8 m
   d   13 m
   e   200 mm
   f   35 m
   g   74 mm
   h   122 mm

2 Change the following lengths to millimetres.
   a   3 cm
   b   6 cm
   c   22 cm
   d   40 cm
   e   200 cm
   f   5.4 cm
   g   13.7 cm
   h   5.15 cm

3 Change the following lengths to metres.

  a 6 km
  b 500 cm
  c 20 km
  d 3000 cm
  e 45 km
  f 0.8 km
  g 1.4 km
  h 2.45 km

4 Change the following weights to grams.

  a 2 kg
  b 30 kg
  c 400 kg
  d 250 kg
  e 2000 kg
  f 55 kg
  g 0.12 kg
  h 4.2 kg

5 Change the following capacities to litres.

  a 4000 ml
  b 7000 ml
  c 20 000 ml
  d 45 000 ml
  e 2500 ml
  f 3700 ml
  g 6520 ml
  h 3130 ml

6 Change the following capacities to millilitres.

  a 3 l
  b 20 l
  c 200 l
  d 450 l
  e 35 l
  f 7.5 l
  g 0.4 l
  h 1.43 l

7 Change the following lengths to kilometres.

  a 3000 m
  b 8000 m
  c 30 000 m
  d 68 000 m
  e 4200 m
  f 5600 m
  g 5410 m
  h 2140 m

8 Change the following weights to tonnes.

  a 5000 kg
  b 6000 kg
  c 40 000 kg
  d 57 000 kg
  e 3600 kg
  f 4500 kg
  g 7630 kg
  h 4250 kg

9 Change the following weights to kilograms.

  a 4000 g
  b 2 tonnes
  c 20 000 g
  d 15 tonnes
  e 200 000 g
  f 3.7 tonnes
  g 6400 g
  h 1230 g

10 A large box contains 3 kg of plastic cubes. Each cube weighs 6 grams.
How many cubes are there in the box?

11 A jar contains 450 ml of water. A tank can fill 200 of these jars. What is the capacity, in litres, of the tank?

12 One lap of a running track is 400 m. How many laps must be run to complete a 10 km race?

13 Neejal estimates that a car weighs $\frac{3}{4}$ tonne. How many kilograms is this?

14 Jessica has a 60 cl bottle of medicine. How many 30 ml doses can be poured from it?

15 A group of men can lay 300 m of pipe in one day. Working at the same rate, how long should it take them to lay the pipe for a length of 15 km?

G

A02 F

A02

A02

A02

A02

A02 E

Example 9 Put the following weights in order, with the largest first.

  250 g   25 g   2 kg   250 kg   3000 g

250 g, 25 g, 2000 g, 250 000 g, 3000 g ← First, change all of the weights to grams.

250 000 g, 3000 g, 2000 g, 250 g, 25 g ← Then order, with the largest first.

250 kg, 3000 g, 2 kg, 250 g, 25 g ← Finally, change back to the original units.

F

Exercise 17I

1 Put these lengths in order, with the smallest first.

   4 m   6 mm   3 cm   4 km   30 cm   60 mm

2 Put these capacities in order, with the smallest first.

   400 m*l*   6 *l*   700 m*l*   3000 m*l*   1*l*

3 Put these weights in order, with the smallest first.

   600 g   450 g   0.5 kg   0.62 kg

4 Put these lengths in order, with the smallest first.

   40 cm   0.6 cm   370 mm   1.4 m   600 mm   55 cm

5 Put these lengths in order, with the smallest first.

   6 cm   55 mm   46 cm   0.2 cm   77 mm   0.4 cm   9 mm

6 Put these capacities in order, with the smallest first.

   600 m*l*   450 m*l*   0.3 *l*   260 m*l*   0.08 *l*   75 m*l*

17.4 Imperial units

◎ Objectives

○ You can convert between imperial units.

○ You know the common metric–imperial conversions.

○ You can convert between metric and imperial units.

⏻ Why do this?

The UK is still making things using imperial units, for parts of the world that still use imperial units (parts of Africa, Asia and the US). It is also important that we understand our mathematical heritage.

⬥ Get Ready

1. Measure the heights of a number of objects in feet and inches. Also measure the same heights in centimetres.

🔑 Key Points

Imperial unit conversions

12 inches = 1 foot
3 feet = 1 yard
16 ounces = 1 pound
14 pounds = 1 stone
8 pints = 1 gallon

Metric–imperial approximate equivalent conversions

Metric	Imperial	Metric	Imperial
8 km ⟶	5 miles	1 kg ⟶	2.2 pounds
1 m ⟶	39 inches	25 g ⟶	1 ounce
30 cm ⟶	1 foot	4.5 litres ⟶	1 gallon
2.5 cm ⟶	1 inch	1 litre ⟶	1.75 pints

Example 10 ▶ A man weighs $10\frac{1}{2}$ stones. Change this weight to pounds.

Use 14 pounds = 1 stone
$10\frac{1}{2}$ stones = 10.5×14 = 147 pounds.

Exercise 17J

1   Change the following imperial measurements.
    a   36 inches into feet   **b**   32 pints into gallons
    c   2 ft 4 in into inches   **d**   4 pounds into ounces
    e $5\frac{1}{4}$ feet into inches   **f**   21 feet into yards
    g   4 ft 4 in into inches   **h**   6 gallons into pints
    i $3\frac{1}{2}$ stones into pounds   **j** $2\frac{1}{4}$ gallons into pints
    k   300 pounds into stones and pounds   **l**   4 yards into feet
    m $\frac{1}{4}$ yard into inches   **n**   22 pints into gallons and pints
    o   1 stone into ounces

2   Ben was 4 foot $10\frac{1}{2}$ inches tall. He has grown another $4\frac{1}{2}$ inches. What is his height now?

3   Anna weighs 9 stone 10 pounds. She wants to lose $\frac{1}{2}$ stone. What will she then weigh?

4   A curtain measures 5 foot 8 inches. It has a folded hem of 6 inches. What is its total length of the curtain material?

Example 11 ▶ Change 20 km into miles.

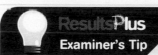

Examiner's Tip

In the exam, you will be expected to know the metric–imperial conversions.

Use 8 km = 5 miles
20 km = $20 \div 8 \times 5$ miles = 12.5 miles.

Exercise 17K

1   Change the following measurements.
    a   15 miles into kilometres   **b**   10 kg into pounds   **c**   4 litres into pints
    d   6 inches into cm   **e**   3 yards into cm   **f**   48 km into miles
    g   11 pounds into kg   **h**   2.5 m into inches   **i**   75 miles into km

2   A house is estimated to be 24 feet high. What is this in metres?

3   A shirt collar measures 14 inches. What is this in centimetres?

F

4 A tank holds 18 litres. How many gallons is this?

A02
A03
5 A baby needs 1 pint of milk a day. His bottle contains 0.2 litres.
How many bottles will his carer have to make for him in one day?

6 Yasmin travels from London to Scotland. The distance is 400 miles. What is the distance in kilometres?

7 A container has a capacity of 5 litres. What is this in pints?

8 A family on holiday in Majorca travel 150 km while touring the island. How many miles do they travel?

E
A02
A03
9 A scuba diver weighs 110 lb. She needs to weigh a total of 57 kg to dive. Her airtank weighs 5 kg.
How many half pound weights does she need to wear?

A02
10 Shop A is selling 4 kg of potatoes for £1.60. Shop B is selling a 5-pound bag of potatoes for £1.60.
Which is the better buy?

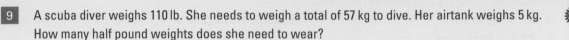

17.5 Speed

◎ Objectives

○ You can calculate speed given distance and time.
○ You can solve problems involving speed.

◈ Why do this?

You need to be able to calculate speed if you are training to improve your times in sports such as running, swimming or athletics.

◈ Get Ready

1. Find out how speed is measured on the speedometer of a car.

◒ Key Points

● **Speed** = $\dfrac{\text{distance}}{\text{time}}$

● Time = $\dfrac{\text{distance}}{\text{speed}}$

● Distance = speed × time

● **Average speed** = $\dfrac{\text{total distance travelled}}{\text{total time taken}}$

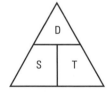

● Units of speed are usually miles per hour (mph), kilometres per hour (km/h) or metres per second (m/s).

⚲ Example 12 A car covers a distance of 160 miles in 4 hours. What is the car's average speed?

Average speed = $\dfrac{\text{total distance travelled}}{\text{total time taken}} = \dfrac{160}{4} = 40$ mph. ⟵

> Units are miles and hours, so speed is given as miles per hour (mph).

Exercise 17L

1 What is the average speed of a car that takes 3 hours to travel 90 miles?

2 Giles ran for 2 hours and covered 16 miles. At what average speed was he running?

3 Gareth was travelling by canal boat. He went 32 miles in 8 hours.
At what average speed was he travelling?

4 Mandy swam 5 miles. It took her $2\frac{1}{2}$ hours. What was her average speed?

5 Ahmed travelled 200 miles on a business trip. He left home at 9 am and arrived at his destination at 1 pm.
What was his average speed?

6 What is the average speed of a train that takes $2\frac{1}{2}$ hours to travel 80 km?

7 After a $3\frac{1}{2}$ hour journey a car had travelled 210 miles. What was its average speed?

8 Rami drives 125 miles in $2\frac{1}{2}$ hours. Work out his average speed.

9 Jon travels 40 km in 30 minutes. Calculate his average speed in km/h.

10 An aeroplane flies 1400 km in $3\frac{1}{2}$ hours. What is its average speed?

Example 13 Jill is a motorcycle courier. She travels at an average speed of 5 mph in the city.
How long will it take her to travel 25 miles?

$$\text{Time} = \frac{distance}{speed} = \frac{25 \text{ miles}}{5 \text{ mph}} = 5 \text{ hours.}$$

Exercise 17M

1 A car travels for 2 hours at 40 mph. How far will the car have gone?

2 Find the time taken to travel 12 km at 3 km/h.

3 A delivery van takes 3 hours to complete a journey at a speed of 35 mph.
What distance will it have covered?

4 How long does it take to travel 60 miles at an average speed of 40 mph?

5 How long does it take to travel 50 km at 20 km/h?

6 Find the time taken to walk 10 miles at an average walking speed of $2\frac{1}{2}$ mph.

7 A man walks at an average speed of 5 km/h. What distance will he cover in 2 hours 30 minutes?

D

8   Find the time taken to travel 10 miles at a speed of 4 mph.

9   An aeroplane flew at 400 mph. How far did it travel in $3\frac{1}{2}$ hours?

10 Rajesh travels for $2\frac{1}{2}$ hours at 64 mph. Calculate the distance he travelled.

Example 14   Martin cycles for 2 hours 18 minutes at a speed of 12 mph.
How far will he have travelled in this time?

$2\,h\,18\,min = 2\frac{18}{60}\,hour = 2.3\,h$

The time needs to be converted into hours.

ResultsPlus
Examiner's Tip

Remember:
$0.1\,h = 6$ minutes
$\frac{1}{10}\,h = 6$ minutes

$Distance = speed \times time = 12 \times 2.3 = 27.6\,miles.$

Exercise 17N

C

1   How far will you have gone if you travel for 1 hour 45 minutes at 50 mph?

2   Mark runs at an average speed of 5 mph. He goes running for 1 hour and 18 minutes.
What distance does he run in that time?

3   A train travelled a distance of 208 km in 2 hours 36 minutes. What was the average speed of the train?

4   A cyclist rode at a speed of 10 mph. She covered a distance of 32 miles. For how long was she cycling?
(Give your answer in hours and minutes.)

5   A speedboat has an average speed of 20 km/h. It travels around the coast for 4 hours 6 minutes.
What distance does it cover in that time?

A02

6   A train travelled from London to Scotland. It left London at 08:00 hours and arrived at its destination at
13:24 hours. It travelled 324 miles. What was the average speed of the train?

7   A lorry made a journey of 390 km at an average speed of 50 km/h. How long did it take?
Give your answer in hours and minutes.

8   A train was travelling on a track at a speed of 90 mph. It was travelling at this speed for
3 hours 18 minutes. How far did it go in this time?

9   A horse takes 6 minutes to gallop 5 km. What is the average speed of the horse?

10 a A racing car travels at 85 m/s. Work out the distance the car travels in 0.4 seconds.
b Change a speed of 85 m/s into km/h.

17.6 Accuracy of measurements

⊙ Objective

⊙ You can recognise that measurements given to the nearest whole unit may be inaccurate by up to one half in either direction.

⟳ Why do this?

Small differences in the measurement of an object can make a big difference if multiplied over 10 objects.

⟳ Get Ready

1. Round the following numbers to the nearest whole number.
   a 6.8   **b** 52.49   **c** 13.5
2. Change the following lengths to cm.
   a 3 m   **b** 70 mm
3. Change the following lengths to mm.
   a 22 cm   **b** 4.7 cm

🌐 Key Points

◉ Measured with a centimetre stick, the length of a piece of A4 paper is 30 cm.

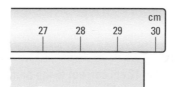

◉ This does not mean that the length of the A4 is exactly 30 cm. It is 30 cm correct to the nearest centimetre.

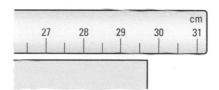

◉ The exact length of the piece of A4 paper is somewhere between 29.5 and 30.5 cm.

◉ Measured with a ruler, the length of a piece of A4 paper is 297 mm.

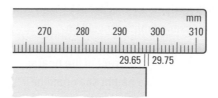

◉ The exact length of the piece of A4 paper is somewhere between 296.5 mm and 297.5 mm.

◉ So 297 mm to the nearest millimetre means that the minimum (least) possible length is 296.5 mm, and the maximum (greatest) possible length is 297.5 mm.

◉ Measurements given to the nearest whole unit may be inaccurate by up to one half of a unit below and one half of a unit above.

Example 15

1. The weight of a cocker spaniel is 14 kg correct to the nearest kilogram.
   Write down   a   the smallest possible weight   b   the greatest possible weight.

2. A small bowl holds 244 millilitres of water correct to the nearest millilitre.
   Write down   a   the smallest possible volume   b   the greatest possible volume.

1. a   13.5 kg   b   14.5 kg
2. a   243.5 ml b   244.5 ml

Example 16

The length of a calculator is 12.8 cm correct to the nearest millimetre.
Write down   a   the minimum possible length   b   the maximum possible length.

12.8 cm = 128 mm   ⟵   Convert the length from centimetres to millimetres. 1 cm = 10 mm.

a   127.5 mm or 12.75 cm
b   128.5 mm or 12.85 cm

Exercise 17O

C

1   The length of a pencil is 12 cm correct to the nearest centimetre.
    Write down the maximum length it could be.

2   The weight of an envelope is 45 grams correct to the nearest gram.
    Write down the minimum weight it could be.

3   The capacity of a jug is 4 litres correct to the nearest litre.
    Write down the minimum capacity of the jug.

4   The radius of a plate is 9.7 cm correct to the nearest millimetre.
    Write down
    a   the least possible length it could be   b   the greatest possible length it could be.

5   Magda's height is 1.59 m correct to the nearest centimetre. Write down in metres
    a   the minimum possible height she could be
    b   the maximum possible height she could be.

6   The length of a pencil is 10 cm correct to the nearest cm.
    The length of a pencil case is 102 mm correct to the nearest mm.
    Explain why the pen might **not** fit in the case.

7   The width of a cupboard is measured to be 82 cm correct to the nearest centimetre.
    There is a gap of 817 mm correct to the nearest mm in the wall.
    Explain how the cupboard might fit in the wall.

Chapter review

- Lines can be measured using a ruler.
- Some dials have a scale. You need to interpret the scale to take an accurate reading from the dial.
- Sometimes you will need to estimate a reading.
- A 12-hour clock goes from 0 to 12, using am for times before noon, and pm for times after noon.
- A 24-hour clock goes from 0 to 24, where 24 is midnight.
- You will need to know the following **conversions** involving time:

60 seconds = 1 minute	366 days = 1 leap year
60 minutes = 1 hour	3 months = 1 quarter
24 hours = 1 day	12 months = 1 year
7 days = 1 week	52 weeks = 1 year
365 days = 1 year	

- The following facts show how to change from one **metric unit** to another metric unit.

Length	Weight	Capacity
10 mm = 1 cm	1000 mg = 1 g	100 cl = 1 litre
100 cm = 1 m	1000 g = 1 kg	1000 ml = 1 litre
1000 mm = 1 m	1000 kg = 1 tonne	1000 l = 1 cubic metre
1000 m = 1 km		1000 cm^3 = 1 litre

- **Imperial unit** conversions

12 inches = 1 foot
3 feet = 1 yard
16 ounces = 1 pound
14 pounds = 1 stone
8 pints = 1 gallon

  Metric–imperial approximate equivalent conversions

Metric		Imperial	Metric		Imperial
8 km	⟶	5 miles	1 kg	⟶	2.2 pounds
1 m	⟶	39 inches	25 g	⟶	1 ounce
30 cm	⟶	1 foot	4.5 litres	⟶	1 gallon
2.5 cm	⟶	1 inch	1 litre	⟶	1.75 pints

- **Speed** $= \dfrac{\text{distance}}{\text{time}}$
- Time $= \dfrac{\text{distance}}{\text{speed}}$
- Distance = speed $\times$ time
- **Average speed** $= \dfrac{\text{total distance travelled}}{\text{total time taken}}$
- Units of speed are usually miles per hour (mph), kilometres per hour (km/h) or metres per second (m/s).
- Measurements given to the nearest whole unit may be inaccurate by up to one half of a unit below and one half of a unit above.

Review exercise

1 **a** Complete the table by writing a sensible **metric** unit for each measurement.
The first one has been done for you.

The length of the river Nile	6700 kilometres
The height of the world's tallest tree	110
The weight of a chicken's egg	70
The amount of petrol in a full petrol tank of a car	40

b Change 4 metres to centimetres.

c Change 1500 grams to kilograms.

June 2008

G

2 This is part of a ruler.

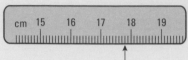

a Write down the length marked with an arrow.

This is a thermometer.

b Write down the temperature shown.

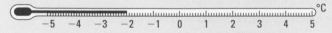

This is a parcel on some scales.

c Write down the weight of the parcel.

F

3 **a** Write down a sensible metric unit that can be used to measure
   i the height of a tree   **ii** the weight of a person.
   b Change 2 centimetres to millimetres.

Nov 2008

4 **a** Write down a sensible **metric** unit for measuring
   i the distance from London to Paris   **ii** the amount of water in a swimming pool.
   b **i** Change 5 centimetres to millimetres.   **ii** Change 4000 grams to kilograms.

Nov 2008

5 **a**

Write down the number marked by the arrow.

b

Find the number 127 on the number line.
Mark it with an arrow (↑).

c

Write down the number marked by the arrow.

d

Find the number 3.18 on the number line.
Mark it with an arrow (↑).

Nov 2008

6

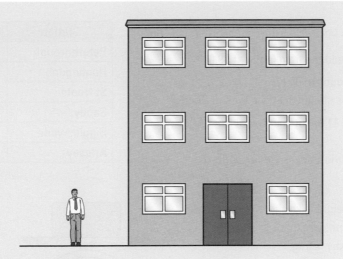

The diagram shows a building and a man.

The man is of normal height.

The man and the building are drawn to the same scale.

a Write down an estimate for the height of the man.

b Write down an estimate for the height of the building.

Nov 2008

7

The picture shows a man standing next to a flagpole.

The man is of normal height.

The man and the flagpole are drawn to the same scale.

a Write down an estimate for the height, in metres, of the man.

b Work out an estimate for the height, in metres, of the flagpole.

June 2008

8 Here is a picture of a woman opening a door that is 2 m high.

Estimate the height of the woman.

Nov 2007

9 Shalim says 1.5 km is less than 1400 m.

Is he right?

Explain your answer.

A03

June 2007

10 Here is part of a train timetable from Peterborough to London.

Station	Time of leaving
Peterborough	08:44
Huntingdon	09:01
St Neots	09:08
Sandy	09:15
Biggleswade	09:19
Arlesey	09:24

a Which station should the train leave at 09:01?

The train arrives in Sandy at 09:12.

b How many minutes should the train wait in Sandy?

The train should take 41 minutes to travel from Arlesey to London.

c What time should the train arrive in London?

Nov 2008

11 a Write down the weight in kg shown on this scale.

b i How many pounds are there in 1 kg?

The weight of a baby is 5 kg.

ii Change 5 kg to pounds.

Nov 2008

12 Here is part of a railway timetable.

Bristol Temple Meads	08:00	08:30	09:00
Bath	08:15	08:45	09:15
Chippenham	08:30	09:00	09:30
Swindon	08:50	09:20	09:50
Didcot	09:15	09:45	10:15
Reading	09:35	10:05	10:35
London Paddington	09:55	10:25	10:55

A train leaves from Bristol Temple Meads at 09:00.

a At what time should the train arrive at Swindon?

Jambaya gets to the station in Chippenham at 08:45.
She waits for the next train to Didcot.

b i How long should she have to wait?

ii At what time should she arrive at Didcot?

All the trains should take the same time to travel from Bath to Reading.

c How long, in minutes, should it take to travel from Bath to Reading?

June 2008

13 Here is part of a bus timetable.

Bus Station	07:00	07:30	08:00
Castle Street	07:10	07:40	08:15
High Street	07:25	07:55	08:25
Station Road	07:37	08:07	08:37
Church Street	07:50	08:20	08:50
Wharf Inn	07:55	08:25	08:55

A bus leaves the Bus Station at 07:00.

a At what time should the 07:00 bus arrive at Station Road?

Jill arrives at High Street at 07:45.
She wants to catch a bus to Wharf Inn.

b How long should she have to wait for the next bus?

A bus leaves Station Road at 08:37.

c How long should this bus take to travel from Station Road to Wharf Inn?

Nov 2007

14 Complete this table.

Write a sensible unit for each measurement.

Three have been done for you.

	Metric	Imperial
Distance from London to Cardiff	km	
Weight of a bag of potatoes		pounds
Volume of fuel in a car's fuel tank		gallons

Nov 2007

15 Zoe is planning a trip to Palma from Sa Pobla on the local train.

This is part of the train timetable.

Sa Pobla	09:23	10:23	11:23	12:23	13:23
Inca	09:41	10:41	11:41	12:41	13:41
Santa Maria	09:57	10:57	11:57	12:57	13:57
Marratxi	10:05	11:05	12:05	13:05	14:05
Palma	10:20	11:20	12:20	13:20	14:20

a How long, in minutes, is the train journey from Sa Pobla to Palma?

The drive from Zoe's hotel to the station will take 30 minutes.

She wants to travel by train from Sa Pobla to Palma, visiting Inca for $1\frac{1}{2}$ hours on the way.

b Complete the table.

	Time
Zoe leaves the hotel	
Zoe leaves Sa Pobla on a train	
Train arrives at Inca (Zoe gets off)	
Zoe leaves Inca on another train	
Zoe arrives at Palma	

16 The distance from London to New York is 3456 miles.

A plane takes 8 hours to fly from London to New York.

Work out the average speed of the plane.

June 2008

17 Car P and Car Q travel from Amfield to Barton.

Car P averages 10 km per litre of petrol.

It needs 48 litres of petrol for this journey.

Car Q averages 4 km per litre of petrol.

Work out the number of litres of petrol Car Q needs for the same journey.

18 There are 14 pounds in a stone.

There are 2.2 pounds in a kilogram.

A man weighs 13 stone 6 pounds.

Work out his weight in kilograms, giving your answer to the nearest kilogram.

You must show all your working.

C

19 Stuart drives 180 km in 2 hours 15 minutes.
Work out Stuart's average speed.

Nov 2008

20 A gold necklace has a mass of 127 grams, correct to the nearest gram.
   a Write down the least possible mass of the necklace.
   b Write down the greatest possible mass of the necklace.

June 2006

PERIMETER AND AREA OF 2D SHAPES

Many people need to be able to calculate both the perimeter and area of various shapes in their day-to-day work. For example, the organisers at Glastonbury need to know the area and perimeter of the festival fields so they can hire enough safety barriers.

Objectives

In this chapter you will:

- learn the difference between the perimeter and area of a shape
- find the perimeter and area of two-dimensional shapes and solve problems involving area.

Before you start

You need to be able to:

- measure the length of a line
- change measurements between millimetres (mm), centimetres (cm), metres (m) and kilometres (km).

18.1 Perimeter

⊙ Objectives

○ You can find the perimeter of simple shapes.
○ You can find the perimeter of shapes made from squares, rectangles and triangles.

⊕ Why do this?

A gardener will need to know the distance around a boundary edge in order to calculate the length of fence needed to surround a garden.

⊕ Get Ready

1. What is the length and width of the shaded rectangle drawn on the centimetre grid?

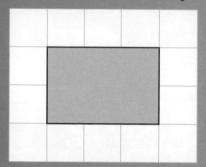

2. Measure the lengths of the sides of the following triangle.

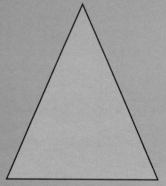

Key Points

⊙ The **perimeter** of a **two-dimensional (2D)** shape is the total distance around the edge of the shape. The examples show you how to work out the perimeter of a variety of shapes.

⊙ To work out the perimeter of a rectangle you can use the following formula.
Perimeter of a rectangle $= l + w + l + w$
$$= 2l + 2w$$

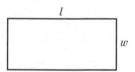

$l =$ the length of the rectangle
$w =$ the width of the rectangle

🔍 Example 1

What is the perimeter of the following shape?

Perimeter $= 3 + 2 + 4 + 5$
$= 14\,cm$ ← Add the lengths of all four sides to find the perimeter.

3 cm
2 cm
4 cm
5 cm

ResultsPlus
Examiner's Tip

Always remember to include the units in your answer.

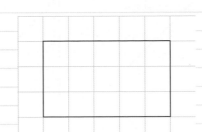 **Example 2**

The diagram shows a rectangle drawn on a centimetre grid.

Work out the perimeter of the rectangle.

Method 1

Perimeter = 5 + 3 + 5 + 3

= 16 cm ← Work out the length of each side and add them together.

Method 2

Perimeter = $2l + 2w$

= $(2 \times 5) + (2 \times 3)$

= 10 + 6

= 16 cm ← Use the formula for the perimeter of a rectangle.

Exercise 18A

Questions in this chapter are targeted at the grades indicated.

1 Here are three shapes drawn on a centimetre grid.

Work out the perimeter of each shape.

G

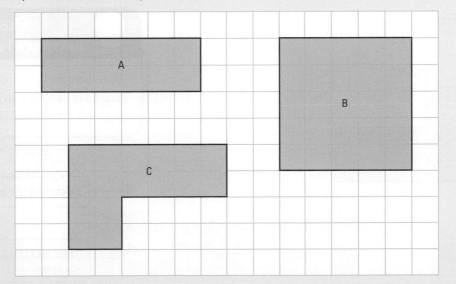

F

2 Work out the perimeters of the following shapes.

a

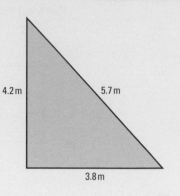

b

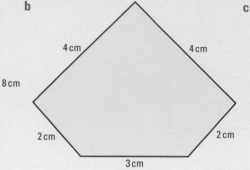

c

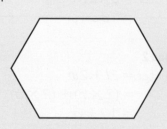

A02 **3** These shapes are drawn accurately. Find the perimeter of each shape.

a **b** **c**

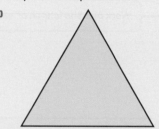

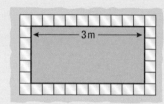

A03 **4** Jenny has a rectangular pond in her garden.
The length of the pond is 3 m.
The width of the pond is half the length of the pond.
She wants to put a low fence around the edge of the pond.
What length of fencing does Jenny need?

Example 3 Work out the perimeter of the following shape.

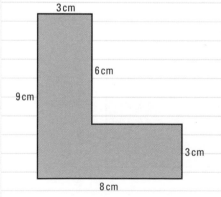

ResultsPlus
Examiner's Tip

Make sure that you know the lengths of all the sides before you work out the perimeter.

Missing length = 8 − 3
= 5 cm ← First, find the missing length.

Perimeter = 3 + 6 + 5 + 3 + 8 + 9
= 34 cm ← Then, add the lengths of all six sides.

Exercise 18B

1   Each side of a regular pentagon has length 8 cm.
Work out the perimeter of the pentagon.

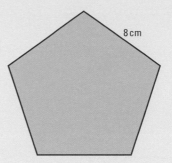

2   Work out the perimeters of the following three shapes.

a
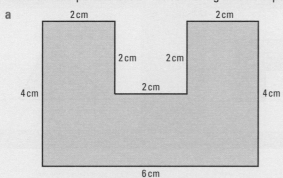

b
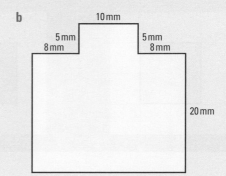

c
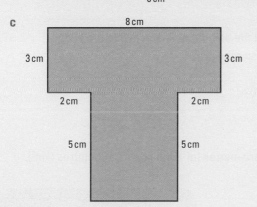

3   Work out the perimeter of this trapezium.

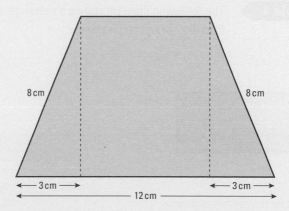

4   The perimeter of an equilateral triangle is 21.6 cm.
Work out the length of each side.

18.2 Area

⊚ **Objectives**

○ You can find the area of simple shapes by counting squares.

○ You can find the area of more complicated shapes by counting squares.

⊘ **Why do this?**

It would be useful to know the area of a wall you were going to paint, so that you could buy the right amount of paint for the job.

⬧ **Get Ready**

1. How many squares do you need to cover each of the shaded shapes?

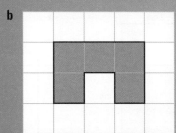

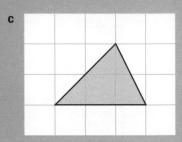

🔍 **Key Points**

⊚ The **area** of a two-dimensional (2D) shape is a measure of the amount of space inside the shape.

⊚ The area of a square with sides of length 1 cm is 1 square centimetre.

1 cm

1 cm

This is written as 1 cm^2.

⊚ Other units of area include square millimetres (mm^2), square metres (m^2) and square kilometres (km^2).

🔍 **Example 4** The following diagram shows a rectangle drawn on a centimetre grid. Find the area of the rectangle.

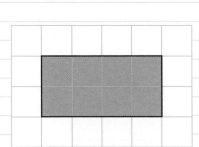

Count the number of squares.

Each square has area of 1 cm^2.

Area = 8 squares = 8 cm^2.

Example 5 Estimate the area of the shaded shape drawn on the centimetre grid.

These cover about 1 square

These cover $1\frac{1}{2}$ squares

Number of whole squares = 6 ← Count the whole squares.

Number of part squares = $1 + 1\frac{1}{2}$ ← Estimate the number of squares covered by the other parts.

$= 6 + 1 + 1\frac{1}{2}$
$= 8\frac{1}{2}\,\text{cm}^2$ ← Add your answers. State the units.

Exercise 18C

1 Find the area of the shape shown on the centimetre grid.
Give the units with your answer.

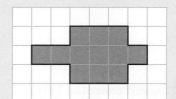

G

2 The diagram shows four shapes drawn on a centimetre grid.
Find the area of each shape.

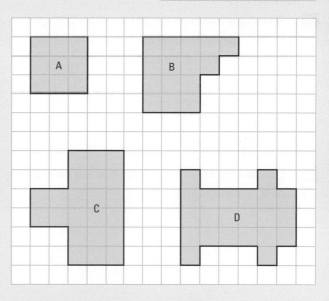

3 Find the area of each of the four triangles, T_1, T_2, T_3 and T_4, drawn on the centimetre grid.

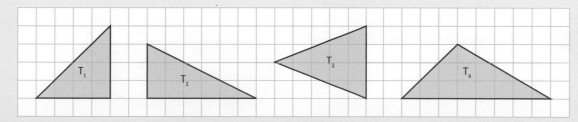

G

4   Find the area of each of the three shapes, F, G and H, drawn on the centimetre grid.

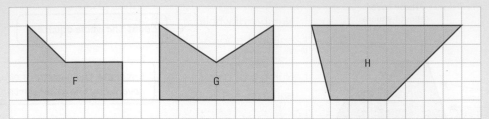

5   Estimate the area of each of the three shapes, P, Q and R, drawn on the centimetre grid.

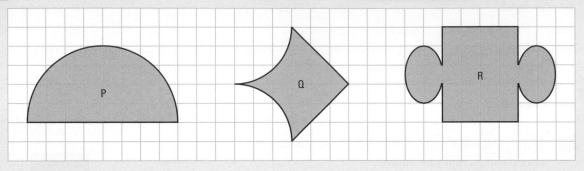

Mixed exercise 18D

G

1   This shape has been drawn on a centimetre grid.
   a   Find the perimeter of the shape.
   b   Find the area of the shape.

2   Estimate the area of this shape that has been drawn
     on a centimetre grid.

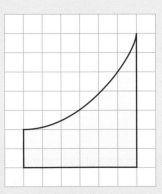

E

3   Work out the perimeter of the shape below.

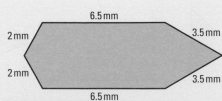

4 The perimeter of triangle ABC is 52 m.
What is the length of side AB?

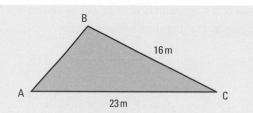

18.3 Finding areas using formulae

◉ Objectives

- You can find the areas of rectangles, triangles and parallelograms.
- You can use the formula to find the area of a trapezium.

◈ Why do this?

It might be necessary for your school to find out the area of a field, so that they can work out if it is big enough for an athletics track.

◈ Get Ready

The diagram shows a triangle drawn on a centimetre grid.
The length of the base is 5 cm.
The vertical height is 3 cm.

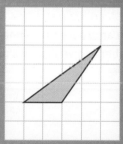

State the length of the base and the vertical height of these triangles.

1.

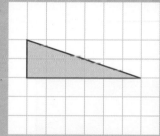

2.

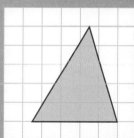

3.

🔍 Key Points

- To work out the area of a rectangle, square, triangle and parallelogram use the following formulae.

Rectangle
Area of a rectangle
= length × width
= $l \times w$

Parallelogram
You can cut a triangle off a parallelogram and put it on the other side to make a rectangle, so:
Area of a parallelogram
= base × vertical height
= $b \times h$

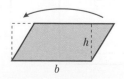

Triangle
The area of a triangle is half the area of a rectangle that surrounds it.
Area of a triangle
= $\frac{1}{2}$ × base × vertical height
= $\frac{1}{2} \times b \times h$

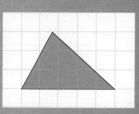

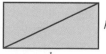

Square
Area of a square
= length × length
= $l \times l$
= l^2

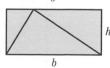

● The area of a trapezium is worked out by finding the average of the lengths of the parallel sides and multiplying by the distance between them.

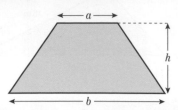

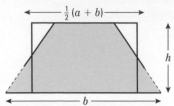

A formula is used to work out the area of a trapezium:

Area = $\frac{1}{2}$ × sum of the lengths of the parallel sides × distance between the parallel sides

$A = \frac{1}{2}(a + b)h$

Example 6 ▸ Work out the area of this rectangle.

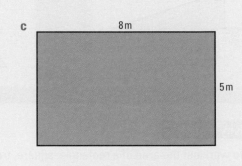

Area of rectangle ← Choose the formula to use.

= length × width

= $l \times w$ ← Put in the values of l and w.

= 6 × 4

= 24 cm² ← Remember to include the units.

Exercise 18E

E

1 Find the area of the following rectangles.
Remember to give the units with your answer.

a 5 cm, 3 cm

b 2 mm, 7 mm

c 8 m, 5 m

2 The following three squares are accurately drawn. Find the area of each one.

a

b

c

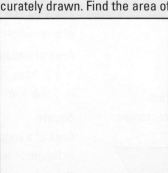

E

3   A decorator wants to paint two rectangular walls.
    The walls are 4.5 m by 2.6 m and 3.1 m by 2.6 m.
    What is the total area of the two walls to be painted?

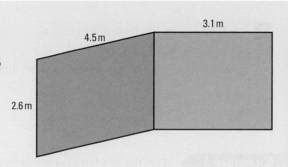

Example 7   Work out the area of this triangle.

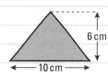

Area of triangle

$= \frac{1}{2} \times$ base $\times$ vertical height   ← Choose the formula to use.

$= \frac{1}{2} \times b \times h$

$= \frac{1}{2} \times 10 \times 6$ ← Put in the values of b and h.

$= 30\,\text{cm}^2$ ← Remember to include the units.

Exercise 18F

D

1   Find the area of the following triangles.

a
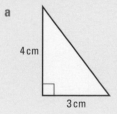
4 cm
3 cm

b
12 m
8 m

c
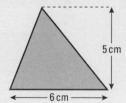
5 cm
6 cm

d
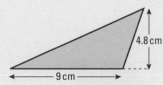
4.8 cm
9 cm

e
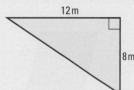
5.4 mm
6.2 mm

2   Find the area of this triangle.

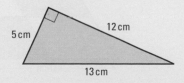
5 cm
12 cm
13 cm

D

3 A company makes flags in this shape.
It makes 50 identical flags.
Work out the area of fabric used to make these flags.

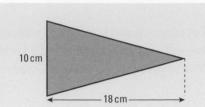

10 cm

18 cm

Example 8 Find the shaded area in the company logo.

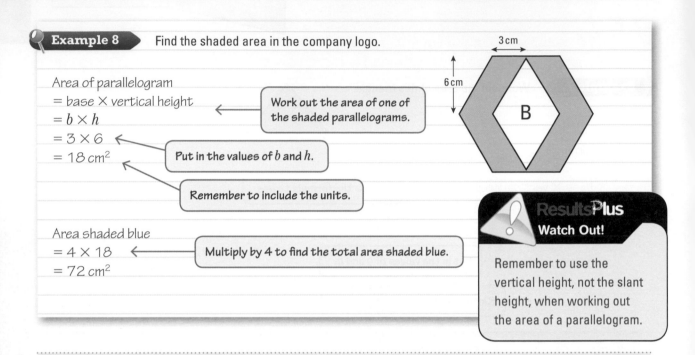

Area of parallelogram
= base × vertical height
= $b \times h$
= 3×6
= $18\,cm^2$

Work out the area of one of the shaded parallelograms.

Put in the values of b and h.

Remember to include the units.

3 cm

6 cm

B

Area shaded blue
= 4×18
= $72\,cm^2$

Multiply by 4 to find the total area shaded blue.

ResultsPlus
Watch Out!

Remember to use the vertical height, not the slant height, when working out the area of a parallelogram.

Exercise 18G

D

1 Find the areas of the parallelograms drawn on the centimetre grid.

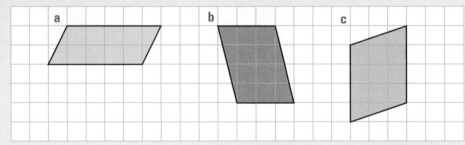

a   b   c

A03

2 A tiler creates the following pattern using
parallelogram-shaped tiles.
Work out the total area covered by the red tiles.

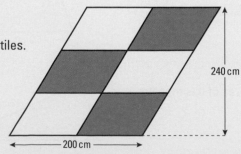

240 cm

200 cm

Example 9   Work out the area of this trapezium.

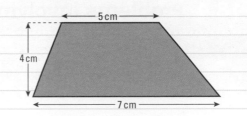

Area of trapezium

$= \frac{1}{2} \times$ sum of parallel sides $\times$ distance between
parallel sides

$= \frac{1}{2}(a + b)h$ ←   Choose the formula to use.

$= \frac{1}{2} \times (5 + 7) \times 4$ ←   Put in the values of a, b and h.

$= \frac{1}{2} \times 12 \times 4$

$= 24 \, cm^2$

Exercise 18H

1   Copy and complete the table to find the area of each trapezium.

	a	b	h	Area
Trapezium 1	4 cm	6 cm	3 cm	
Trapezium 2	10 cm	12 cm	5 cm	
Trapezium 3	9 m	7 m	6 m	
Trapezium 4	5 m	10 m	4 m	

2   Work out the area of each trapezium.

a
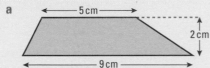

b
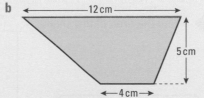

c
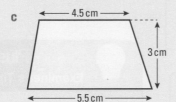

d
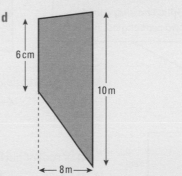

3   A trapezium has an area of 40 cm².
The two parallel sides have lengths 7 cm and 13 cm.
The distance between the two parallel sides is h cm.
Work out the value of h.

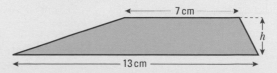

C

18.4 Problems involving areas

◉ Objective

○ You can find the area of a more complicated shape by splitting it up into simple shapes.

⑦ Why do this?

Garden patios are not always simple shapes. A gardener will need to work out the area of a new patio to know how many tiles to buy.

⬆ Get Ready

1. Here are some shapes. Copy them and draw lines to show how each one can be split up into squares, rectangles and triangles.

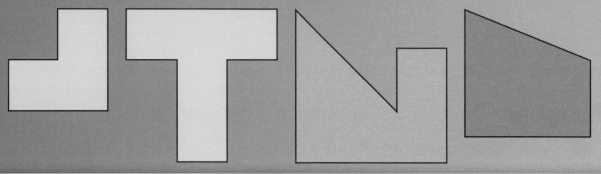

🔍 Key Point

◉ To find the area of more complicated shapes you will need to split the shape into a number of simpler shapes such as rectangles, squares, triangles or parallelograms. You can then find the area of each part and add these areas together to find the total area.

The following examples will show you how to do this.

Example 10 Find the area of this shape.

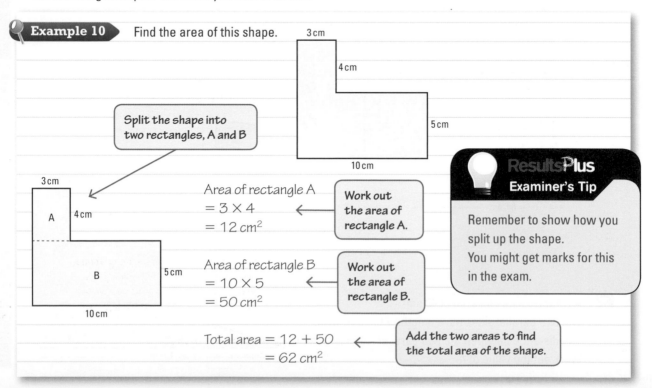

Split the shape into two rectangles, A and B

Area of rectangle A
= 3 × 4
= 12 cm²

Work out the area of rectangle A.

Area of rectangle B
= 10 × 5
= 50 cm²

Work out the area of rectangle B.

Total area = 12 + 50
= 62 cm²

Add the two areas to find the total area of the shape.

Results Plus
Examiner's Tip

Remember to show how you split up the shape. You might get marks for this in the exam.

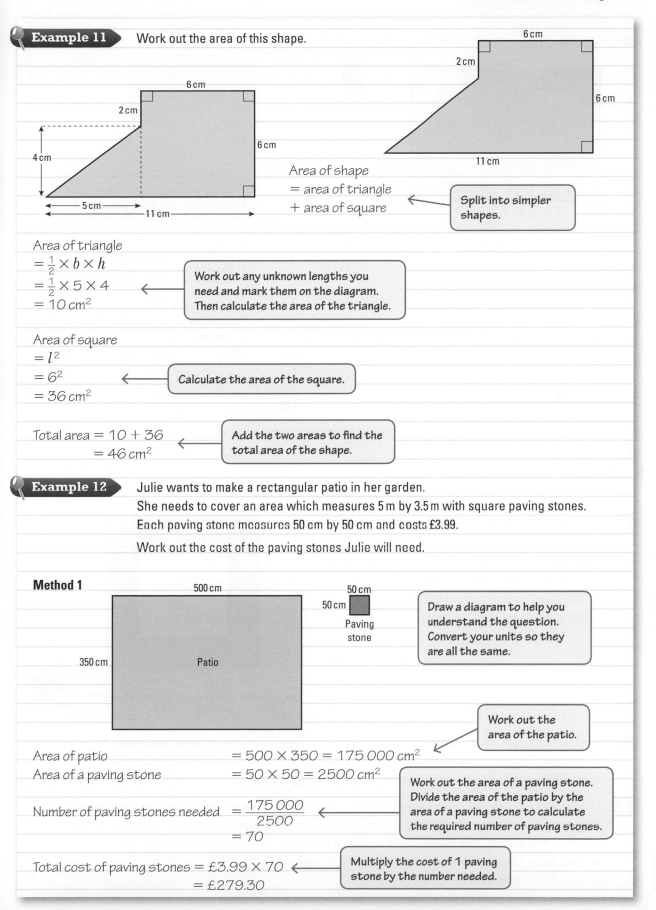

Example 11 ▶ Work out the area of this shape.

Area of shape
= area of triangle
+ area of square

Split into simpler shapes.

Area of triangle
$= \frac{1}{2} \times b \times h$
$= \frac{1}{2} \times 5 \times 4$
$= 10 \, cm^2$

Work out any unknown lengths you need and mark them on the diagram. Then calculate the area of the triangle.

Area of square
$= l^2$
$= 6^2$
$= 36 \, cm^2$

Calculate the area of the square.

Total area = 10 + 36
= 46 cm²

Add the two areas to find the total area of the shape.

Example 12 ▶ Julie wants to make a rectangular patio in her garden.
She needs to cover an area which measures 5 m by 3.5 m with square paving stones.
Each paving stone measures 50 cm by 50 cm and costs £3.99.

Work out the cost of the paving stones Julie will need.

Method 1

Draw a diagram to help you understand the question. Convert your units so they are all the same.

Work out the area of the patio.

Area of patio   = 500 × 350 = 175 000 cm²
Area of a paving stone   = 50 × 50 = 2500 cm²

Work out the area of a paving stone. Divide the area of the patio by the area of a paving stone to calculate the required number of paving stones.

Number of paving stones needed $= \frac{175\,000}{2500}$
= 70

Total cost of paving stones = £3.99 × 70
= £279.30

Multiply the cost of 1 paving stone by the number needed.

Method 2

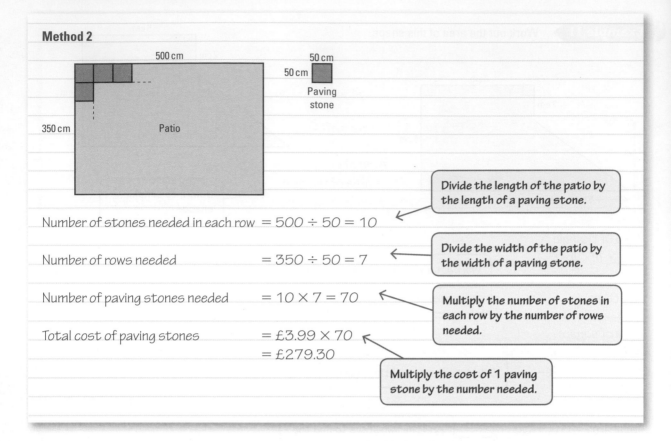

Number of stones needed in each row $= 500 \div 50 = 10$

> Divide the length of the patio by the length of a paving stone.

Number of rows needed $= 350 \div 50 = 7$

> Divide the width of the patio by the width of a paving stone.

Number of paving stones needed $= 10 \times 7 = 70$

> Multiply the number of stones in each row by the number of rows needed.

Total cost of paving stones $= £3.99 \times 70$
$= £279.30$

> Multiply the cost of 1 paving stone by the number needed.

Exercise 18I

E

1 A square piece of card has sides of length 10 cm.
A hole is cut from the card.
The hole is a square of side 6 cm.
a Work out the area of the large square.
b Work out the area of the small square.
c Work out the area of the card left.

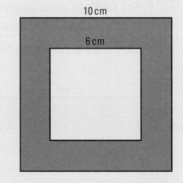

D

A02

2 The floor of the hall in a house is a 225 cm by 150 cm rectangle. Tiles which are squares of side 15 cm are used to tile the floor. Work out how many tiles are needed.

A02

3 Liam wants to replace the carpet in his room. The floor of the room is a rectangle measuring 4 metres by 3 metres. The carpet he wants to buy costs £8.65 per square metre.
Work out how much it will cost Liam to buy enough carpet to cover the floor.

A03

4 Libby wants to buy some grass seed so that she can sow a new lawn in her garden. She wants the lawn to be a rectangle measuring 3.2 metres by 2.5 metres. She needs 35 grams of lawn seed for every square metre of lawn. One box of lawn seed contains 250 g.
a How many boxes of lawn seed will Libby need to buy?
b How much lawn seed will be left over?

5 Find the area of the following shapes.

a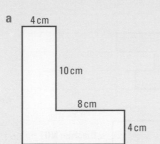

b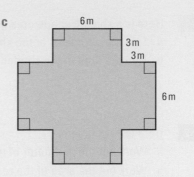

c

6 Work out the shaded area in each diagram.

a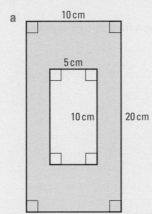

b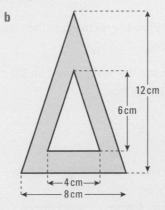

c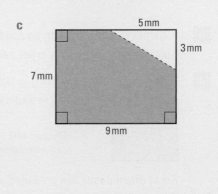

Chapter review

- The **perimeter** of a **2D shape** is the total distance around the edge of the shape.
- The perimeter of a rectangle can be found using the formula
  Perimeter of a rectangle $= l + w + l + w$
  $$= 2l + 2w$$
- The **area** of a 2D shape is the amount of space inside the shape.
- The area of a 2D shape can be found by counting squares or by using the formulae:
  - Area of rectangle = length × width
    $$= l \times w$$
  - Area of square = length × length
    $$= l^2$$
  - Area of triangle $= \frac{1}{2} \times$ base × vertical height
    $$= \frac{1}{2} \times b \times h$$
  - Area of parallelogram = base × vertical height
    $$= b \times h$$
  - Area of trapezium $= \frac{1}{2} \times$ sum of the lengths of the parallel sides × distance between the parallel sides
    $$= \frac{1}{2}(a + b)h$$
- To find the area of more complicated shapes you will need to split the shape into a number of simpler shapes such as rectangles, squares, triangles or parallelograms. You can then find the area of each part and add these areas together to find the total area.

⚙ **Review exercise**

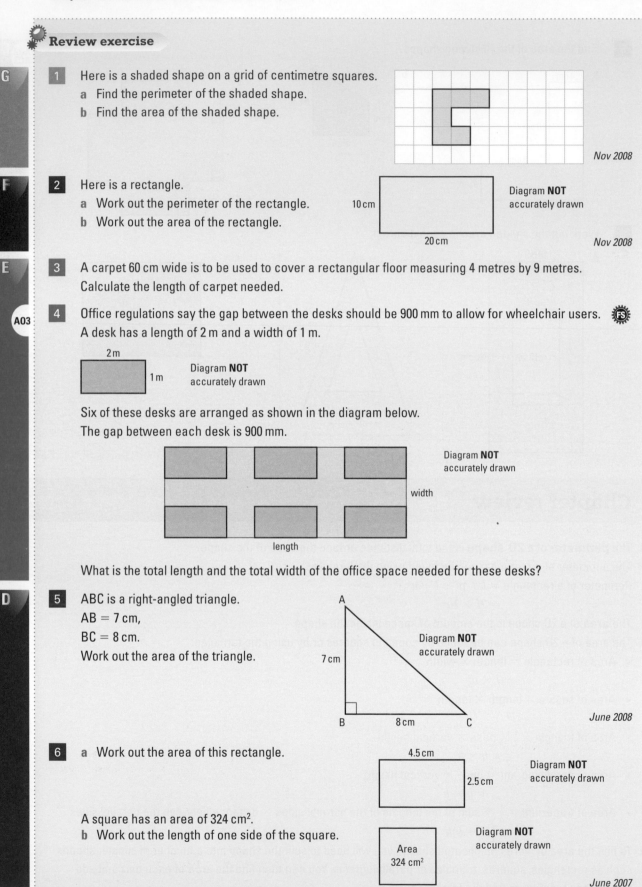

G

1 Here is a shaded shape on a grid of centimetre squares.
   a Find the perimeter of the shaded shape.
   b Find the area of the shaded shape.

Nov 2008

F

2 Here is a rectangle.
   a Work out the perimeter of the rectangle.
   b Work out the area of the rectangle.

10 cm

20 cm

Diagram **NOT** accurately drawn

Nov 2008

E

3 A carpet 60 cm wide is to be used to cover a rectangular floor measuring 4 metres by 9 metres. Calculate the length of carpet needed.

A03

4 Office regulations say the gap between the desks should be 900 mm to allow for wheelchair users. Ⓕ
A desk has a length of 2 m and a width of 1 m.

2 m
1 m

Diagram **NOT** accurately drawn

Six of these desks are arranged as shown in the diagram below.
The gap between each desk is 900 mm.

Diagram **NOT** accurately drawn

width

length

What is the total length and the total width of the office space needed for these desks?

D

5 ABC is a right-angled triangle.
AB = 7 cm,
BC = 8 cm.
Work out the area of the triangle.

A

7 cm

B   8 cm   C

Diagram **NOT** accurately drawn

June 2008

6 **a** Work out the area of this rectangle.

4.5 cm

2.5 cm

Diagram **NOT** accurately drawn

A square has an area of 324 cm².
   b Work out the length of one side of the square.

Area
324 cm²

Diagram **NOT** accurately drawn

June 2007

7 The diagram shows Rob's patio.
All the corners are right angles.
The patio is made up of square paving
stones each 50 cm by 50 cm.
Work out how many of these paving
stones are needed to tile Rob's patio.

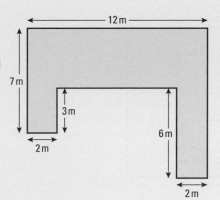

Diagram **NOT**
accurately drawn

8 A room has four interior walls.

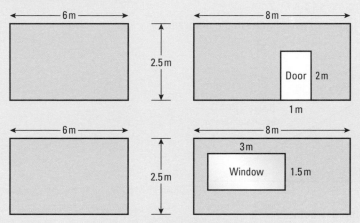

Diagram **NOT**
accurately drawn

Alesha paints the walls with emulsion paint. She does not paint the door.
A 3 litre tin of emulsion paint covers 30 m² of wall.
Work out how many 3 litre tins she needs to buy. Show all your working.

9

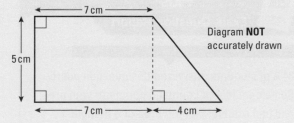

Diagram **NOT**
accurately drawn

Work out the area of the shape.

Nov 2008

10 The diagram shows a rectangle inside a triangle.
The triangle has a base of 12 cm and a height of 10 cm.
The rectangle is 5 cm by 3 cm.
Work out the area of the region shown shaded in the
diagram.

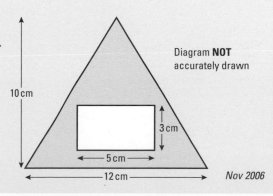

Diagram **NOT**
accurately drawn

Nov 2006

C A02 A03

11

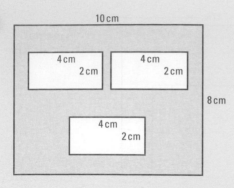

Diagram **NOT** accurately drawn

The diagram shows 3 small rectangles inside a large rectangle.

The large rectangle is 10 cm by 8 cm.

Each of the 3 small rectangles is 4 cm by 2 cm.

Work out the area of the region shown shaded in the diagram.

June 2007

12

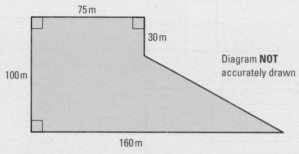

Diagram **NOT** accurately drawn

The diagram shows the plan of a field.

The farmer sells the field for £3 per square metre.

Work out the total amount of money the farmer should get.

March 2007

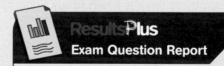

ResultsPlus
Exam Question Report

84% of students answered this question poorly. Some candidates confused perimeter with area, and had trouble with the area of a triangle.

The photo shows a work of art by the artists Christo and Jeanne-Claude in which they wrapped the Pont Neuf Bridge in Paris in $40\,876$ m^2 ($454\,178$ sq ft) of silky golden fabric. In order to wrap the bridge, they needed to work out the surface area so they could calculate the amount of fabric required.

Objectives

In this chapter you will:
- learn how to recognise and draw 3D shapes
- find the volumes of shapes made from cuboids
- find the volumes and surface areas of prisms.

Before you start

You need to be:
- familiar with two-dimensional shapes.

19.1 Recognising three-dimensional shapes

Objectives

○ You can recognise 3D shapes.
○ You can count the vertices, edges and faces of 3D shapes.

Why do this?

A designer or architect needs to be able to describe the shape they want to build. The correct names for 3D shapes make it easier to explain what the product will look like.

Get Ready

1. Can you remember the names of these two-dimensional shapes?

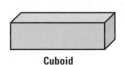

Key Points

● Here are some **three-dimensional** shapes.

Cube

Cylinder

Cuboid

ResultsPlus
Examiner's Tip

You need to know the names of all of these shapes.

Sphere **Cone**

● **Pyramids** have a base, which can be any shape, and sloping triangular sides.

Triangular pyramid called a tetrahedron   **Rectangular-based pyramid**

● A **prism** has two parallel faces and a number of rectangular sides joining them.

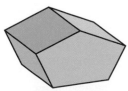
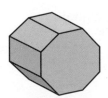

Triangular prism   **Pentagonal prism**   **Octagonal prism**

● The flat surfaces of a 3D shape are called **faces**.
● The lines where two faces meet are called **edges**.
● The point (corner) at which edges meet is called a **vertex**. The plural of vertex is vertices.

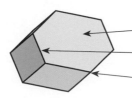

Face (two-dimensional surface).

Edge (where two faces meet).

Vertex (corner where three or more edges meet).

Example 1
a Name this shape.
b How many faces does it have?
c How many edges does it have?
d How many vertices does it have?

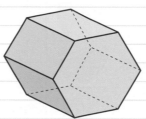

Two parallel hexagons
Rectangular sides

a Hexagonal prism
b 8 faces
c 18 edges
d 12 vertices

The parallel faces are called the cross-section of the prism. The cross-section of this prism is a hexagon.

Exercise 19A

Questions in this chapter are targeted at the grades indicated.

1 Name this shape.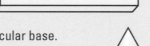

2 This 3D shape has a circular base.
Name the 3D shape.

3 A prism has a base which has five sides. What type of prism is it?

4 This pyramid has a special name. Name this pyramid.

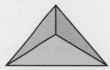

5 Look at the picture. Copy and complete the table with as many 3D shapes as you can find.

	Shape	Object
1	sphere	football
2		
3		
4		
5		
6		
7		
8		
9		
10		
11		
12		

G

⚙ **Exercise 19B**

1 Copy and complete the table below.

	Shape	Faces	Edges	Vertices
A	Cube			
B	Pentagonal prism			
C	Triangular prism			
D	Square-based pyramid			
E	Cuboid			
F	Tetrahedron			
G	Octagonal prism			

2 What is the shape of the cross-section of this prism?

3 Draw a sketch of a prism with a pentagonal cross-section.

4 A pyramid has six triangular faces.
   a What is the shape of its other face?   **b** What type of pyramid is it?

19.2 Isometric paper

◎ **Objective**

● You can draw 3D shapes using isometric paper.

⊘ **Why do this?**

A designer making a 3D container must be able to make an accurate 3D drawing of it.

⬦ **Get Ready**

Draw these shapes.

1. A cube   **2.** A triangular prism   **3.** A square-based pyramid

Key Points

● **Isometric paper** will help you to make scale drawings of three-dimensional objects.
● Isometric paper must be the right way up – **vertical** lines down the page and no **horizontal** lines.

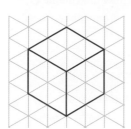

This cube has sides
of 2 cm.

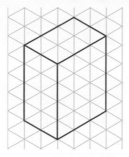

This cuboid has height
4 cm, length 3 cm and
width 2 cm.

This prism has a
triangular face.

Shapes can be
joined together.

ResultsPlus
Examiner's Tip

Draw the shape at an angle
as if you are looking from the
bottom-right corner.

Exercise 19C

1 On isometric paper draw a cube of side 3 cm.

2 Use isometric paper to draw a cuboid with height 2 cm, width 4 cm and length 3 cm.

3 The diagram shows a shape made up of three cubes.
   On isometric paper draw a different shape made up of the same three cubes.

4 Use isometric paper to make full-sized drawings of these prisms.

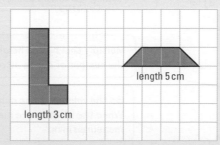

length 3 cm

length 5 cm

19.3 Volume of a prism

Objectives

- You can find the volumes of shapes made from cuboids.
- You can find the volumes of prisms.

Why do this?

Food manufacturers need to know the volume of food containers, for example, cereal packets or drink cartons.

Get Ready

1. What is the area of this rectangle? Each small square has side 1 cm.

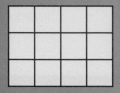

2. What is the area of this shape? Each small square has side 1 cm.

3. Use the formula
area = length × width
to find the area of this shape.

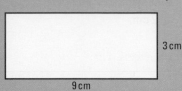

3 cm

9 cm

4. Find the area of these two-dimensional shapes.

a

4 cm

5 cm

b

7 cm

3 cm

c

3 cm

4 cm

2 cm

Key Points

- The **volume** of a 3D shape is the amount of space it takes up. The diagram shows a cube of side 1 cm. Its volume is 1 cm³.

- The volume of a 3D shape with measurements in centimetres is the number of centimetre cubes it contains.
- The volume of a prism is the area of the **cross-section** × its length.

Example 2

The diagram shows a cuboid made from centimetre cubes. Find the volume of the cuboid.

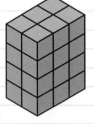

Remember some of the cubes are hidden. Try building the shape using multilink.

Multiply the length by the width (3 × 2) to find the number of cubes (6) in each layer.

There are 6 centimetre cubes in each layer.

Multiply this by the height (4) to give the number of cubes altogether (24).

There are 4 layers.

So the volume of the cuboid = length × width × height. When the lengths are measured in metres, volume is measured in m³.

The volume of the cuboid is 6 × 4 = 24 cm³.

Example 3

Work out the volume of the cuboid shown below.

Volume of the cuboid = length × width × height
Volume = 8 × 3 × 2 m³ = 48 m³

ResultsPlus
Examiner's Tip

Remember, you need to specify the units in all of your answers.

Exercise 19D

G

1 The diagrams below show prisms that have been made from centimetre cubes.
   Find the volume of each prism.

   a   b   c

2 Work out the volumes of these cuboids.

   a   b

E

3 Work out the volume of the following cuboids.
   a length 7 m, width 8 m and height 4 m
   b length 17 mm, width 12 mm and height 3 mm

4 Work out the volume of a cube of side 7 cm.

5 A cuboid has volume 150 cm³. Its width is 3 cm and its height is 10 cm.
   Find the length of the cuboid.

6 A cereal packet was measured and found to have height 35 cm, width 20 cm and depth 10 cm.
   Give the dimensions of a box that will hold 24 cereal packets.

A03

Example 4   Work out the volume of this prism.

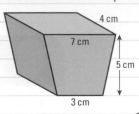

ResultsPlus
Examiner's Tip

If you can't remember how to use the formula, you can split the trapezium into a rectangle and two triangles.

The cross-section is a trapezium. ← **Decide on the shape of the cross-section.**

area of a trapezium $= \frac{1}{2}(a + b) \times l$

$$= \frac{1}{2} \times (3 + 7) \times 5 \quad ←\quad \text{Put in the lengths you know.}$$

$$= 25\,\text{cm}^2 \quad ←\quad \text{Don't forget the units.}$$

$$\text{Volume} = 25 \times 4$$
$$= 100\,\text{cm}^3$$

 Exercise 19E

1   The diagram shows a prism.
The cross-section of this prism is a hexagon with an area of $5\,\text{m}^2$.
If the length of the prism is 3 m, what is the volume of the prism?

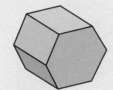

2   The diagram shows the cross-section of a prism of length 8 cm.
Work out the volume of the prism.

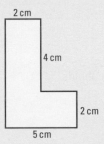

3   The diagram shows a prism with a right-angled triangle as its cross-section.
Work out the volume of the prism.

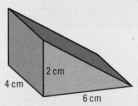

A02
A03   **4**   A metal bar of length 30 cm has a square cross-section of side 5 cm.
The metal is to be recast into triangular rods of length 10 cm. The cross-section of the rods is a right-angled triangle as shown.
   a   Work out the volume of the metal bar.
   b   Work out the volume of each rod.
   c   Work out the maximum number of rods that can be made from the bar.
   d   Work out how much metal is left over.

5   A triangular prism has length 12 cm. The triangular face has base 8 cm and height 9 cm.
Calculate the volume of the prism.

19.4 Surface area of a prism

Objective

⦿ You can find the surface area of a prism.

Why do this?

The surface area is the amount of space on an object available for design, information or advertising.

Get Ready

1. Find the area of these two-dimensional shapes.

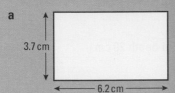

a
3.7 cm
6.2 cm

b
3.7 m
4.8 m

Key Points

⦿ The **surface area** of a prism is the area of the net that can be used to build the shape.

⦿ Area is measured in square units — mm^2, cm^2, m^2 and km^2 are common.

⦿ Each of the six faces of a cube is a square, so the surface area of a cube is 6 times the area of one face.

⦿ The area of each square face of this cube is $2 \times 2 = 4\ cm^2$.
So the surface area of the cube is $6 \times 4 = 24\ cm^2$.

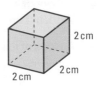
2 cm
2 cm
2 cm

Example 5 Work out the surface area of this cuboid.

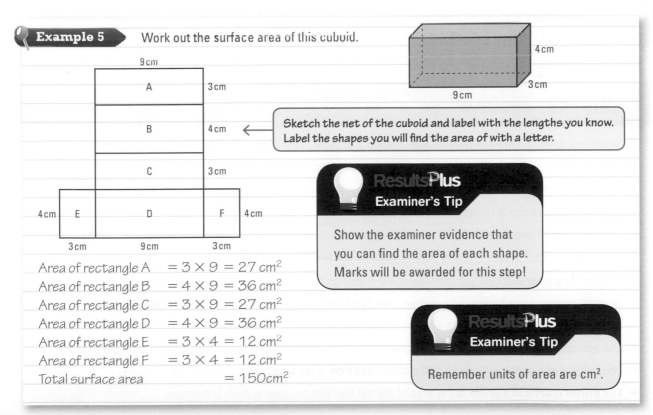

4 cm
9 cm
3 cm

9 cm

A 3 cm

B 4 cm

C 3 cm

4 cm E D F 4 cm

3 cm 9 cm 3 cm

Sketch the net of the cuboid and label with the lengths you know. Label the shapes you will find the area of with a letter.

Area of rectangle A $= 3 \times 9 = 27\ cm^2$
Area of rectangle B $= 4 \times 9 = 36\ cm^2$
Area of rectangle C $= 3 \times 9 = 27\ cm^2$
Area of rectangle D $= 4 \times 9 = 36\ cm^2$
Area of rectangle E $= 3 \times 4 = 12\ cm^2$
Area of rectangle F $= 3 \times 4 = 12\ cm^2$
Total surface area $= 150 cm^2$

ResultsPlus
Examiner's Tip

Show the examiner evidence that you can find the area of each shape. Marks will be awarded for this step!

ResultsPlus
Examiner's Tip

Remember units of area are cm^2.

Exercise 19F

E

1 Find the surface area of this cuboid.

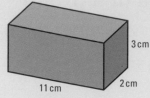

3 cm
2 cm
11 cm

A02 **2** The diagram shows a piece of cheese.
Work out the surface area of the cheese.

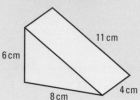

11 cm
6 cm
8 cm
4 cm

A02 **3** A cereal packet in the shape of a cuboid has height 35 cm, width 20 cm and depth 20 cm.
Work out the surface area of the cereal packet.

D
A02 **4** The diagram shows a plastic part from a child's toy.
Work out the surface area of the part.

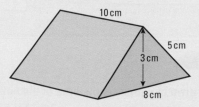

10 cm
5 cm
3 cm
8 cm

C
A02
A03 **5** A cube has a surface area of 24 cm².
Work out the length of the side of the cube.

A03 **6** A block of icing is sold in the shape of a prism.
The cross section of the prism is in the shape of a square.
The length of the prism is twice the length of the side of the square.
The surface area of the prism is 90 cm².
Work out the length of the side of the square.

Chapter review

● Some examples of **three-dimensional** shapes are: **cube, cylinder, cuboid, sphere** and **cone**.

● **Pyramids** have a base which can be any shape, and sloping triangular sides.

● A **prism** has two parallel faces and a number of rectangular sides joining them.

● The flat surfaces of a 3D shape are called **faces**.

● The lines where two faces meet are called **edges**.

● The point (corner) at which edges meet is called a **vertex**. The plural of vertex is vertices.

● **Isometric paper** will help you to make scale drawings of three-dimensional objects.

● Isometric paper must be the right way up – **vertical** lines down the page and no **horizontal** lines.

● The **volume** of a 3D shape is the amount of space it takes up.

● The volume of a 3D shape with measurements in centimetres is the number of centimetre cubes it contains.

● The volume of a prism is the area of the **cross-section** × its length.

● The **surface area** of a prism is the area of the net that can be used to build the shape.

Review exercise

1 Here is a solid prism made from centimetre cubes.

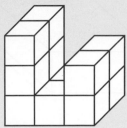

Work out the volume of the solid prism.

Nov 2008

2 Write down the mathematical name of each of these two 3D shapes.

i   ii

Nov 2008

3 Find the volume of this prism.

Diagram **NOT** accurately drawn

represents 1 cm³

June 2008

4 Here is a diagram of a cuboid.

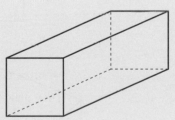

Write down the number of
i faces   ii edges   iii vertices.

June 2008

5 Here is a diagram of a 3D prism.

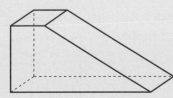

Write down the number of
i faces   ii edges   iii vertices.

Nov 2007

E

6   Work out the volume of the cuboid.

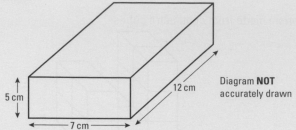

Diagram **NOT**
accurately drawn

5 cm

7 cm

12 cm

Nov 2008

7   The diagram shows a solid cuboid.
On a triangular isometric grid, make an accurate
full-size drawing of the cuboid.

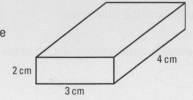

2 cm

3 cm

4 cm

June 2007

D

8   Cereal boxes are packed into cartons.
A cereal box measures 4 cm by 6 cm by 10 cm.
A carton measures 20 cm by 30 cm by 60 cm.

The carton is completely filled with cereal boxes.
Work out the number of cereal boxes that
will completely fill **one** carton.

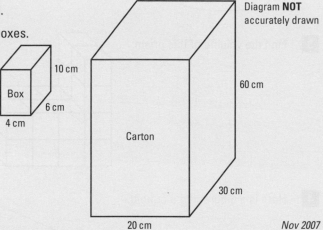

Diagram **NOT**
accurately drawn

10 cm

Box

6 cm

4 cm

60 cm

Carton

30 cm

20 cm

Nov 2007

C

9   Here is a triangular prism.
Calculate the volume of the prism.

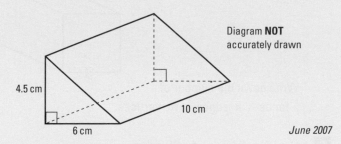

Diagram **NOT**
accurately drawn

4.5 cm

6 cm

10 cm

June 2007

10   A solid cube has sides of length 5 cm.
Work out the total surface area of the cube.
State the units of your answer.

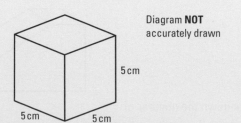

Diagram **NOT**
accurately drawn

5 cm

5 cm

5 cm

Nov 2009

11   Work out the total surface area of the triangular prism.

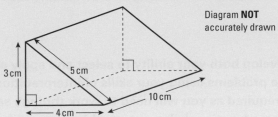

Diagram **NOT**
accurately drawn

3 cm

5 cm

10 cm

4 cm

May 2008

MULTIPLICATION

The following question helps you to develop both your ability to select and apply a method (AO2) and your ability to solve problems using your skills of interpretation (AO3). Your AO3 skills are particularly required as you will need to work through several steps to solve this problem. There are also some functional elements as this is a real-life situation and there is a problem to solve.

Example

Adam runs a coach company. He has 6 small coaches, 4 medium coaches, 3 large coaches and 1 double-decker coach.

The table gives information on how many passengers each coach can seat, the cost of hiring the coach and a driver for a day, and how many of these coaches Adam owns.

Adam's Coach Company			
Coach type	Number of seats	Cost of hire	Number owned
Small	25	£100	6
Medium	38	£110	4
Large	54	£120	3
Double-decker	78	£140	1

Rachel wants to hire some coaches from Adam to take 222 people out for the day.
What is the cheapest way for Rachel to do this?

> As the number of seats increases, the cost goes down proportionally. Therefore you need to use the largest coach, the double-decker, first.

Solution

1 double-decker	£140	78
3 large	£360 +	162 +
	£500	240 seats

← This leaves 144 people to fit in. This could be done with three large coaches but would leave 18 empty seats.

1 double-decker	£140	78
2 large	£240	108
1 medium	£110 +	38 +
	£490	224 seats

← If two large coaches were used then this would leave 36 people to fit in, so a medium coach would be needed as well.

The cheapest way is £490 and there are only two spare seats.

Now try these

1   Sam is a salesman. He gets paid expenses when he drives his car on
company business.

He gets paid 45p for each mile he drives.

He also gets paid a meal allowance.

Here is Sam's time and mileage sheet for one week.

> **Meal Allowance**
> *Lunch £8.50*
> *Dinner £22*
>
> *Only paid if Sam arrives home after 8 pm

Day	Miles driven	Lunch claimed	Time arrived home
Monday	180	Yes	9 pm
Tuesday	48		5 pm
Wednesday	64	Yes	8.30 pm
Thursday	33		5 pm
Friday	75	Yes	7.30 pm

Work out Sam's total expenses for the week.

2   Lynsey took part in a sponsored swim. Her target was to raise £100 for charity. Her nan promised her
that she would make up the £100 if Lynsey did not raise enough.

Here is Lynsey's sponsor form.

Lynsey swam 32 lengths in a pool of length 40 m.

Will her nan have to give her any money?

You must explain your answer.

Sponsor	Amount
Ali	£5
Rob	25p for each length
Will	30p for each length
Mum	50p for each length
Jade	2p for each metre

3   Here are the rates charged for Mr Pitkin's telephone.

Line rental	£29.36
Daytime cost	4p for each minute
Evening and weekend	3p for each minute
To mobiles	11p for each minute
National rate	8p for each minute

Here are the details of calls made by Mr Pitkin one quarter.

Type of call	Minutes
Daytime	78
Evening	312
To mobiles	42
National rate	25

Calculate Mr Pitkin's telephone bill for that quarter.

The following question helps you develop your ability to select and apply a method (AO2) and your ability to analyse and interpret problems (AO3).

Example ▶ The side of a shed is the shape of a trapezium as shown in the diagram.

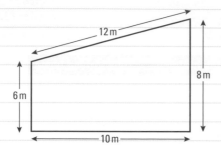

The side is to be given two coats of paint. The paint is sold in 1 litre cans costing £3 each.
1 litre of paint covers 15 square metres.
How much will it cost to paint the wall?

Solution ▶

Using the formula

Area (trapezium) $= \frac{1}{2}(a + b)h$

$$= \frac{1}{2}(6 + 8)10$$

$$= 70 \, m^2$$

> You need to find the area of the side of the shed.
> You may choose to use the formula, or divide the shape into a rectangle and a triangle.
> The formula is given on the formulae sheet.

Dividing the shape

Area of rectangle $= 6 \times 10 = 60 \, m^2$

Area of triangle $= \frac{1}{2} \times 10 \times 2 = 10 \, m^2$

$\qquad$ Total area $= 70 \, m^2$

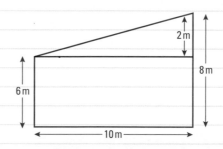

The number of tins of paint required for one coat is $70 \div 15$.

> Total area ÷ area covered by 1 tin

The number of tins needed for two coats is
$140 \div 15 = 9.3$.
So 10 tins will be needed.
The cost of the paint will be £30.

1 This shape is made by joining six squares.

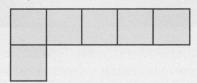

Find two shapes which have the same area but different perimeters.

2 The diagram shows a wall which is to be built with bricks.
The bricks measure 200 mm × 100 mm.
They are sold in packs of 100. One pack costs £35.
Find the cost of the bricks.

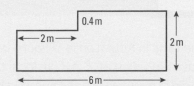

3 The diagram shows a rectangular path around a lawn. The path is 1 m wide.

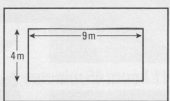

Gravel costs £124 per tonne.
1 tonne of gravel covers 15 m².
Work out the cost of covering the path with gravel.

4 Find the perimeter of three different rectangles which each have an area of 36 cm².

5 The diagram shows a bathroom wall in the shape of a trapezium. The wall is to be painted.

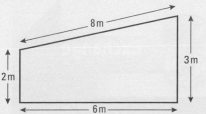

The paint chosen is sold in 1 litre cans costing £4 each. 1 litre covers 12 square metres.
How much will it cost to paint the wall?

2012 OLYMPICS

The Olympic Games is held every four years in different countries around the world. The Olympic committee picked London to host the 2012 games because campaigners promised to improve local communities and encourage involvement in sport.

People around the world live in different time zones. Time zones are based on Greenwich Mean Time (GMT). They state how many hours in front or behind of British time they are.

QUESTION

1. The time in Kingston is 5 hours behind the time in London. If the Jamaican sprint team leave Kingston at 2.32 pm local time and fly to London on a flight that takes 11 hours and 45 minutes, what time will they arrive in London?

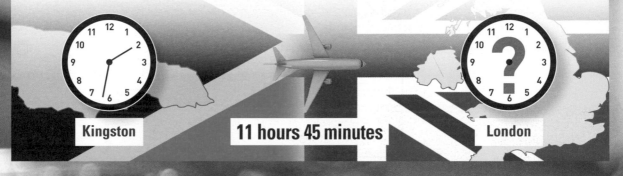

Kingston   **11 hours 45 minutes**   **London**

QUESTION

2. An American visitor to the Olympics plans to travel to France after the Games. He needs to change $500 into Euros before flying to Paris. How many euros will he get?

$500   **Exchange**   **€ ?**

The exchange rate is £1 = $1.10 £1 = €1.65

MODERN PENTATHLONS

Fencing – 35 bouts
Baseline 25 victories = 1000 points
Each victory over 25 gains
an extra 24 points.
Each victory below 25 loses
24 points.

Running – 3000 m
Baseline 10 minutes = 1000 points
Each second faster gains an
extra 4 points.
Each second slower loses
4 points.

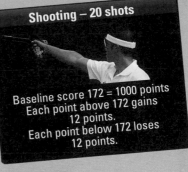

Shooting – 20 shots
Baseline score 172 = 1000 points
Each point above 172 gains
12 points.
Each point below 172 loses
12 points.

Show jumping – 15 jumps
Baseline is a clean round
(no mistakes)
Within the time limit = 1200 points.
Athletes are penalised 28 points for
each mistake.

Swimming – 200 m
Baseline 2 min 30 sec = 1000 points
Every 0.1 second faster gains
an extra point.
Every 0.1 second slower loses
a point.

QUESTION

3. Modern Pentathlons consist of five events: shooting, fencing, swimming, running and show jumping. For each event competitors are scored points above or below a baseline. Which of the three athletes below has the highest total score?

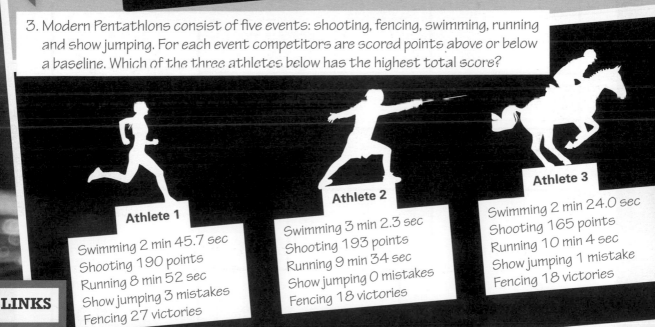

Athlete 1

Swimming 2 min 45.7 sec
Shooting 190 points
Running 8 min 52 sec
Show jumping 3 mistakes
Fencing 27 victories

Athlete 2

Swimming 3 min 2.3 sec
Shooting 193 points
Running 9 min 34 sec
Show jumping 0 mistakes
Fencing 18 victories

Athlete 3

Swimming 2 min 24.0 sec
Shooting 165 points
Running 10 min 4 sec
Show jumping 1 mistake
Fencing 18 victories

LINKS

- For **Question 1** you need to understand how to use time. You learnt about this in **Chapter 17**.

- You learnt about decimals in **Chapter 3**. You will need to be able to use decimals in calculations for **Question 2**.

- For **Question 3** you need to be able to add and subtract numbers from a baseline. You learnt how to do this in **Chapter 1**.

FS UNIVERSITY

Moving away to university is the first time many teenagers leave home. The majority of university students rent houses with small groups of friends.

QUESTION

1. Elaine and 4 friends decide to pay £1122 a month to rent a five-bedroom house near campus. Some rooms are bigger than others so the rent for each room is calculated as a proportion of the floor space of the house. How much can they each expect to pay?

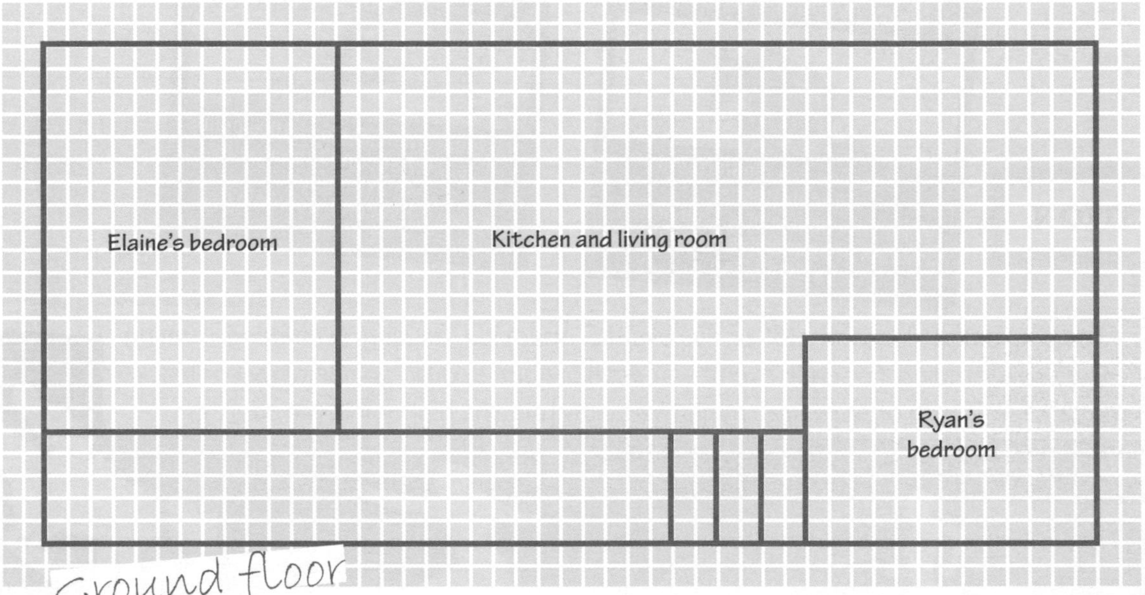

Elaine's bedroom
Kitchen and living room
Ryan's bedroom
Ground floor

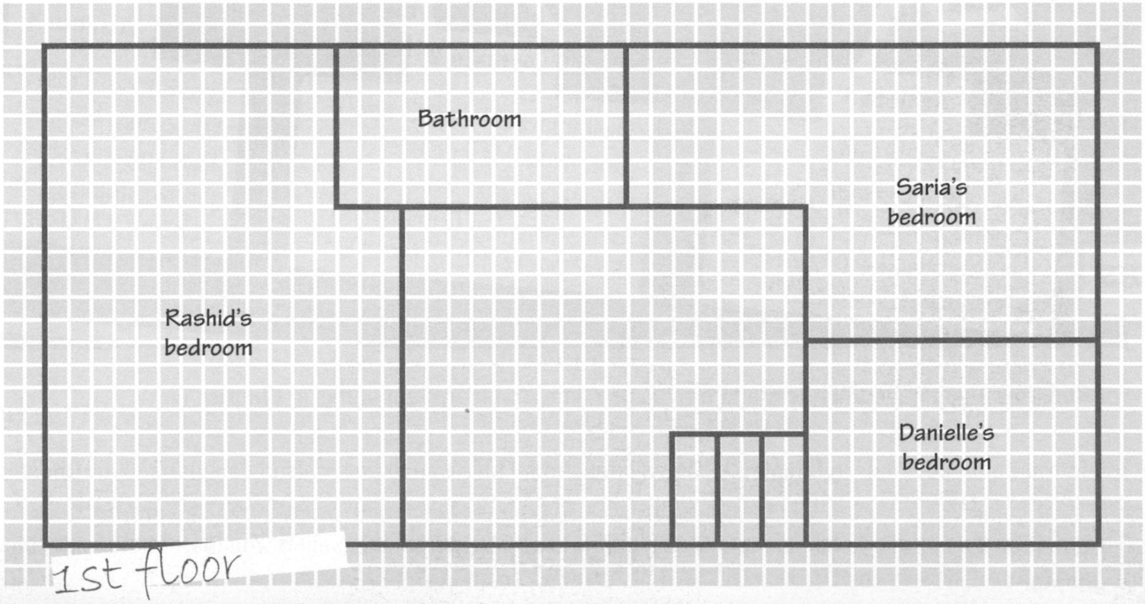

Bathroom
Saria's bedroom
Rashid's bedroom
Danielle's bedroom
1st floor

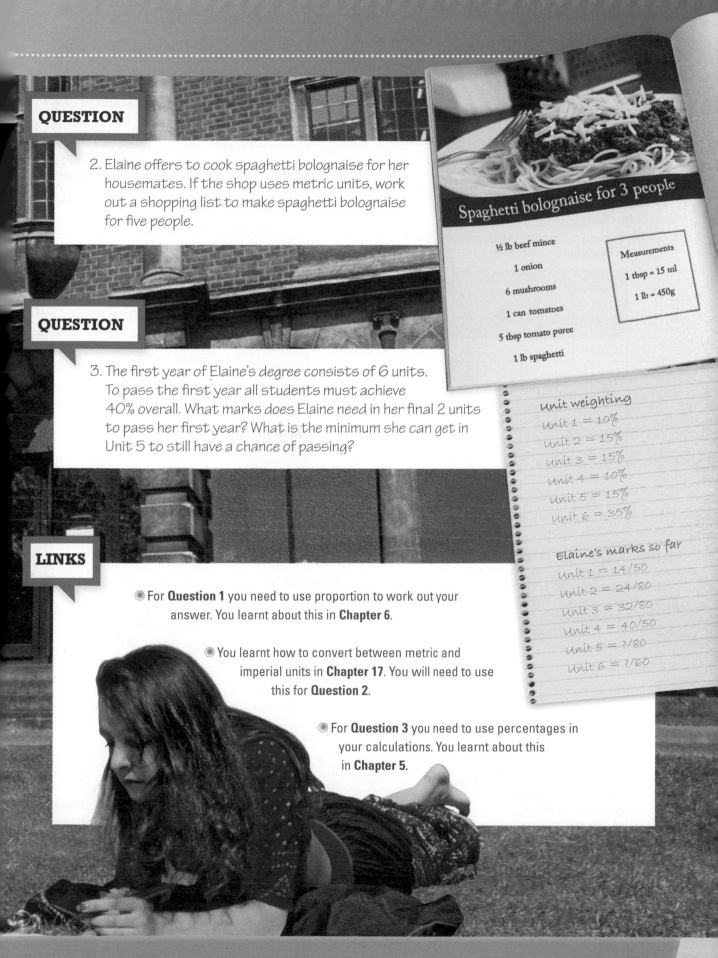

2. Elaine offers to cook spaghetti bolognaise for her housemates. If the shop uses metric units, work out a shopping list to make spaghetti bolognaise for five people.

Spaghetti bolognaise for 3 people

½ lb beef mince

1 onion

6 mushrooms

1 can tomatoes

5 tbsp tomato puree

1 lb spaghetti

Measurements

1 tbsp = 15 ml

1 lb = 450g

3. The first year of Elaine's degree consists of 6 units. To pass the first year all students must achieve 40% overall. What marks does Elaine need in her final 2 units to pass her first year? What is the minimum she can get in Unit 5 to still have a chance of passing?

Unit weighting

Unit 1 = 10%

Unit 2 = 15%

Unit 3 = 15%

Unit 4 = 10%

Unit 5 = 15%

Unit 6 = 35%

Elaine's marks so far

Unit 1 = 14/50

Unit 2 = 24/80

Unit 3 = 32/80

Unit 4 = 40/50

Unit 5 = ?/80

Unit 6 = ?/60

LINKS

⦿ For **Question 1** you need to use proportion to work out your answer. You learnt about this in **Chapter 6**.

⦿ You learnt how to convert between metric and imperial units in **Chapter 17**. You will need to use this for **Question 2**.

⦿ For **Question 3** you need to use percentages in your calculations. You learnt about this in **Chapter 5**.

Answers

Chapter 1 Answers

1.1 Get ready

1 508, 510
2 13
3 56

Exercise 1A

1 Students' numbers; the digits that must be as in the examples below are shown in **bold**.

	Ten thousand	Thousands	Hundreds	Tens	Units
a		**4**	1	2	3
b				**3**	9
c	8	9	**1**	5	5
d			5	3	**9**
e			2	4	**0**	8
f	6	7	**4**	5	5
g		**7**	7	3	**7**
h	3	**6**	5	9	**6**

2 For example:
Teacher on left: 51, 52, 53, 54, 55
Teacher in middle: 134, 154, 384, 494, 514
Teacher on right: 1100, 2135, 4189, 8176, 3144
3 a 60, sixty   b 600, six hundred
  c 60 000, sixty thousand
  d 6, six   e 6000, six thousand

1.2 Get ready

1 a Forty   b Four   c Four hundred

Exercise 1B

1 a 325   b 1718   c 6204   d 19 420
2 a two hundred and thirty-seven
  b three hundred and twenty-one
  c one thousand seven hundred and ninety-two
  d six thousand five hundred and two
  e one thousand and fifty-three
3 a 73, 179, 183, 190, 235
  b 970, 2015, 2105, 2439, 2510
  c 2998, 3000, 3003, 3033, 30 300
  d 56 321, 56 745, 56 762, 59 342
4 a 69, 1010, 2306   b 76 152, 70 363, 151 400
5 a ten million four hundred and sixty-seven thousand five hundred and forty-two

b seven hundred and ninety-three thousand nine hundred and sixty-three
c one million three hundred and forty thousand four hundred and fifteen
d sixty-four million three hundred and fifty-one thousand
e ten million six hundred and twenty-seven thousand two hundred and fifty

6 a

Peugeot 505	seven thousand nine hundred and ninety-five pounds
Focus	eleven thousand four hundred and ninety-five pounds
Ka	four thousand eight hundred and thirty-five pounds
Mini	six thousand five hundred and forty-nine pounds
Sharan	thirteen thousand two hundred and five pounds

b

Sharan	£13 205
Focus	£11 495
Peugeot 505	£7995
Mini	£6549
Ka	£4835

1.3 Get ready

1 £3645, £4190, £5250, £5490
2 462 690, 10 348 276, 10 524 145, 40 280 780, 60 424 213
3 70 363, 76 152, 150 400

Exercise 1C

1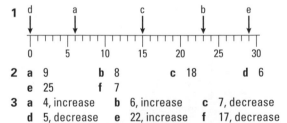

2 a 9   b 8   c 18   d 6
  e 25   f 7
3 a 4, increase   b 6, increase   c 7, decrease
  d 5, decrease   e 22, increase   f 17, decrease

1.4 Get ready

1 a 62   b 7   c 17

Exercise 1D

1 43   2 80   3 331   4 270 marks
5 45   6 96 fish   7 84 passengers   8 216 songs
9 401   10 416 people

Exercise 1E

1 305   2 3166   3 1003   4 12 CDs
5 a 19   b 17   c 36
6 a 88   b 59   c 133

1.5 Get ready

1 **a** 169 **b** 0

Exercise 1F

1 **a** i 270   ii 2700   iii 27 000
   b i 80   ii 800   iii 8000
   c i 3010   ii 30 100   iii 301 000
   d i 600   ii 6000   iii 60 000
   e i 50 200   ii 502 000   iii 5 020 000
2 **a** 700   **b** 5200   **c** 3660
   d 21 500   **e** 4600   **f** 642 000
3 **a** 455   **b** 159   **c** 1884
   d 4184   **e** 1185   **f** 1596
4 **a** 408   **b** 975   **c** 1749
   d 5024   **e** 24 581   **f** 14 144
   g 7738   **h** 1102
5 923   6 4494
7 782 miles   8 1449 supporters
9 1548 matches   10 384 tins

Exercise 1G

1 **a** 366   **b** 43   **c** 900   **d** 870
2 **a** 24   **b** 23   **c** 14   **d** 16
   e 160   **f** 113   **g** 18   **h** 89
   i 92   **j** 91   **k** 204   **l** 340
3 **a** 16   **b** 44   **c** 16   **d** 41
   e 21   **f** 34   **g** 40   **h** 300
4 **a** 21   **b** 15   **c** 40   **d** 20
   e 14
5 £480
6 **a** 10 finalists   **b** 30 finalists   **c** 36 finalists
7 **a** 7 trips   **b** 13 trips   **c** 36 trips
8 **a** 12 cases   **b** 41 cases   **c** 80 cases

1.6 Get ready

1 £11
2 £5.40

Exercise 1H

1 **a** 60   **b** 60   **c** 190
   d 190   **e** 990   **f** 2410
2 **a** 300   **b** 700   **c** 2400
   d 3100   **e** 8800   **f** 29 500
3 **a** 2000   **b** 36 000   **c** 29 000
   d 322 000   **e** 717 000   **f** 2 247 000
4

	Length (ft)	Cruising speed (mph)	Takeoff weight (lb)
Airbus A310	150	560	36 100
Boeing 737	90	580	130 000
Saab 2000	90	400	50 300
Dornier 228	50	270	12 600
Lockheed L1011	180	620	496 000

1.7 Get ready

1 266, 403, 557, 577, 615

Exercise 1I

1 **a** i 5, −10   ii −10, −3, 0, 4, 5
   b i 0, −13   ii −13, −9, −7, −2, 0
   c i 13, −15   ii −15, −6, −3, 6, 13
   d i −2, −21   ii −21, −20, −13, −5, −2
2 **a** 0, −1   **b** −2, −5   **c** 3, 7   **d** −7, −12
   e −15, −24   **f** −1, 2
3 **a** 7   **b** −3   **c** −1   **d** −5
   e −6   **f** 3   **g** 0   **h** −7
   i −5   **j** −2
4 **a** −40   **b** −70   **c** 30   **d** −30
   e −160   **f** 270   **g** −30   **h** 0
   i 120   **j** −400
5 **a** Minsk   **b** Minsk, 49°C   **c** Tripoli, 26°C
6 22°C
7 6°C

1.8 Get ready

1 Civetta
2 Val d'Isere

Exercise 1J

1 **a** 5°C   **b** 3°C   **c** 6°C   **d** 10°C
   e 10°C   **f** 8°C   **g** 5°C   **h** 13°C
2 **a** −2°C   **b** −3°C   **c** −7°C   **d** 3°C
   e 3°C   **f** −6°C   **g** −9°C

1.9 Get ready

1 **a** −2°C   **b** −10°C

Exercise 1K

1 **a** −7   **b** 4   **c** 10   **d** 9
   e −1   **f** 2   **g** −2   **h** −10
2 11 metres
3 −12°C

Exercise 1L

1 **a** −3   **b** −3   **c** 4
   d 12   **e** −4   **f** −12
2 **a** −90   **b** 4   **c** 5
   d −14   **e** −2   **f** 12
3 **a** −20   **b** 2   **c** −20
   d 6   **e** 9   **f** −42
4 **a** 24   **b** −15   **c** −4
   d 27   **e** −40   **f** 3
5 **a** 10   **b** −56   **c** 36
   d 21   **e** −3   **f** −42

Review exercise

1 For example, 4**2** 983
2 **a** Fifty   **b** Five thousand
   c Fifty thousand   **d** Fifty   **e** Five
3 **a** Three thousand seven hundred and twenty-three
   b One hundred and seven
   c Two thousand and seven
   d Fifteen thousand and seventy-one

4 a 21 231   **b** 507   **c** 70 203
5 a 2005   **b** 2001   **c** 2003   **d** 2005
6 £92 534
7 583 girls
8

Number of packs	Number of cans
1	24
2	48
3	72
4	96
5	120

9 a 30   **b** 300   **c** 2000
   d 78 940   **e** 8 000 000   **f** 80
10 644
11 £103.74
12 £738
13 a 9°C   **b** −5°C
14 a −4   **b** 2   **c** −15   **d** 10   **e** −2
15 a −21   **b** 20   **c** −8   **d** 5   **e** −7
16 For example, 9, 8 and 3
17 7 cm
18 77
19 £3.50
20 a −4°C   **b** 7°C   **c** Leeds
21 63 (both 9)
22 a 152   **b** 7 coaches
23 For example, the carton could have dimensions 12 cm by 20 cm by 20 cm

Chapter 2 Answers

2.1 Get ready

1 a 42   **b** 63   **c** 64   **d** 18

Exercise 2A

1 42, 18, 1110, 73 536, 500 000
2 105, 537, 811, 36 225
3 a 14, 16   **b** 30, 32   **c** 198, 200
4 a 3   **b** 29   **c** 199
5 5674, 6574, 7564, 5764, 6754, 7654, 4576, 4756, 5476, 5746, 7456, 7546
6 a 8732   **b** 2387
7 4, 8, 12, 12, 14, 14, 16, 16, 16, 22, 22, 42, 31, 23, 17, 17, 15, 15, 9, 9, 3, 3, 1

Exercise 2B

1 a 1, 3, 5, 15   **b** 1, 2, 4, 5, 10, 20
   c 1, 2, 3, 4, 6, 8, 12, 24   **d** 1, 2, 3, 6, 9, 18
   e 1, 13
   f 1, 2, 3, 5, 6, 9, 10, 15, 18, 30, 45, 90
2 a 2, 4, 6, 8, 10   **b** 5, 10, 15, 20, 25
   c 10, 20, 30, 40, 50   **d** 7, 14, 21, 28, 35
   e 13, 26, 39, 52, 65
3 Students' multiples of 10 greater than 50, e.g. 60, 70, 80
4 a 1, 4, 6, 8, 12   **b** 5, 20   **c** 1, 4, 8, 16   **d** 6, 9, 12
5 31, 37
6 97

7 5432
8 a 12, 18, 24, 27, 36, 54, 108, 216 – although a bunch of roses is unlikely to have more than 24 roses.
   b 14, 15, 21, 30, 42, 70 – although a bunch of roses is unlikely to have more than 21 roses.

Exercise 2C

1 a 1, 2   **b** 1, 5   **c** 1, 2, 3, 4, 6, 12
   d 1, 2, 5, 10   **e** 1, 2, 4, 8   **f** 1, 5
   g 1, 2, 4   **h** 1, 2, 3, 6   **i** 1
2 a 2, 3, 5   **b** 5   **c** 2, 3, 7
   d 3, 13   **e** 3, 5, 7
3 a $3 \times 3 \times 5$ **b** $2 \times 2 \times 3 \times 3$
   c $2 \times 2 \times 7$ **d** $2 \times 2 \times 2 \times 2 \times 5$
   e $2 \times 2 \times 2 \times 3 \times 3$

2.2 Get ready

1 a $3 \times 3 \times 11$ **b** $2 \times 2 \times 3 \times 3 \times 3 \times 3$
   c $5 \times 5 \times 7$

Exercise 2D

1 a 4   **b** 3   **c** 6   **d** 2   **e** 7
2 a 12   **b** 12   **c** 60   **d** 144   **e** 850
3 a 36, 6   **b** 360, 60   **c** 168, 12
   d 910, 13   **e** 288, 24   **f** 120, 20
4 120 seconds or 2 minutes
5 20 seconds

2.3 Get ready

1 a 121   **b** 169   **c** 27

Exercise 2E

1 a 9   **b** 36   **c** 100   **d** 4   **e** 400
2 a 8   **b** 125   **c** 343   **d** 8000   **e** 1728
3 a 25   **b** 216   **c** 81   **d** 27

Review exercise

1 a 6, 12   **b** 4, 16   **c** 3, 4, 6, 12   **d** 8, 27
2 a 5, 9, 27, 35, 37   **b** 9, 12, 27, 36   **c** 4, 8, 12, 16
   d 5, 37   **e** 4, 9, 16, 36   **f** 8, 27
3 a Any of number using all the digits once only is a multiple of 3
   b 7416, 4716, 7164 or 1764
4 factor
5 a 8   **b** 7
6 15, 30, 45, 60, 75, 90
7 2 multiplied by another prime number will give an even number.
8 1000
9 a 48   **b** 17   **c** 13
10 6 times (every 12 seconds)
11 a $2 \times 2 \times 3 \times 3 \times 3$
   b 12
12 a i $2 \times 2 \times 3 \times 5$ **ii** $2 \times 2 \times 2 \times 2 \times 2 \times 3$
   b 12   **c** 480
13 3 packs of doughnuts, 4 packs of cakes
14 120 000 miles
15 a 5   **b** 2   **c** 2

Chapter 3 Answers

3.1 Get ready

1 a two hundred and seventy three
   b four thousand and seventy six
   c three thousand seven hundred and fifty three

Exercise 3A

1

		Hundreds	Tens	Units	•	Tenths	Hundredths	Thousandths
a			4	1	•	6		
b				4	•	1	6	
c		7	3	4	•	6		
d				1	•	4	6	3
e				0	•	6	4	3
f				1	•	0	0	5
g				5	•	0	1	
h				0	•	0	8	6

2 a 5 units   b 5 tenths   c 4 tenths
   d 4 hundredths   e 1 ten   f 9 thousandths
   g 3 hundreds   h 7 ten-thousandths
   i 0 hundredths   j 3 tenths   k 1 thousandth
   l 2 hundredths

3.2 Get ready

1 a One tenth   b One hundredth
   c One   d Ten   e One thousandth

Exercise 3B

1 £1.09, £1.13, £1.18, £1.20, £1.29, £1.31
2 a 0.9, 0.76, 0.71, 0.68, 0.62
   b 3.75, 3.4, 3.12, 2.13, 2.09
   c 0.42, 0.407, 0.3, 0.09, 0.065
   d 6.52, 6.08, 3.7, 3.58, 3.0
   e 0.13, 0.105, 0.06, 0.024, 0.009
   f 2.2, 2.09, 1.3, 1.16, 1.1087, 1.08

3

Rascini	52.037 seconds
Killim	53.027 seconds
Ascarina	53.072 seconds
Bertollini	53.207 seconds
Silverman	53.702 seconds
Alloway	54.320 seconds

3.3 Get ready

1 a 61   b 45

Exercise 3C

1	6.1	2	3.25	3	68.9
4	26.02	5	1.0	6	18.725
7	19.8	8	11.001	9	1.914
10	118.17	11	31.97	12	28.71
13	19.122	14	18.326	15	11.064
16	31.006	17	15.0976	18	178.585

Exercise 3D

1 a £6.20   b £4.14   c £14.10   d £97.30
   e £0.11   f £0.35   g £6.19   h £11.14
   i £0.90   j £4.07   k £14.03   l £13.25
2 a 1.225   b 11.649   c 2.254   d 168.58

3.4 Get ready

1 a 1856   b 21 425   c 2856

Exercise 3E

1 a £13.50   b £5.48   c £5.20   d £1.20
2 a 4.5   b 45   c 450
   d 2.03   e 20.3   f 203
3 a 30.4   b 3.04   c 0.304
   d 11.25   e 1.125   f 0.1125
   g 1.125   h 0.01125   i 0.001 125
4 a 64.2   b 642   c 6.42
   d 562.3   e 56.23   f 0.5623
5 a 172.2 kg   b 0.0945 seconds   c 0.0144 m
   d 0.04 miles   e 0.9 litres   f 0.0012 hours
6 a £116.25   b £167.40   c £255.75
7 a £117.75   b £196.25   c £337.55
8 a 68.25 litres   b 113.75 litres
   c 295.75 litres

3.5 Get ready

1 a 9   b 36   c 121   d 64

Exercise 3F

1 a 9.61 b 17.64 c 28.09 d 4.1209 e 0.16
2 a 3.375 b 15.625 c 32.768 d 0.008 e 0.125

3.6 Get ready

1 a 41   b 65   c 6

Exercise 3G

1 a 3.45   b 0.345   c 0.0345
   d 207.1   e 0.2701   f 0.027 01
   g 6.5   h 0.65   i 0.065
2 a 16.12   b 0.633   c 14.84
   d 34.221   e 5.027   f 0.0046
3 a 7.5 b 5.75 c 1.125 d 1.75
   e 1.2 f 0.85 g 6.2 h 1.8
4 £15.40
5 0.625 kg or 625 g
6 a 15.5 b 13.2 c 21 d 3.5
   e 1.07 f 6.2 g 41.4 h 32

Answers

3.7 Get ready

1 4
2 5
3 a 56   **b** 3

Exercise 3H

1 a 8   **b** 13   **c** 14   **d** 6
   e 11   **f** 20   **g** 1   **h** 20
   i 1   **j** 100   **k** 20   **l** 2
2 a 3.6   **b** 5.3   **c** 0.1   **d** 9.3
   e 10.7   **f** 8.0   **g** 2.1   **h** 0.5
   i 2.5   **j** 125.7   **k** 0.1   **l** 9.9
3 a 14 mm   **b** 80 m   **c** 1 kg   **d** £204
   e 4 lb   **f** 0 tonne   **g** 11 g   **h** 8 min

Exercise 3I

1 a i 4.226   **ii** 4.23   **b i** 9.787   **ii** 9.79
   c i 0.416   **ii** 0.42   **d i** 0.058   **ii** 0.06
2 a i 10.517   **ii** 10.52   **b i** 7.503   **ii** 7.50
   c i 21.730   **ii** 21.73   **d i** 9.089   **ii** 9.09
3 a i 15.598   **ii** 15.60   **b i** 0.408   **ii** 0.41
   c i 7.247   **ii** 7.25   **d i** 6.051   **ii** 6.05
4 a i 29.158 cm   **ii** 29.16 cm   **b i** 0.055 kg   **ii** 0.05 kg
   c i 13.379 km   **ii** 13.38 km   **d i** £5.998   **ii** £6.00
5 a 5.617   **b** 0.0   **c** 0.9240
   d 0.9   **e** 9.7   **f** 1.01

3.8 Get ready

1 a 20   **b** 50   **c** 200

Exercise 3J

1 a 40   **b** 700   **c** 300   **d** 0.3
   e 20 000   **f** 0.007   **g** 1000   **h** 5
   i 10   **j** 20
2 100 medals
3 30 000 spectators

3.9 Get ready

1 Manchester United 80 000
   Chelsea 40 000
   Real Madrid 80 000
   Newcastle United 50 000
   Liverpool 40 000
   Barcelona 70 000

Exercise 3K

1 a i 0.062   **ii** 0.0618   **b i** 0.16   **ii** 0.165
   c i 96   **ii** 96.3   **d i** 41   **ii** 41.5
2 a i 730   **ii** 735   **b i** 0.079   **ii** 0.0795
   c i 5.7   **ii** 5.69   **d i** 590   **ii** 586
3 a i 0.015   **ii** 0.0148   **b i** 2200   **ii** 2220
   c i 76   **ii** 76.2   **d i** 0.38   **ii** 0.380
4 a i 8.4   **ii** 8.38   **b i** 36   **ii** 36.0
   c i 190   **ii** 187   **d i** 0.067   **ii** 0.0666
5 a i 220 000   **ii** 219 000   **b i** 4 000 000   **ii** 3 990 000
   c i 310 000   **ii** 307 000   **d i** 26 000   **ii** 25 600

6 £1.61
7 83.3 seconds

3.10 Get ready

1 600   **2** 1200   **3** 0.8

Exercise 3L

1 a $60 \times \dfrac{60}{30} = 120$

   b $\dfrac{200 \times 300}{200} = 300$

   c $\dfrac{10 \times 30 \times 100}{300} = 100$

   d $\dfrac{2000}{10 \times 100} = 2$

   e $\dfrac{500}{10 \times 50} = 1$

   f $\dfrac{100 \times 90}{20 \times 30} = 15$

2 $50 \times 100 = 5000$ seats
3 $70 \times 50 = 3500$ tins
4 $30 \times 5 = £150$
5 a $20 \times \dfrac{0.2}{4} = 1$

   b $6 \times \dfrac{30}{1} = 180$

   c $\dfrac{900}{20} \times 0.5 = 22.5$

3.11 Get ready

1 a 60   **b** 600   **c** 6000
2 a 30   **b** 3   **c** 0.3

Exercise 3M

1 a 1792   **b** 1792   **c** 17.92
2 a 146.4   **b** 1.464   **c** 0.1464
3 a 726   **b** 0.726   **c** 7.26
4 a 0.64   **b** 0.64   **c** 64

Review exercise

1 a 4.09, 4.85, 5.16, 5.23, 5.9
   b 0.021, 0.07, 0.34, 0.37, 0.4
   c 5, 5.007, 5.01, 7.07, 7.23
   d 0.06, 0.23, 1.001, 1.08, 1.14
2 Thiamin   0.0014 g
   Riboflavin   0.0015 g
   Vitamin B6   0.002 g
   Iron   0.014 g
   Sodium   0.02 g
   Fibre   1.5 g
3 a 8.5   **b** 18.57   **c** 0.1172
   d 11.1432   **e** 58.4   **f** 15.58
   g 46.4   **h** 0.0118
4 Yes, it weighs 19.9 kg.
5 211.6 kg
6 a 11.7   **b** 1.44   **c** 0.12
   d 40.96   **e** 5.1   **f** 20.35

7 360 000

8 1.35 m

9 a 1000   **b** 1

  c 1000   **d** 0.1

10   **a** £2.45   **b** £2.85   **c** £2.45

11 0.85 kg

12 146.2 km

13 She buys 13 bottles and has 16p left.

14   **a** 57   **b** 0.103   **c** 600

15   **a** 23.5   **b** 1.8   **c** 0.3   **d** 150.0

16   **a** 7.26   **b** 73.04   **c** 0.042   **d** 0.721

17   **a** 8300   **b** 20 100   **c** 0.5   **d** 20.9

18 $40 \times 8 = £320$

19 $6 \div 20 = £0.30$ or 30p

20   **a** £153.90

     b £59.85

21 $30 \times 6 = £180$

22   **a** $(800 \times 5000)/3000 = 1333$

     b $(4 \times 5)/10 = 2$

     c $(2 \times 8)/(4 \times 2) = 2$

23   **a** 1632   **b** 16.32   **c** 3.4

24   **a** 15 456   **b** 0.15456   **c** 3220

25 $2000 \div 50 \div 0.4 = 100$ miles

26 £3.74

27   **a** Energy consumption is less for A (0.95 kWh/cycle).

     b 160 kWh, £19.20

     c Yes, because a washing machine should last at least 3 years.

Chapter 4 Answers

4.1 Get ready

1 a 4   **b** 20   **c** 7   **d** 6

Exercise 4A

1

Shape	Fraction shaded	Fraction not shaded
circle	$\frac{1}{2}$	$\frac{1}{2}$
square	$\frac{1}{4}$	$\frac{3}{4}$
pentagon	$\frac{2}{5}$	$\frac{3}{5}$
rectangle grid	$\frac{3}{10}$	$\frac{7}{10}$
octagon	$\frac{5}{8}$	$\frac{3}{8}$
triangle	$\frac{4}{9}$	$\frac{5}{9}$

2 a Any single box shaded, e.g.   **b** Any 3 boxes shaded, e.g.

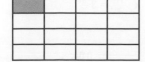

c Any 8 boxes shaded, e.g.   **d**

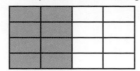 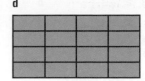

3 a Any single sector shaded, e.g.

b Any 3 sectors shaded, e.g.

c Any 3 sectors shaded, e.g.

4 a $\frac{15}{28}$ **b** $\frac{13}{28}$

5 a 75 surfers   **b** $\frac{47}{75}$ **c** $\frac{28}{75}$

4.2 Get ready

1 a 8   **b** 9   **c** 6

Exercise 4B

1 Students' fractions, e.g.

   a e.g. $\frac{6}{8}, \frac{3}{4}$ **b** e.g. $\frac{2}{6}, \frac{1}{3}$ **c** e.g. $\frac{3}{12}, \frac{1}{4}$ **d** e.g. $\frac{4}{8}, \frac{2}{4}, \frac{1}{2}$

2 a $\frac{3}{4} = \frac{6}{8} = \frac{9}{12} = \frac{12}{16} = \frac{15}{20} = \frac{18}{24}$

   b $\frac{2}{7} = \frac{4}{14} = \frac{6}{21} = \frac{8}{28} = \frac{10}{35} = \frac{12}{42}$

   c $\frac{4}{5} = \frac{8}{10} = \frac{12}{15} = \frac{16}{20} = \frac{20}{25} = \frac{24}{30}$

   d $\frac{1}{3} = \frac{3}{9} = \frac{6}{18} = \frac{9}{27} = \frac{12}{36} = \frac{15}{45}$

3 a $\frac{3}{18}$ **b** $\frac{6}{14}$ **c** $\frac{18}{48}$ **d** $\frac{12}{21}$

   e $\frac{30}{36}$ **f** $\frac{6}{9}$ **g** $\frac{24}{54}$ **h** $\frac{40}{56}$

   i $\frac{90}{100}$ **j** $\frac{84}{144}$ **k** $\frac{49}{56}$ **l** $\frac{18}{81}$

4 a $\frac{3}{6}, \frac{2}{6}$

   b   **i** $\frac{12}{30}, \frac{15}{30}$ **ii** $\frac{7}{70}, \frac{10}{70}$ **iii** $\frac{3}{12}, \frac{10}{12}$

      iv $\frac{5}{10}, \frac{6}{10}$ **v** $\frac{16}{24}, \frac{3}{24}$ **vi** $\frac{15}{20}, \frac{12}{20}$

4.3 Get ready

1 a 20   **b** 70   **c** 12

Exercise 4C

1 a $\frac{8}{20}, \frac{5}{20}, \frac{1}{4}$ **b** $\frac{10}{20}, \frac{16}{20}, \frac{2}{4}$

   c $\frac{8}{12}, \frac{9}{12}, \frac{2}{3}$ **d** $\frac{6}{10}, \frac{7}{10}, \frac{3}{5}$

2 a $\frac{3}{6}$ **b** $\frac{1}{7}$ **c** $\frac{5}{6}$

   d $\frac{3}{5}$ **e** $\frac{2}{3}$ **f** $\frac{3}{4}$

Answers

3 a $\frac{1}{2}, \frac{2}{3}, \frac{3}{4}$ **b** $\frac{7}{15}, \frac{4}{5}, \frac{5}{6}$

  c $\frac{1}{2}, \frac{3}{4}, \frac{4}{5}$ **d** $\frac{5}{14}, \frac{3}{7}, \frac{1}{2}, \frac{4}{7}$

4 $\frac{7}{8}, \frac{3}{4}, \frac{1}{2}, \frac{2}{5}, \frac{2}{10}$

4.4 Get ready

1 a $5\frac{3}{4}$ **b** 18   **c** 35

Exercise 4D

1 a $2\frac{1}{2}$ **b** $1\frac{3}{4}$ **c** $1\frac{2}{7}$ **d** $1\frac{3}{8}$

  e $1\frac{1}{8}$ **f** $3\frac{1}{5}$ **g** $2\frac{3}{10}$ **h** $4\frac{4}{5}$

  i $2\frac{2}{7}$ **j** $2\frac{2}{5}$ **k** $6\frac{2}{3}$ **l** $1\frac{7}{9}$

  m $9\frac{3}{4}$ **n** $5\frac{2}{5}$ **o** $2\frac{8}{9}$ **p** $1\frac{7}{10}$

2 a $\frac{3}{2}$ **b** $\frac{11}{2}$ **c** $\frac{11}{4}$ **d** $\frac{5}{3}$

  e $\frac{13}{4}$ **f** $\frac{22}{5}$ **g** $\frac{37}{10}$ **h** $\frac{26}{5}$

  i $\frac{31}{4}$ **j** $\frac{9}{4}$ **k** $\frac{19}{10}$ **l** $\frac{28}{3}$

  m $\frac{17}{6}$ **n** $\frac{43}{8}$ **o** $\frac{29}{8}$ **p** $\frac{109}{100}$

4.5 Get ready

1 a 78   **b** 133   **c** 105

Exercise 4E

1 a £4   **b** £5   **c** £7   **d** £9

  e 7 cm   **f** 21 cm   **g** 44 kg   **h** 18 kg

2 $82\frac{1}{2}$ hours

3 a $\frac{3}{8}$ **b** $\frac{3}{32}$ **c** $\frac{8}{25}$ **d** $\frac{9}{32}$

  e $\frac{5}{36}$ **f** $\frac{21}{40}$ **g** $\frac{9}{50}$ **h** $\frac{4}{9}$

  i $\frac{3}{16}$ **j** $\frac{8}{15}$ **k** $\frac{4}{21}$ **l** $\frac{4}{15}$

  m $\frac{2}{35}$ **n** $\frac{10}{21}$ **o** $\frac{9}{8} = 1\frac{1}{8}$ **p** $\frac{1}{5}$

4 a $\frac{2}{5}$ **b** $\frac{3}{5}$ **c** $\frac{1}{2}$ **d** $\frac{6}{25}$

  e $\frac{5}{8}$ **f** $\frac{1}{8}$ **g** $\frac{4}{15}$ **h** $\frac{4}{7}$

  i $\frac{2}{7}$ **j** $\frac{5}{14}$ **k** $\frac{1}{5}$ **l** $\frac{1}{6}$

  m $\frac{1}{7}$ **n** $\frac{2}{5}$ **o** $\frac{7}{2} = 3\frac{1}{2}$ **p** $\frac{13}{20}$

5 a $\frac{7}{2} = 3\frac{1}{2}$ **b** $\frac{10}{3} = 3\frac{1}{3}$ **c** $\frac{24}{5} = 4\frac{4}{5}$ **d** 6

  e 14   **f** 6   **g** 4   **h** 10

6 a $\frac{8}{9}$ **b** $\frac{14}{15}$ **c** $\frac{3}{8}$ **d** $\frac{13}{8} = 1\frac{5}{8}$

  e $\frac{17}{6} = 2\frac{5}{6}$ **f** $\frac{10}{9} = 1\frac{1}{9}$ **g** $\frac{7}{4} = 1\frac{3}{4}$

7 a $\frac{21}{4} = 5\frac{1}{4}$ **b** $\frac{133}{24} = 5\frac{13}{24}$ **c** $\frac{21}{5} = 4\frac{1}{5}$ **d** $\frac{21}{4} = 5\frac{1}{4}$

  e $\frac{5}{8}$ **f** $\frac{28}{15} = 1\frac{13}{15}$ **g** 16   **h** 3

4.6 Get ready

1 a 24   **b** 30   **c** 70

Exercise 4F

1 a $\frac{4}{3} = 1\frac{1}{3}$ **b** $\frac{3}{4}$ **c** $\frac{3}{2} = 1\frac{1}{2}$ **d** $\frac{5}{7}$

  e $\frac{10}{3} = 3\frac{1}{3}$ **f** $\frac{15}{8} = 1\frac{7}{8}$ **g** $\frac{10}{9} = 1\frac{1}{9}$ **h** $\frac{7}{8}$

  i $\frac{4}{9}$ **j** $\frac{8}{15}$ **k** $\frac{9}{16}$ **l** 2

2 a 16   **b** 16   **c** 10   **d** $\frac{64}{7} = 9\frac{1}{7}$

  e 5   **f** $\frac{12}{7} = 1\frac{5}{7}$ **g** 15   **h** 24

3 a 5   **b** $\frac{26}{20}$ **c** $\frac{5}{3} = 1\frac{2}{3}$ **d** $\frac{39}{76}$

  e $\frac{1}{2}$ **f** 2   **g** $\frac{17}{27}$ **h** $\frac{21}{40}$

4 a $\frac{3}{32}$ **b** $\frac{5}{12}$ **c** $\frac{1}{10}$ **d** $\frac{4}{25}$

  e $\frac{1}{3}$ **f** $\frac{13}{24}$ **g** $\frac{17}{60}$ **h** $\frac{1}{6}$

  i $\frac{2}{9}$ **j** $\frac{1}{9}$ **k** $\frac{7}{6} = 1\frac{1}{6}$ **l** $\frac{7}{4} = 1\frac{3}{4}$

4.7 Get ready

1 a 56   **b** 15   **c** 6

Exercise 4G

1 a $\frac{7}{8}$ **b** $\frac{7}{9}$ **c** 1   **d** $\frac{8}{9}$

  e $\frac{1}{2}$ **f** $\frac{6}{7}$ **g** 1   **h** $\frac{8}{5} = 1\frac{3}{5}$

  i $3\frac{2}{9}$ **j** $2\frac{2}{3}$ **k** $1\frac{3}{4}$ **l** $1\frac{7}{8}$

2 a $\frac{3}{4}$ **b** $\frac{5}{8}$ **c** $\frac{11}{8} = 1\frac{3}{8}$ **d** $\frac{5}{6}$

  e $\frac{7}{6} = 1\frac{1}{6}$ **f** $\frac{7}{10}$ **g** $\frac{4}{3} = 1\frac{1}{3}$ **h** $\frac{11}{10} = 1\frac{1}{10}$

3 a $\frac{11}{8} = 1\frac{3}{8}$ **b** $\frac{17}{20}$ **c** $\frac{31}{36}$ **d** $\frac{71}{40} = 1\frac{31}{40}$

  e $\frac{17}{30}$ **f** $\frac{13}{20} = 1\frac{1}{12}$ **g** $\frac{23}{24}$ **h** $\frac{19}{18} = 1\frac{1}{18}$

  i $\frac{7}{8}$ **j** $\frac{19}{24}$ **k** $\frac{17}{16} = 1\frac{1}{16}$

4 a $\frac{5}{6}$ **b** $\frac{17}{30}$ **c** $\frac{33}{40}$ **d** $\frac{31}{36}$

  e $\frac{53}{42} = 1\frac{11}{42}$ **f** $\frac{83}{70} = 1\frac{13}{70}$ **g** $\frac{41}{30} = 1\frac{11}{30}$ **h** $\frac{31}{35}$

  i $\frac{23}{40}$ **j** $\frac{11}{30}$ **k** $\frac{20}{21}$

5 a $\frac{29}{8} = 3\frac{5}{8}$ **b** $\frac{53}{8} = 6\frac{5}{8}$ **c** $\frac{65}{16} = 4\frac{1}{16}$ **d** $\frac{35}{8} = 4\frac{3}{8}$

  e $\frac{67}{16} = 4\frac{3}{16}$ **f** $\frac{89}{30} = 2\frac{29}{30}$ **g** $\frac{145}{42} = 3\frac{19}{42}$ **h** $\frac{167}{42} = 3\frac{41}{42}$

  i $\frac{88}{15} = 5\frac{13}{15}$ **j** $\frac{26}{9} = 2\frac{8}{9}$

6 $7\frac{1}{2}$ miles

7 a $\frac{23}{4} = 5\frac{3}{4}$ **b** $\frac{19}{6} = 3\frac{1}{6}$ **c** $\frac{33}{8} = 4\frac{1}{8}$ **d** $\frac{109}{12} = 9\frac{1}{12}$

  e $\frac{83}{16} = 5\frac{3}{16}$ **f** $\frac{44}{12} = 3\frac{2}{3}$ **g** $\frac{43}{6} = 7\frac{1}{6}$ **h** $\frac{109}{15} = 7\frac{4}{15}$

Exercise 4H

1 a $\frac{2}{11}$ **b** $\frac{2}{9}$ **c** $\frac{3}{4}$ **d** $\frac{1}{6}$

  e $\frac{1}{2}$ **f** $\frac{1}{4}$ **g** $\frac{1}{2}$ **h** $\frac{3}{7}$

2 $\frac{3}{5}$

3 a $\frac{1}{4}$ **b** $\frac{1}{8}$ **c** $\frac{1}{8}$ **d** $\frac{5}{8}$

  e $\frac{1}{2}$ **f** $\frac{1}{4}$ **g** $\frac{1}{2}$ **h** $\frac{1}{5}$

  i $\frac{1}{8}$ **j** $\frac{3}{8}$ **k** $\frac{1}{6}$

4 a $\frac{1}{6}$ **b** $\frac{7}{24}$ **c** $\frac{1}{30}$ **d** $\frac{13}{30}$

  e $\frac{2}{15}$ **f** $\frac{3}{20}$ **g** $\frac{11}{30}$ **h** $\frac{3}{20}$

  i $\frac{103}{20} = 5\frac{3}{20}$ **j** $\frac{43}{6} = 7\frac{1}{6}$

5 $\frac{9}{16}$

6 a $\frac{19}{8} = 2\frac{3}{8}$ **b** $\frac{5}{4} = 1\frac{1}{4}$ **c** $\frac{11}{5} = 2\frac{1}{5}$ **d** $\frac{27}{10} = 2\frac{7}{10}$

  e $\frac{9}{10}$ **f** $\frac{3}{4}$ **g** $\frac{14}{5} = 2\frac{4}{5}$ **h** $\frac{77}{24} = 3\frac{5}{24}$

  i $\frac{22}{9} = 2\frac{4}{9}$ **j** $\frac{137}{40} = 3\frac{17}{40}$ **k** $\frac{111}{35} = 3\frac{6}{35}$

4.8 Get ready

1 a 0.375   **b** 0.5   **c** 0.04

Exercise 4I

1 **a** 0.6 **b** 0.5 **c** 0.7 **d** 0.35
e 0.16 **f** 0.06 **g** 0.875 **h** 0.45
i 0.76 **j** 0.3125 **k** 0.125 **l** 0.54
m 0.09 **n** 0.065 **o** $0.\dot{6}$ **p** 0.95

2 **a** $\frac{3}{10}$ **b** $\frac{37}{100}$ **c** $\frac{93}{100}$ **d** $\frac{137}{1000}$
e $\frac{293}{1000}$ **f** $\frac{7}{10}$ **g** $\frac{59}{100}$ **h** $\frac{3}{1000}$
i $\frac{3}{100\,000}$ **j** $\frac{13}{10\,000}$ **k** $\frac{77}{100}$ **l** $\frac{77}{1000}$
m $\frac{39}{100}$ **n** $\frac{41}{10\,000}$ **o** $\frac{19}{1000}$ **p** $\frac{31}{1000}$

3 **a** 0.8 **b** 0.75 **c** 1.125 **d** 0.19
e 3.6 **f** 0.52 **g** 0.625 **h** 3.425
i 0.14 **j** 4.1875 **k** 3.15 **l** 4.3125
m 0.007 **n** 1.28 **o** 15.9375 **p** 2.35

4 **a** $\frac{12}{25}$ **b** $\frac{1}{4}$ **c** $1\frac{7}{10}$ **d** $3\frac{203}{500}$
e $4\frac{3}{1000}$ **f** $2\frac{1}{40}$ **g** $\frac{49}{1000}$ **h** $4\frac{7}{8}$
i $3\frac{3}{4}$ **j** $10\frac{101}{1000}$ **k** $\frac{5}{8}$ **l** $2\frac{64}{125}$
m $\frac{13}{16}$ **n** $14\frac{7}{50}$ **o** $9\frac{3}{16}$ **p** $60\frac{13}{200}$

Review exercise

1 **a** $\frac{1}{4}$ **b** $\frac{2}{3}$ **c** $\frac{3}{4}$ **d** $\frac{2}{5}$

2 Students' equivalent fractions, e.g.
a $\frac{8}{10}, \frac{12}{15}, \frac{16}{20}$ **b** $\frac{4}{14}, \frac{6}{21}, \frac{8}{28}$ **c** $\frac{3}{4}, \frac{6}{8}, \frac{9}{12}$
d $\frac{4}{5}, \frac{8}{10}, \frac{12}{15}$ **e** $\frac{3}{10}, \frac{6}{20}, \frac{9}{30}$ **f** $\frac{2}{3}, \frac{4}{6}, \frac{6}{9}$

3 **a** 0.25 **b** 0.375 **c** 0.7
d 0.6 **e** 0.12 **f** 0.74

4 **a** $\frac{17}{50}$ **b** $\frac{1}{8}$ **c** $\frac{3}{10}$
d $\frac{1}{40}$ **e** $\frac{3}{20}$ **f** $3\frac{1}{10}$

5 $\frac{3}{20}$

6 **a** $\frac{3}{8}$ **b** $\frac{1}{3}$ **c** $\frac{1}{12}$ **d** $\frac{1}{24}$ **e** $\frac{1}{6}$

7 **a** $\frac{2}{3}$ **b** $\frac{4}{5}$ **c** $\frac{3}{4}$ **d** $\frac{4}{9}$

8 $\frac{1}{3}, \frac{3}{10}, \frac{29}{100}, \frac{2}{7}, \frac{4}{15}$

9 $\frac{4}{5}$

10 $\frac{450}{1000}, 0.6, \frac{7}{10}, \frac{3}{4}$

11 **a** $\frac{2}{15}$ **b** $\frac{1}{9}$ **c** $\frac{3}{4}$

12 $\frac{7}{25}$

13 **a** $\frac{2}{3}$ **b** $\frac{5}{9}$ **c** $\frac{39}{40}$ **d** $1\frac{1}{36}$ **e** $\frac{4}{9}$

14 $3\frac{2}{15}$ m

15 **a** $\frac{18}{24} = \frac{3}{4}$ of the bracelet is gold
b $\frac{1}{3}$
c 30 g
d 50 g
e 10 carat

16 **a** $8\frac{1}{2}$ **b** $1\frac{5}{16}$ **c** $\frac{98}{125}$

17 $3\frac{1}{3}$

18 **a** $1\frac{5}{16}$ **b** $1\frac{3}{4}$ **c** $2\frac{7}{8}$ **d** $2\frac{11}{15}$

19 $\frac{4}{15}$

20 Yes, it is $6\frac{2}{16}$ cm long

21 15 glasses

Chapter 5 Answers

5.1 Get ready

1 $\frac{4}{5}$ **2** $\frac{2}{5}$ **3** 0.27

Exercise 5A

1 **a** 70% **b** 36% **c** 50% **d** 75%
e 40% **f** 60%
2 **a** 30% **b** 64% **c** 50% **d** 25%
e 60% **f** 40%
3 **a** any 7 squares shaded
b any 2 squares shaded
4 40%
5 75%
6 27%
7 65%

Exercise 5B

1 **a** 0.5 **b** 0.45 **c** 0.62 **d** 0.95
e 0.29 **f** 0.3 **g** 0.03 **h** 0.07
2 1.25
3 0.125
4 0.032

Exercise 5C

1 **a** $\frac{3}{5}$ **b** $\frac{3}{4}$ **c** $\frac{7}{20}$ **d** $\frac{9}{10}$
e $\frac{1}{20}$ **f** $\frac{4}{5}$ **g** $\frac{21}{25}$ **h** $\frac{8}{25}$
2 $\frac{16}{25}$
3 $\frac{6}{25}$
4 $\frac{11}{20}$
5 **a** $\frac{1}{8}$ **b** $\frac{1}{40}$ **c** $\frac{3}{8}$ **d** $\frac{7}{40}$

Exercise 5D

1 **a** 0.23 **b** 0.25 **c** $\frac{1}{4}$
2 **a** 0.74 **b** 0.7 **c** 74%
3 **a** $0.45, 48\%, \frac{1}{2}$ **b** $0.53, 55\%, \frac{6}{10}$
c $68\%, 0.7, \frac{3}{4}$ **d** $0.2, 27\%, \frac{3}{10}$
4 $\frac{7}{40}$
5 **a** $30\%, \frac{1}{3}, 0.4, 45\%, \frac{1}{2}$ **b** $\frac{1}{20}, 10\%, 0.12, 15\%, \frac{1}{5}$
c $0.6, 0.63, \frac{13}{20}, \frac{2}{3}, 68\%$ **d** $\frac{3}{5}, 62\%, 0.65, \frac{27}{40}, 70\%$

Mixed Exercise 5E

1 20%
2 85%
3 **a** $\frac{3}{10}$ **b** 30% **c** 5 more squares shaded
4 0.7
5 $\frac{37}{100}$
6 $\frac{7}{25}$
7 $0.25, 0.3, \frac{1}{3}, 35\%, \frac{3}{8}$
8 Ryan, as Sam only got 40%

5.2 Get ready

1 8 **2** 16 **3** 4.5

Exercise 5F

1 a	£12	**b**	40 kg	**c**	8 m	**d**	38 p
e	£50	**f**	9 cm	**g**	$30	**h**	£23
2 a	£6	**b**	7 km	**c**	14 km	**d**	£36
e	£16	**f**	3 kg	**g**	210 ml	**h**	£14

3 No, he has saved £3540

4 150 g	**5** £5400	**6** 34p	**7** £72
8 £185	**9** £42	**10** £16 280	

Exercise 5G

1 a £4 **b** 36 kg **c** £8.10 **d** 720 km

5.3 Get ready

1 a 15 **b** 180 **c** 6.5 **d** 2700

Exercise 5H

1 a	1.1	**b**	1.2	**c**	1.5	**d**	1.03
2 a	£360	**b**	117 kg	**c**	50 km	**d**	£2400
3 a	0.9	**b**	0.8	**c**	0.5	**d**	0.97
4 a	£360	**b**	160 kg	**c**	49 m	**d**	£300

Review exercise

1 a 0.1 **b** 0.04 **c** $\frac{13}{50}$
2 a 0.25 **b** $\frac{1}{4}$
3 a $\frac{3}{5}$ **b** 45%
4 a £30 **b** 5 m
5 a 0.92 **b** $\frac{3}{100}$ **c** 20 g
6 a 20 g **b** 60 g
7 45
8 84
9 £52.50
10 Emma, Majda got 67.5%
11 £67
12 2
13 £1.44
14 a £63 **b** £162 **c** £12.60
15 81 600
16 £46

Chapter 6 Answers

6.1 Get ready

1 $\frac{5}{8}$ **2** $\frac{3}{4}$ **3** $\frac{4}{5}$

Exercise 6A

1 a 3 : 7 **b** 8 : 7 **c** 4 : 11
2 a 16 : 13 **b** 13 : 16
3 a 2 : 3 **b** 2 : 5 **c** 5 : 3 **d** 2 : 3 : 5

Exercise 6B

1 $\frac{1}{3}$
2 a $\frac{2}{5}$ **b** $\frac{3}{5}$
3 a $\frac{5}{8}$ **b** $\frac{3}{8}$
4 a $\frac{1}{10}$ **b** $\frac{7}{10}$

5 1 : 1
6 1 : 2

Exercise 6C

1 a 1 : 2 **b** 1 : 3 **c** 7 : 5 **d** 2 : 3
e 4 : 3 **f** 1 : 3 **g** 8 : 15 **h** 1 : 4
2 2 : 3
3 4 : 3
4 5 : 9 : 6
5 3 : 4

Exercise 6D

1 a 1 : 3 **b** 2 : 5 **c** 3 : 10 **d** 4 : 1
2 a 3 : 10 **b** 20 : 3 **c** 8 : 3 **d** 1 : 4
3 9 : 20
4 3 : 16
5 a 3 : 1 **b** 1 : 3

Exercise 6E

1 a 1 : 2.5 **b** 1 : 2.4 **c** 1 : 0.3 **d** 1 : 1.25
2 1 : 18
3 1 : 14.5
4 a 1 : 5 **b** 1 : 20 **c** 1 : 0.2 **d** 1 : 6

Mixed Exercise 6F

1 1 : 2
2 a $\frac{4}{9}$ **b** $\frac{5}{9}$
3 a 8 : 3 **b** 1 : 4 **c** 3 : 8 : 2
4 a 3 : 1 **b** 2 : 5
5 3 : 20
6 a 2 : 3 **b** 1 : 1.5
7 1 : 2.25
8 7 adults

6.2 Get ready

1 4 **2** $\frac{2}{8}, \frac{3}{12}$ **3** 0.0365 m

Exercise 6G

1 20
2 a 4 **b** 6 **c** 20
3 a 6 kg **b** 60 kg
4 a i 40 g **ii** 200 g **iii** 300 g
b i 90 g **ii** 150 g **iii** 375 g
5 a 200 ml **b** 750 ml
6 15
7 1.25 litres

Exercise 6H

1 400 cm
2 720 cm
3 240 m
4 25 cm
5 a 360 cm **b** 26 cm
6 6 km
7 1.75 km
8 12 cm

6.3 Get ready

1 $\frac{4}{7}$ **2** $2:3$ **3** $1:15$

Exercise 6I

1 £16, £64
2 £9, £15
3 £18, £27
4 Alex 30 sweets, Ben 10 sweets
5 copper 680 g, nickel 120 g
6 15
7 £5, £15, £20
8 flour 450 g, sugar 150g, butter 300 g
9 £6
10 40 l

Mixed Exercise 6J

1 1440 cm
2 100 g
3 160 cm
4 25
5 13.5 km
6 £6
7 manganese 25 kg, nickel 5 kg
8 60 cm : 24 cm : 36 cm

6.4 Get ready

1 £1.35
2 17p
3 £1.95

Exercise 6K

1 £1.20
2 £2.24
3 £57.50
4 96 m
5 £3.75
6 £74.90
7 17.50 g
8 500 sheets

Exercise 6L

1 200 g flour, 100 g margarine, 150 g cheese
2 150 g pastry, 150 g bacon, 112.5 g cheese, 3 eggs, 225 ml milk
3 a 350 g   **b** 75 g   **c** 75 g

Exercise 6M

1 216 euros
2 26 970 rubles
3 a $426   **b** £250
4 £75
5 £40

Mixed Exercise 6N

1 300 g
2 £43.75
3 £42.50

4 a 472 euros   **b** £200
5 250 g margarine, 250 g caster sugar, 5 eggs, 562.5 g flour, 75 ml milk
6 £80
7 7 people
8 Francs

Review exercise

1 $2:3$
2 $21:4$
3 $5:2$
4 45 apples
5 10 British cars
6 a 50 cm   **b** 16 m
7 110 green counters
8 900 g of sugar, 18 g of butter, 720 g of condensed milk, 135 ml of milk
9 a €280   **b** £8
10 5:03 pm
11 Paris
12 45 litres

Chapter 7 Answers

7.1 Get ready

1 $5a$ **2** $3a + 2b$

Exercise 7A

1 $4a$ **2** $6a$ **3** $3p$
4 $5x$ **5** $6j$

Exercise 7B

1 a $p + 3$ **b** $x + 4$ **c** $q - 5$ **d** $g - 5$
   e $h + 4$ **f** $k - 6$ **g** $j - 6$ **h** $a + 3$
   i $y - 4$ **j** $m - 3$ **k** $p + 6$ **l** $h + 7$
2 $c + 12$
3 $a - 3$
4 $d + 12$
5 $g - 7$
6 $x + y$

Exercise 7C

1 $4f + 12t$ **2** $4f + 10t$ **3** $4a + 9b$
4 $50g + 20s$ **5** $5p + 3m + 4n$ **6** $6x + 12y$
7 $12x + 6y$ **8** $90r + 10d$

7.2 Get ready

1 $s + b$ **2** $h - 12$ **3** $r + g$

Exercise 7D

1 a a, b **b** x, y **c** a, t **d** x, y
   e t, d **f** a, s **g** b **h** g
   i t **j** a, b
2 a $3a, 4b$ **b** $x, 4y$ **c** $5a, 4t$ **d** x, y
   e $2t, 5d$ **f** $2a, 5s, 8$ **g** $4b, 6h$ **h** $9g, 6r, 4$
   i $5t, 7s, 3$ **j** $2a, 5b$

Answers

3
a a	**b** y	**c** t	**d** x
e d	**f** a, s	**g** b	**h** g
i t	**j** a, b		

4 Students' expressions, e.g. $3a + 4y$, $x + 9g + 6$
5 Students' terms, e.g. $8a$, $7b$, $5x$

7.3 Get ready

1 **a** 8 apples   **b** 5 bananas − 1 pear
   c 6 apples + 5 bananas

Exercise 7E

1 $5t$	2 $2c$	3 $4x$
4 $3a$	5 $6y$	6 $8a$

Exercise 7F

1 $5a$	2 $2p$	3 $8s$
4 $3x$	5 $6b$	6 $5k$
7 $9a$	8 $6x$	9 $9b$
10 p	11 $10n$	12 $4p$
13 $4x$		

Exercise 7G

1 $7a + 9b$	2 $9m + 7n$	3 $7p + 8q$
4 $2e + 2f$	5 $2g + 7h$	6 $2p + 5r$
7 $3j + 3k$	8 $6m + 8n$	9 $8a + 5b$
10 $3a + 3b$	11 $3m + 3n$	12 $7p + 2q$
13 $6e + f$	14 $2g + 3h$	15 $6p + r$
16 $9j + 3k$	17 $8n$	18 $3a + 2b$
19 $3p + 3j$	20 $4t$	21 $6x$
22 0	23 $4g$	24 m

Exercise 7H

1 $7a + 9$	2 $9m + 7$	3 $5p + 3q + 7$
4 $2e + 2$	5 $7h + 2$	6 $g + 4$
7 $3j + 3$	8 $6m + 1$	9 $8a + 5b + 1$
10 $3a + c + 3$	11 $6m + 5n + 9$	12 $3p + 6q + 2$
13 $4e + f + 2$	14 $3p + 4r + 4$	15 $p + r$

Exercise 7I

1 $2x^2$	2 $5y^2$	3 $2a^2$
4 $7a^2 + 9b^3$	5 $9m^2 + 7n$	6 $7p^3 + 2pq$
7 $4ef$	8 $2g^2 + 7h^3$	9 $2pq + 7r^3 - 2r^2$
10 $6jk$	11 $6m^3 + 8n$	12 $7a^2 + 4b^2$
13 $2a^3 + 3b^2$	14 $2m + 3n^2$	15 $7pq + 5p^3 - 3q^2$
16 $6e^3 + f^2$	17 $gh + 2h^3$	18 $7pqr$

Exercise 7J

1 $-3a$	2 $-2m + 5n$	3 $-3p - 4q$
4 $-5e - 1$	5 $-2g - 3h$	6 $2p^2 - 5r^3$
7 $-3k$	8 $-m^3 - 8n$	9 $-2a - 5b$
10 $-ab$	11 $-2m - 2$	12 $-3p - 6$
13 $-e - f$	14 $-2g^2 - 2$	15 $-4p + 2r + 2$
16 $-2j + 3k$	17 $-8n - 5$	18 $3a - 8b$
19 $-3p - 7j$	20 -8	21 $-4x + 1$
22 0	23 $-3g^3$	24 $-8mn + 2m^2$

7.4 Get ready

1 **a** 144	**b** 80	**c** 800

Exercise 7K

1 ab	2 xy	3 b^2
4 d^3	5 rst	6 abc
7 g^3	8 $2ef$	9 $3jk$
10 h^2	11 $5s^2$	12 $6t^3$
13 rt	14 xyt	15 $3mn$
16 $7abc$		

Exercise 7L

1 $6ab$	2 $20xy$	3 $6b^2$
4 $12d^3$	5 $35rs$	6 $12bc$
7 $15g^2$	8 $14ef$	9 $24jk$
10 $20h^2$	11 $25s^2$	12 $12t^3$
13 $12rt$	14 $35xy$	15 $18mn$
16 $30abc$	17 $4g^2$	18 $49h^2$
19 $8x^3$	20 $25n^2$	21 $30fgh$
22 $24jk$	23 $36hi$	24 $4a^2b$

7.5 Get ready

1 **a** 3	**b** $\frac{3}{2}$	**c** $\frac{3}{4}$

Exercise 7M

1 $4p$	2 p	3 4	4 $4n$
5 $2t$	6 15	7 $6k$	8 $2a$
9 $2x$	10 3	11 $5x$	12 8
13 $2p$	14 2	15 $4c$	16 2

7.6 Get ready

1 $2a + 6$	2 $3p + 6$

Exercise 7N

1 $2a + 8$	2 $3b + 6$	3 $4c + 24$
4 $5a - 20$	5 $3b - 15$	6 $5x + 15$
7 $2y - 4$	8 $6n + 12$	9 $15 + 3g$
10 $10 - 2x$	11 $6 - 3y$	12 $20 - 5h$
13 $10a + 50$	14 $3g + 21$	15 $4s - 20$
16 $21 - 3w$		

Exercise 7O

1 $6a + 8$	2 $15b + 12$	3 $20c + 24$
4 $6a - 15$	5 $15b - 21$	6 $10x + 25$
7 $6y - 8$	8 $12n + 42$	9 $15 + 9g$
10 $10 - 4x$	11 $6 - 15y$	12 $20 - 15h$
13 $40a + 30$	14 $15g + 21$	15 $12s - 20$
16 $21 - 12w$		

Exercise 7P

1 $a^2 + 4a$	2 $b^2 + 2b$	3 $ac + 6a$
4 $2a^2 - 4a$	5 $b^2 - 5b$	6 $x^2 + 3x$
7 $2y^2 - 2y$	8 $3n^2 + 2n$	9 $5g + g^2$
10 $5x - 2x^2$	11 $2y - 3y^2$	12 $4h - 5h^2$

13 $2a^2 + 2ab$ **14** $2g^2 + 14g$ **15** $4s^2 + 4st$
16 $21w - 3w^2$ **17** $10p^2 + 15p$ **18** $15x^2 - 5xy$
19 $2gh + 6h^2$ **20** $20p - 10p^2$

7.7 Get ready

1 a $6p + 38$ **b** $26 - 14s$ **c** $55 + 5g$

Exercise 7Q

1 $2(a + 3)$ **2** $2(n + 4)$ **3** $2(a - 6)$
4 $3(k + 2)$ **5** $3(f - 3)$ **6** $5(p - 2)$
7 $5(r + 4)$ **8** $3(x - 4)$ **9** $7(w + 2)$
10 $3(m - 5)$ **11** $4(q + 2)$ **12** $2(s + 1)$
13 $5(a - 5)$ **14** $6(x + 5)$ **15** $8(p - 5)$
16 $5(y - 1)$

Exercise 7R

1 $a(a + 2)$ **2** $a(a + 8)$ **3** $y(y^2 + 2)$
4 $j(j^2 - 3)$ **5** $s(s - 9)$ **6** $x(x^2 - 5)$
7 $p(p + 6)$ **8** $a(a - 1)$ **9** $p(p^2 + 1)$
10 $m^2(m - 1)$ **11** $c(c^2 + 8)$ **12** $a(2 + a)$
13 $x^2(x - 2)$ **14** $x(x + 7)$ **15** $p(p^2 - 1)$
16 $y(y - 5)$

Exercise 7S

1 $2a(3a + 1)$ **2** $3a(a + 3)$ **3** $2y(2y^2 + 1)$
4 $3j(2j^2 - 1)$ **5** $3s(s - 3)$ **6** $5x(2x^2 - 1)$
7 $3p(p + 2)$ **8** $2a(a - 1)$ **9** $5p(p^2 + 2)$
10 $3m^2(2m - 1)$ **11** $4c(c^2 + 2)$ **12** $6a(2 + a)$
13 $2x^2(3x - 1)$ **14** $5x(x + 6)$ **15** $4p(2p^2 - 1)$
16 $5y(5y - 1)$

7.8 Get ready

1 Students' expressions

Exercise 7T

1 Formula   **2** Expression
3 Equation   **4** Equation
5 Formula   **6** Equation
7 Expression   **8** Formula
9 Formula   **10** Formula

7.9 Get ready

1 a 11   **b** 1   **c** 14

Exercise 7U

1 4   **2** 15   **3** 12
4 3   **5** 6   **6** 15
7 20   **8** 10   **9** 9
10 18 **11** 13 **12** 8
13 19 **14** 7   **15** 10
16 6   **17** 27 **18** 10
19 0   **20** 3

Exercise 7V

1 10   **2** −6   **3** 12

4 7   **5** −6   **6** 15
7 20   **8** −5   **9** 28
10 11   **11** 7   **12** −7
13 31   **14** 8   **15** 20
16 −11 **17** 16 **18** 6
19 24   **20** 5

Exercise 7W

1 14   **2** 24   **3** 20
4 15   **5** 36   **6** 40
7 −8   **8** −5   **9** 24
10 78   **11** 5   **12** 22
13 35   **14** 2   **15** 76
16 4   **17** 102 **18** 20
19 −36 **20** −20

Review exercise

1 If Luke has x pounds, the total is $4x + 9$ pounds.
2 a

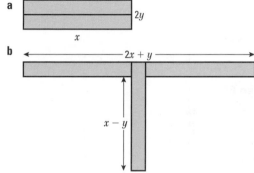

b

3 $8a + 4x$
4 a $10x - 20y$ **b** 210
5 a $3c$ **b** $3e + 2f$ **c** $5a$
    d $4xy$ **e** $2a + 7b + 8$
6 a $3bc$ **b** $2x + 5y$ **c** m^3 **d** $6np$
7 $x(x + 4)$
8 $12a - 28$

Chapter 8 Answers

8.1 Get ready

1 a 4   **b** 125   **c** 512   **d** 1 000 000

Exercise 8A

1 a 32   **b** 256   **c** 1   **d** 10 000
    e 625   **f** 7776
2 a 2^4 **b** 4^5 **c** 1^6 **d** 8^3
    e $3^2 \times 8^4$ **f** $2^3 \times 4^4$
3 a 16   **b** 243   **c** 216   **d** 25
    e 512   **f** 11 664 **g** 65 536 **h** 10 125
    i 31 104 **j** 256
4

Power of 10	Index	Value	Value in words
10^3	3	1000	One thousand
10^2	2	100	One hundred
10^6	6	1 000 000	One million
10^1	1	10	Ten
10^5	5	100 000	One hundred thousand

Answers

5 **a** 6400 **b** 400 **c** 6000 **d** 4
   e 125 **f** 16
6 **a** 3 **b** 4 **c** 6 **d** 4
   e 2 **f** 3 **g** 4 **h** 2

8.2 Get ready

1 **a** 4 **b** 64 **c** 2401 **d** 10 000

Exercise 8B

1 **a** 6^{11} **b** 8^8 **c** 2^6
2 **a** 4 **b** 6^3 **c** 7^4
3 **a** 4^5 **b** 5^2 **c** 3
4 **a** 5^{13} **b** 2^{11}
5 **a** 10^5 **b** $9^0 = 1$
6 **a** 6^{11} **b** 5^6
7 **a** 3^8 **b** 4^6
8 **a** 6^9 **b** 5^5 **c** 4^{12}
9 **a** 5^6 **b** 7^8

8.3 Get ready

1 **a** 7^7 **b** 8^2 **c** 5^6

Exercise 8C

1 **a** x^{10} **b** y^{11} **c** x^{14}
2 **a** a^8 **b** b^6 **c** d^{11}
3 **a** p^3 **b** q^{10} **c** t^4
4 **a** j^6 **b** k **c** n^2
5 **a** x^9 **b** y^9 **c** z^{10}
6 **a** $6x^5$ **b** $15y^{29}$ **c** $24z^{10}$
7 **a** $3p^5$ **b** $5q^2$ **c** $2r^3$
8 **a** d^{12} **b** e^{10} **c** f^9 **d** g^{63}
9 **a** g^{24} **b** h^4 **c** 1 **d** 1
10 **a** $2187d^{14}$ **b** $64e^3$ **c** 1
11 **a** 1 **b** b^4 **c** c^{10}
12 **a** $8d^{10}$ **b** $2e^4$ **c** $16f^4$

8.4 Get ready

1 **a** 9 **b** 11 **c** 17

Exercise 8D

1 **a** 9 **b** 1 **c** 25 **d** 13
   e 21 **f** 23 **g** 6 **h** 4
   i 8 **j** 5 **k** 6 **l** 2
   m 1 **n** 5 **o** 4
2 **a** $4 + 5 = 9$ **b** $4 \times 5 = 20$
   c $(2 + 3) \times 4 = 20$ **d** $(3 - 2) \times 5 = 5$
   e $(5 - 2) \times 3 = 9$ **f** $4 \div 2 + 8 = 10$
   g $5 \times 4 + 5 + 2 = 27$ **h** $5 \times 4 + 5 - 2 = 23$
3 **a** 49 **b** 25 **c** 243 **d** 123
   e 72 **f** 17 **g** 22 **h** 7
   i 7 **j** 0 **k** 7 **l** 0

8.5 Get ready

1 **a** 48 **b** 3 **c** 120

Exercise 8E

1 $5x + 14$ **2** $20x + 17$
3 $23x + 14$ **4** $-8x + 12$ or $12 - 8x$
5 $9 - 21x$ **6** $15 - 23x$
7 $12x - 22y$ **8** $22y - 2x$
9 $-4x - 21y$ **10** $x + 4y$
11 $7y + 2$ **12** 27
13 $x - 3$ **14** $7y - 16x + 6$
15 $21x - 19y$ **16** $21y - 27x$
17 $10x - 7y$ **18** $20x - 32y$
19 $8xy + 3x$ **20** $8xy + 2x + y$
21 $6x - 4y - 3xy$ **22** $10xy - 20x^2 - 2y^2$

8.6 Get ready

1 **a** 8 **b** 4 **c** ab

Exercise 8F

1 **a** $2(x + 3)$ **b** $2(3y + 1)$ **c** $5(3b - 1)$
   d $2(2r - 1)$ **e** $x(3 + 5y)$ **f** $4(3x + 2y)$
   g $4(3x - 4)$ **h** $3(3 - x)$ **i** $3(3 + 5g)$
2 **a** $x(3x + 4)$ **b** $y(5y - 3)$ **c** $a(2a + 1)$
   d $b(5b - 2)$ **e** $c(7 - 3c)$ **f** $d(d + 3)$
   g $m(6m - 1)$ **h** $x(4y + 3)$ **i** $n^2(n - 8)$
3 **a** $4x(2x + 1)$ **b** $3p(2p + 1)$ **c** $3x(2x - 1)$
   d $3b(b - 3)$ **e** $3a(4 + a)$ **f** $5c(3 - 2c)$
   g $7x^3(3x + 2)$ **h** $4y^2(4y - 3)$ **i** $2d^2(3d^2 - 2)$
4 **a** $ax(x + 1)$ **b** $pr(r - 1)$ **c** $ab(b - 1)$
   d $q(r^2 + q)$ **e** $ax(a + x)$ **f** $by(b - y)$
   g $3a^2(2a - 3)$ **h** $4x^3(2 - x)$ **i** $6x^3(3 + 2x^2)$
5 **a** $6ab(2a + 3b)$ **b** $2xy(2x - y)$
   c $4ab(a + 2b + 3)$ **d** $2xy(2x + 3y - 1)$
   e $3ax(4x + 2a - 1)$ **f** $abc(a + b + c)$
6 **a** $5(x + 4)$ **b** $2(6y - 5)$
   c $x(3x + 5)$ **d** $y(4 - 3y)$
   e $2a(4 + 3a)$ **f** $4b(3b - 2)$
   g $cy(y + 1)$ **h** $3dx(x - 2)$
   i $3cd(3c + 5d)$

Review exercise

1 **a** 6^3 **b** 11^2 **c** 2^6
2 **a** 625 **b** 128 **c** 1000 **d** 100 000
3 **a** 144 **b** 784 **c** 400 **d** 30 000
4 **a** $2^2 \times 3^3$ **b** $5^2 \times 7^2$ **c** $4^2 \times 8^4$ **d** $2^3 \times 6^3$
5 **a** 512 **b** 10 000 **c** 125
   d 144 **e** 800
6 **a** 5 **b** 5 **c** 3
7 **a** $5(x + 3y)$ **b** $3(5p - 3q)$ **c** $c(d + e)$
8 6 or −6
9 2
10 2^{14}
11 $5(p - 4)$
12 **a** 2^7 **b** 5^5 **c** 3^5 **d** 7^3
   e 9^4 **f** 8^2 **g** 7^3 **h** 6^5
13 **a** x^9 **b** x^3 **c** x^{15} **d** x
   e x^5 **f** x^{12} **g** 1 **h** x^6
   i x^{11} **j** x^9 **k** x^3 **l** x^{11}
14 **a** $4x^8$ **b** $15x^8$ **c** $21x^5$ **d** $4x^4$
   e $8x^5$ **f** $9x$ **g** x^{10} **h** x^9

15 a a^7 **b** $15x^3y^4$
16 a $9ab - 6a^2 - 4b^2$ **b** $11pq + 18p^2$
   c $13c^2 + 12cd$ **d** $a^2 + 2ab + b^2$
   e $5ab + ac - bc$ **f** $-4ab - 4ac - 9bc$
17 a $x(x - 7)$ **b** $t(t + a)$ **c** $x(bx - 1)$
   d $p(3p + y)$ **e** $a(q^2 - t)$
18 a $x(x - 5)$
   b 10 500

Chapter 9 Answers

9.1 Get ready

1 2, 4, 6, 8, 10, 12, 14, 16, 18, 20
2 1, 3, 5, 7, 9, 11, 13, 15, 17, 19
3 Check

Exercise 9A

1 a 12, 15; add 3 **b** 15, 19; add 4 **c** 25, 30; add 5
   d 22, 27 ; add 5 **e** 13, 16; add 3 **f** 13, 15; add 2
   g 23, 28; add 5 **h** 16, 19; add 3 **i** 17, 22; add 4
   j 40, 50; add 10
2 a, b i 21, 25; add 4 **ii** 17, 20; add 3 **iii** 23, 27; add 4
      iv 24, 28; add 4 **v** 20, 23; add 3 **vi** 29, 35; add 6
      vii 22, 26; add 4 **viii** 31, 37; add 6 **ix** 43, 51; add 8
      x 25, 29; add 4
3 question 1
   a 30 **b** 39 **c** 50 **d** 47
   e 28 **f** 23 **g** 48 **h** 31
   i 38 **j** 100
   question 2
   i 37 **ii** 29 **iii** 39 **iv** 40
   v 32 **vi** 59 **vii** 38 **viii** 55
   ix 75 **x** 41
4 a

Week	1	2	3	4	5
Money in piggy bank (£)	2	4	6	8	10

   b 10 weeks

Exercise 9B

1 a 12, 10; subtract 2 **b** 9, 7; subtract 2
   c 35, 30; subtract 5 **d** 22, 17; subtract 5
   e 10, 7; subtract 3 **f** 11, 9; subtract 2
   g 17, 10; subtract 7 **h** 13, 10; subtract 3
   i 13, 9; subtract 4 **j** 50, 40; subtract 10
2 a, b i 25, 21; subtract 4 **ii** 15, 12; subtract 3
      iii 43, 39; subtract 4 **iv** 22, 19; subtract 3
      v 18, 15; subtract 3 **vi** 37, 31; subtract 6
      vii 14, 12; subtract 2 **viii** 31, 26; subtract 5
      ix 36, 29; subtract 7 **x** $-4, -6$; subtract 2
3 question 1
   a 2 **b** -1 **c** 10 **d** -3
   e -5 **f** 1 **g** -18 **h** -2
   i -7 **j** -10
   question 2
   i 5 **ii** 0 **iii** 23 **iv** 7
   v 3 **vi** 7 **vi** 4 **viii** 6
   ix 1 **x** -10

4

Day	M	Tu	W	Th	F
Money left at end of day (£)	17	14	11	8	5

£5

Exercise 9C

1 a 16, 32; multiply by 2 **b** 256; multiply by 4
   c 25, 625; multiply by 5 **d** 1000, 10 000; multiply by 10
   e 48, 96; multiply by 2 **f** 54, 162; multiply by 3
   g 128, 512; multiply by 4 **h** 20 000, 200 000; multiply by 10
   i 250, 1250; multiply by 5 **j** 375; multiply by 5
2 a, b i 32, 64; multiply by 2
      ii 243, 729; multiply by 3
      iii 1024, 4096; multiply by 4
      iv 3125, 15 625; multiply by 5
      v 80, 160; multiply by 2
      vi 324, 972; multiply by 3
      vii 810, 2430; multiply by 3
      viii 50 000, 500 000; multiply by 10
      ix 160, 320; multiply by 2
      x 7776, 46 656; multiply by 6
3 question 1
   a 512 **b** 262 144
   c 1 953 125 **d** 1 000 000 000
   e 1536 **f** 39 366
   g 524 288 **h** 2 000 000 000
   i 3 906 250 **j** 5 859 375
   question 2
   i 1024 **ii** 59 049
   iii 1 048 576 **iv** 9 765 625
   v 2560 **vi** 78 732
   vii 196 830 **viii** 5 000 000 000
   ix 5120 **x** 60 466 176
4 a

Month	1	2	3	4	5
Number of rabbits	2	4	8	16	32

   b 1024 rabbits

Exercise 9D

1 a 4, 2; divide by 2 **b** 16; divide by 4
   c 25, 5; divide by 5 **d** 100, 10; divide by 10
   e 12, 6; divide by 2 **f** 6; divide by 3
   g 64, 32; divide by 2 **h** 300, 30; divide by 10
   i 50, 10; divide by 5 **j** 2, 0.2; divide by 10
2 a, b i 4, 2; divide by 2
      ii 3, 1; divide by 3
      iii 8, 4; divide by 2
      iv $1, \frac{1}{5}$ (0.2); divide by 5
      v 5, 2.5; divide by 2
      vi 12, 4; divide by 3
      vii 30, 10; divide by 3
      viii 5, 0.5; divide by 10
      ix 10, 5; divide by 2
      x $1, \frac{1}{6}$; divide by 6
3 question 1
   a 0.5 **b** 0.0625 **c** 0.04 **d** 0.01
   e 1.5 **f** $\frac{2}{9}$ **g** 8 **h** 0.03
   i 0.08 **j** 0.0002

Answers

question 2

i 0.5   **ii** $\frac{1}{9}$ **iii** 1   **iv** 0.008

v 0.625   **vi** $\frac{4}{9}$ **vii** $\frac{10}{9}$ **viii** 0.005

ix 1.25   **x** $\frac{1}{216}$

4 a

Years	0	10	20	30	40
Number of atoms	2560	1280	640	320	160

b 2.5 atoms

Exercise 9E

1 a i

    ii Add 1 cross to each row

    iii

b

    ii Add 2 crosses to each row

    iii

c i

    ii Add 6 matches

    iii

d i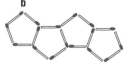

    ii Add 2 matches

    iii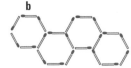

e

    ii Add 1 cross to each row

    iii

2 a 5, 9, 13

b

c Add 4 matches   **d** 41 matches

3 a 6, 11, 16

b

c Add 5 matches   **d** 51 matches

9.2 Get ready

1 a 8   **b** 6   **c** 14   **d** 20

2 a 4   **b** 0   **c** −4

Exercise 9F

1 a

Term number	Term
1	3
2	6
3	9
4	12

b Multiply the term number by 3   **c** Add 3

2 a

Term number	Term
1	7
2	14
3	21
4	28

b Multiply the term number by 7   **c** Add 7

3 a

Term number	Term
1	4
2	8
3	12
4	16

b Multiply the term number by 4   **c** Add 4

4 a

Term number	Term
1	2
2	4
3	6
4	8

b Multiply the term number by 2   **c** Add 2

5 a

Term number	Term
1	8
2	16
3	24
4	32

b Multiply the term number by 8   **c** Add 8

6 a

Term number	Term
1	10
2	20
3	30
4	40

b Multiply the term number by 10   **c** Add 10

7 a

Term number	Term
1	12
2	24
3	36
4	48

b Multiply the term number by 12   **c** Add 12

8 a

Term number	Term
1	50
2	100
3	150
4	200

b Multiply the term number by 50
c Add 50

Exercise 9G

1 a

Term number	1	2	3	4	5
Term	5	8	11	14	17

b Multiply by 3 and add 2
c Add 3

2 a

Term number	1	2	3	4	5
Term	3	5	7	9	11

b Multiply by 2 and add 1
c Add 2

3 a

Term number	1	2	3	4	5
Term	2	5	8	11	14

b Multiply by 3 and subtract 1
c Add 3

4 a

Term number	1	2	3	4	5
Term	1	5	9	13	17

b Multiply by 4 and subtract 3
c Add 4

5 a

Term number	1	2	3	4	5
Term	3	8	13	18	23

b Multiply by 5 and subtract 2
c Add 5

6 a

Term number	1	2	3	4	5
Term	7	10	13	16	19

b Multiply by 3 and add 4
c Add 3

7 a

Term number	1	2	3	4	5
Term	8	11	14	17	20

b Multiply by 3 and add 5
c Add 3

8 a

Term number	1	2	3	4	5
Term	4	9	14	19	24

b Multiply by 5 and subtract 1
c Add 5

9 a

Term number	1	2	3	4	5
Term	7	11	15	19	23

b Multiply by 4 and add 3
c Add 4

10 a

Term number	1	2	3	4	5
Term	1	6	11	16	21

b Multiply by 5 and subtract 4
c Add 5

Exercise 9H

1

Term number	Term
1	4
2	7
3	10
4	13
5	16
↓	↓
10	31
↓	↓
11	34

2

Term number	Term
1	1
2	3
3	5
4	7
5	9
↓	↓
10	21
↓	↓
13	25

3

Term number	Term
1	8
2	13
3	18
4	23
5	28
↓	↓
10	53
↓	↓
15	78

4

Term number	Term
1	1
2	5
3	9
4	13
5	17
↓	↓
10	37
↓	↓
12	45

Answers

5

Term number	Term
1	11
2	21
3	31
4	41
5	51
↓	↓
10	101
↓	↓
15	151

6

Term number	Term
1	2
2	7
3	12
4	17
5	22
↓	↓
10	47
↓	↓
14	67

7 a 39   b 12
8 a 49   b 14
9 a 35   b 15

9.3 Get ready

1 a 5   b −5   c 3
   d 4   e −3
2 a 50   b −5   c 31
   d 43   e 23

Exercise 9I

1 (1) $3n + 1$ (2) $2n − 1$ (3) $5n + 3$
2 a $2n − 1$; 39   b $2n + 1$; 41   c $3n − 1$; 59
   d $3n + 2$; 62   e $4n − 3$; 77   f $4n − 2$; 78
   g $5n − 3$; 97   h $5n − 1$; 99   i $5n + 3$; 103
   j $2n + 3$; 43   k $45 − 5n$; − 55   l $40 − 2n$; 0
   m $38 − 3n$; −22   n $22 − 2n$; −18   o $21 − 2n$; −19
   p $200 − 10n$; 0
3 a

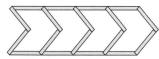

   b

Pattern number	1	2	3	4	5	6
Number of sticks	6	10	14	18	22	26

   c $4n + 2$
   d 82 sticks

9.4 Get ready

1 a $3n + 1$, 61   b $5n − 2$, 98   c $2n + 11$, 51

Exercise 9J

1 The sequence is of all the odd numbers, so 21 is a member of the number pattern because it is an odd number.
34 is not in the pattern as it is not an odd number.
2 The sequence is of all the odd numbers above 1, so 63 is a member of the number pattern because it is an odd number higher than 1.
86 is not in the pattern as it is not an odd number.
3 The nth term is $3n − 1$. If 50 is in the pattern, $50 = 3n − 1$, giving $n = 17$, so 50 is in the pattern.
None of the terms are a multiple of 3. 66 is a multiple of 3, so 66 is not in the pattern.
4 The nth term is $3n + 2$. If 50 is in the pattern, $50 = 3n + 2$, giving $n = 16$, so 50 is in the pattern.
The nth term is $3n + 2$. If 62 is in the pattern, $62 = 3n + 2$, giving $n = 20$, so 62 is in the pattern.
5 The nth term is $4n − 3$. If 101 is in the pattern, $101 = 4n − 3$, giving $n = 26$, so 101 is in the pattern.
The nth term is $4n − 3$. If 150 is in the pattern, $150 = 4n − 3$, giving $n = 38.25$, so 150 is not in the pattern.
6 All the terms are even, so 101 is not in the pattern.
The nth term is $4n − 2$. If 98 is in the pattern, $98 = 4n − 2$, giving $n = 25$, so 98 is in the pattern.
7 The sequence contains all the numbers ending in 2 or 7, so 97 is in the pattern.
All the numbers end in 2 or 7, so 120 is not in the pattern.
8 All the numbers end in 4 or 9, so 168 is not in the pattern.
The sequence contains all the numbers ending in 4 or 9, so 169 is in the pattern.
9 The sequence contains all the numbers below 45 ending in 0 or 5, so 85 is not in the pattern, as it is higher than 40.
All the numbers end in 0 or 5, so 4 is not in the pattern.
10 All the terms are even, so 71 is not in the pattern.
The sequence contains all the even numbers below 40, so 82 is not in the pattern as it is higher than 38.
11 All the terms are odd, so 46 is not in the pattern.
The nth term is $4n − 1$. If 79 is in the pattern, $79 = 4n − 1$, giving $n = 20$, so 79 is in the pattern.
12 The nth term is $6n − 1$. If 119 is in the pattern, $119 = 6n − 1$, giving $n = 20$, so 119 is in the pattern.
All the terms are odd, so 72 is not in the pattern.

Review exercise

1 a

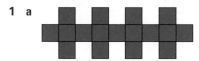

   b 41 squares
2 a 116
   b 112
   c No. 9 is an odd number and all the terms in the sequence are even.
3

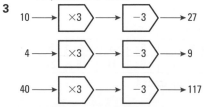

4 a

Input	Output
4	−2
2	−6
−3	−16

b $2(n − 5)$

5 No. When $n = 9$, $n^2 + 4 = 85$, which is not a prime number.

6 a 8, 13
b 55

7 Dylan could be right if you are doubling the previous term.
Evie could be right if the difference between the terms was increasing by one.

8 The nth term is $3n + 2$. If 140 is in the sequence,
$140 = 3n + 2$, giving $n = 46$, so 140 is in the sequence.

9 $x = 4.5$

10 a 31
b $4n − 1$

Chapter 10 Answers

10.1 Get ready

1 a One square to the right and one down
b Two squares to the right
c One square up

Exercise 10A

1 a i Ghost Ride **ii** Big Dipper **iii** Helter Skelter
iv Exit **v** Dodgems
b i (3, 0) **ii** (4, 3) **iii** (0, 6)
iv (1, 2) **v** (4, 5)
2 a i B **ii** F **iii** A
iv E **v** O
b i (0, 0) **ii** (6, 2) **iii** (3, 0)
iv (2, 7) **v** (0, 4)
3 a i U **ii** T **iii** S
iv P **v** V
b i (3, 1) **ii** (6, 2) **iii** (5, 4)
iv (5, 1) **v** (3, 7)

4

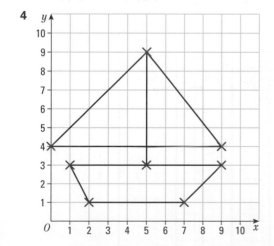

5

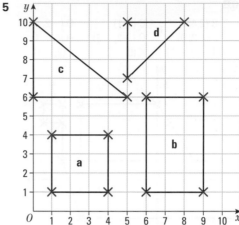

6 a, b

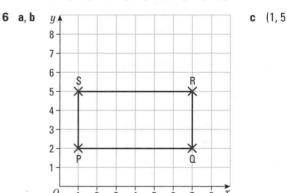

c (1, 5

7 a, b

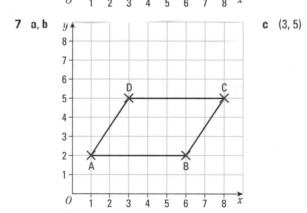

c (3, 5)

10.2 Get ready

1

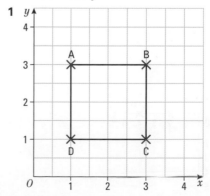

2 (2, 2)

Exercise 10B

1 A (5, 2)   B (2 4)   C (0, 1)   D (−2, 4)
E (−4, 2)   F (−3, 0)   G (−2, −3)   H (4, −1)
I (5, 0)   J (−4, −2)   K (0, −2)   L (2, −1)

2

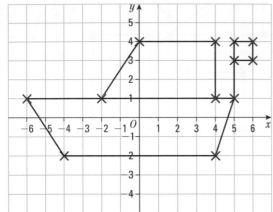

3 Students' pictures and coordinates

10.3 Get ready

1 a 25   **b** 35   **c** 5   **d** 6
   e 1   **f** 2   **g** 1   **h** 1.5

Exercise 10C

1 a (2, 3)   **b** (4, 3)   **c** (7, 2.5)
   d (3.5, 3.5)   **e** (2.5, 2)
2 a (4, 4)   **b** (4, 6)   **c** (5.5, 5)
   d (4.5, 4.5)
3 a (5, 5)   **b** (4, 6.5)   **c** (3.5, 4.5)
   d (4, 4.5)   **e** (4, 5.5)   **f** (4.5, 4.5)

Exercise 10D

1 a (1, 2)   **b** (3.5, −0.5)   **c** (−1.5, 3)
   d (−2.5, −0.5)   **e** (−4, 0.5)   **f** (−0.5, −2)
   g (2.5, −0.5)   **h** (0.5, 2.5)   **i** (−2, 3)
   j (4.5, −0.5)
2 a (1.5, 0.5)   **b** (4, 1)
   c (1.5, 1)   **d** (3.5, 2.5)
3 a (4, 4)   **b** (−2, 2.5)   **c** (1.5, −3.5)
   d (−3, 4.5)   **e** (2, −2.5)   **f** (2.5, −1.5)

Review exercise

1 a (3, 3)   **b** (4, 0)
   c

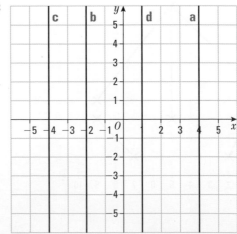

   d (−3, 0)

2 a i (−6, 4)   **ii** (3, 4)   **iii** (3, −2)
   b i

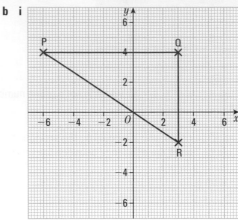

   ii right-angled triangle
   c i (3, 1)   **ii** (1.5, 4)   **iii** (−1.5, 1)
   d (−6, −2)
3 a i (2, 3)   **ii** (2, −1)
   b (−4, −1)
   c (2, 1)

Chapter 11 Answers

11.1 Get ready

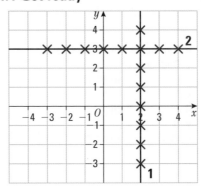

3 (2, 3)

Exercise 11A

1 a $y = 3$ **b** $y = 1$ **c** $y = -1$ **d** $y = -3$
2 a $x = -4$ **b** $x = -2$ **c** $x = 1$ **d** $x = 4$
3

4

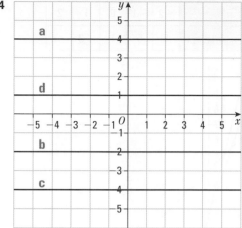

iii

x	−3	−2	−1	0	1	2	3
$y = 3x + 1$	−8	−5	−2	1	4	7	10

iv

x	−3	−2	−1	0	1	2	3
$y = x + 4$	1	2	3	4	5	6	7

v

x	−2	−1	0	1	2
$y = 4x + 1$	−7	−3	1	5	9

b

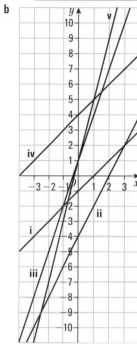

5 a, b

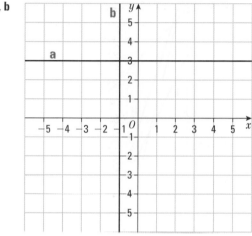

c $(-1, 3)$

11.2 Get ready

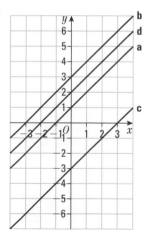

2

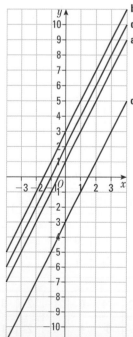

3

Exercise 11B

1 a i

x	−3	−2	−1	0	1	2	3
$y = x - 1$	−4	−3	−2	−1	0	1	2

ii

x	−3	−2	−1	0	1	2	3
$y = 2x - 4$	−10	−8	−6	−4	−2	0	2

Answers

4

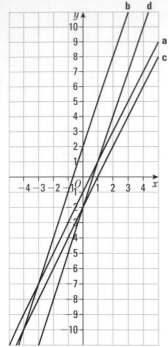

5

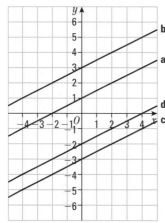

Exercise 11C

1 a i

x	−3	−2	−1	0	1	2	3
$y = -x - 1$	2	1	0	−1	−2	−3	−4

ii

x	−3	−2	−1	0	1	2	3
$y = -2x + 4$	2	0	−2	−4	−6	−8	−10

iii

x	−3	−2	−1	0	1	2	3
$y = -3x + 1$	10	7	4	1	−2	−5	−8

iv

x	−3	−2	−1	0	1	2	3
$y = -x + 4$	7	6	5	4	3	2	1

v

x	−2	−1	0	1	2
$y = -4x + 1$	9	5	1	−3	−7

b

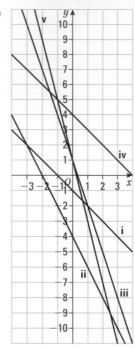

2

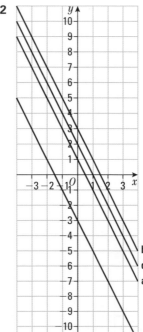

3

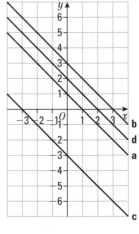

4

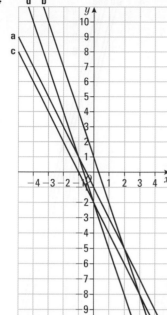

5

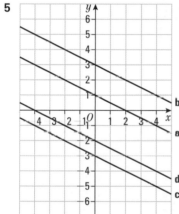

11.3 Get ready

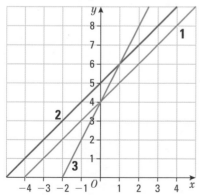

They both cross the y-axis at $y = 4$.

Exercise 11D

1

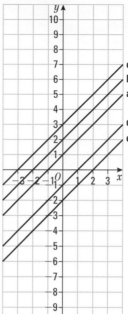

2

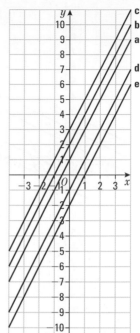

3

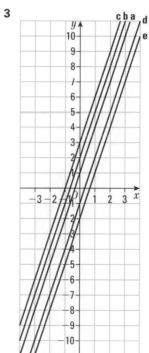

4

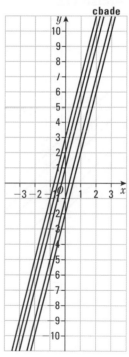

5

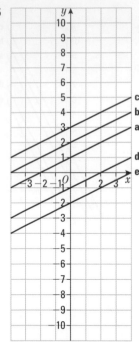

3

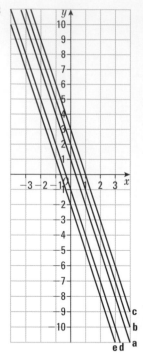

4

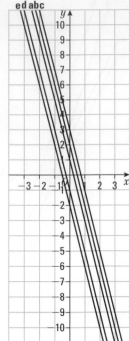

For each question, the lines are parallel.

Exercise 11E

1

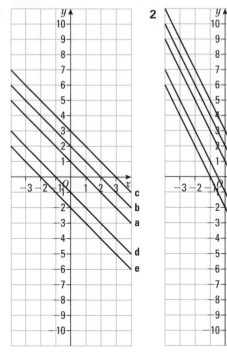

2

5

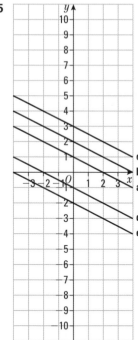

For each question, the lines are parallel.

Exercise 11F

1

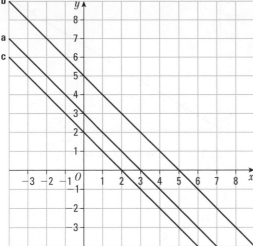

2

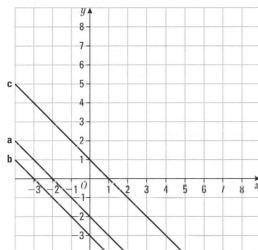

3

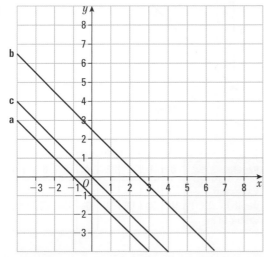

11.4 Get ready

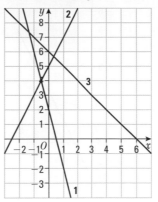

Exercise 11G

1 a $y = x + 3$
  b $y = 2x + 1$
  c $x + y = 4$
  d $y = -\frac{4}{3}x + 1$

2 a $y = 3$
  b $y = 2x + 3$
  c $y = -3x + 1$
  d $x = -3$
  e $x + y = 2$

3 a $y = -1$
  b $y = 3x + 2$
  c $y = -\frac{1}{2}x + 1$
  d $x = 3$
  e $y = -2x - 1$

4 a $y = 1$
  h $y = 2x + 1$
  c $y = -4x + 1$
  d $x = 1$
  e $x + y = 1$

Review exercise

1 i B   **ii** C   **iii** D   **iv** A

2 a

x	−1	0	1	2	3
$y = 2x - 1$	8	6	4	2	0

b

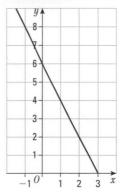

c $y = 1.2$ **d** $x = -0.3$

Answers

3
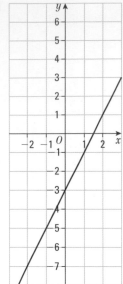

4
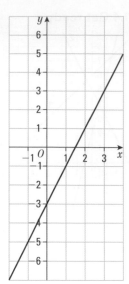

5
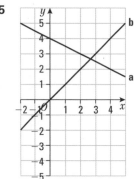

c $x = 2.67$

6 A $y = 2x + 1$ B $y = -3$ C $y = 3x - 3$

Chapter 12 Answers

12.1 Get ready

1

Number of bars	1	2	3	4	5	6	7	8	9	10
Cost in pence	20	40	60	80	100	120	140	160	180	200

Exercise 12A

1 a
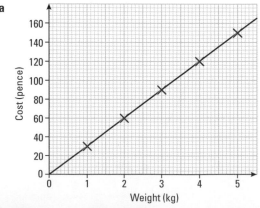

b 75p **c** 180p or £1.80

2 a
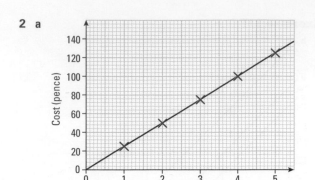

b i 200p or £2 **ii** 150p or £1.50

3 a
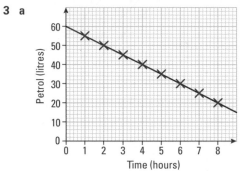

b 60 litres
c 32.5 litres

4 a

Distance travelled in km	0	5	10	15	20	25
Petrol used in litres	0	2	4	6	8	10

b
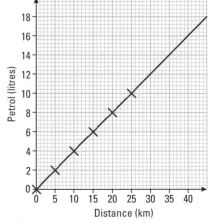

c 1.6 km
d 37.5 km

5 a

Weeks	0	1	2	3	4	5	6	7	8
Expected depth of water in m	144	140	136	132	128	124	120	116	112

b

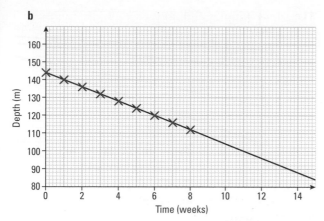

c 104 m   **d** 12 weeks

6 a

Usage time in minutes	Cost in pounds
0	0
5	2
10	4
15	6
20	8
25	10
30	12
35	14
40	16
45	18
50	20

b

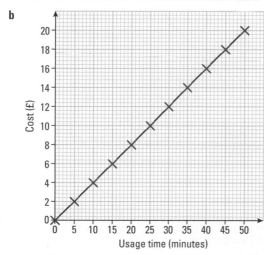

c £12.80
d 21 minutes

7 a

Usage time in minutes	Cost in pounds
0	15
5	15.5
10	16
15	16.5
20	17
25	17.5
30	18
35	18.5
40	19
45	19.5
50	20

b

c £18.20
d 20 minutes

8 a

Units used	Cost in pounds
0	0
10	5
20	10
30	15
40	20
50	25
60	30
70	35
80	40
90	45
100	50

b

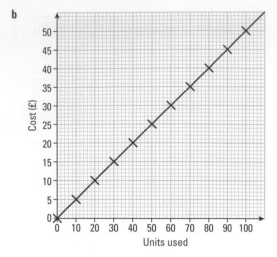

c £16

d 90 units

9 a

Units used	Cost in pounds
0	20
10	22.5
20	25
30	27.5
40	30
50	32.5
60	35
70	37.5
80	40
90	42.5
100	45

b

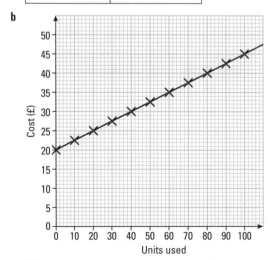

c £28

d 72 units

10 a Tariff 1   B

Tariff 2   A

Tariff 3   C

b Tariff 2 would be cheapest because the cost of £24 is the lowest of the three tariffs.

11

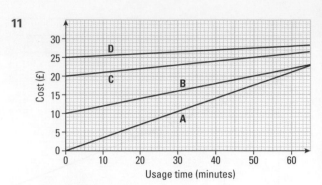

Jodie should choose Tariff A.

12.2 Get ready

1 a £4   **b** £8   **c** £16   **d** £80

2 a €2   **b** €10   **c** €30   **d** €20

Exercise 12B

1 a i HK$120   **ii** HK$60   **iii** HK$96

iv HK$1200   **v** HK$2400

b i £5   **ii** £2.50   **iii** £7.50

iv £50   **v** £100

2

°C	5	20	27	28	10	38	35	80	93	40
°F	41	68	80	82	50	100	95	176	200	104

3 a

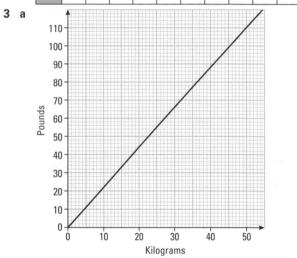

b

Kilograms	0	4.5	9	45	30	15	23	6	35	50
Pounds	0	10	20	99	66	33	51	13	77	110

4

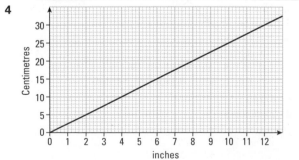

Inches	0	1	2	4	6	8	9	8	10	12
Centimetres	0	2.5	5	10	15	20	22.5	20	25	30

5

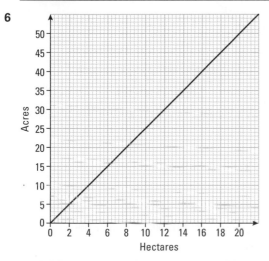

Miles	0	5	10	40	22.5	30	45	12.5	24	50
Kilometres	0	8	16	64	36	48	72	20	38.4	80

6

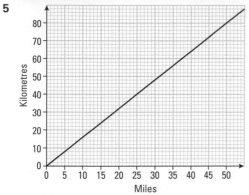

Hectares	0	8	12	12	15	17	9.6	18	3	20
Acres	0	20	30	30	37.5	42.5	24	45	7.5	50

12.3 Get ready

1 30 mph **2** 100 miles **3** 2 hours

Exercise 12C

1 a 10 minutes **b** 300 m **c** 15 minutes
   d 15 minutes **e** 30 m per minute = 1.8 km per hour
   f 20 m per minute = 1.2 km per hour
2 a Students' stories, e.g.
   David travelled for 1 hour, stopped for a cup of tea for
   half an hour and then drove faster for another hour,
   when he reached his aunt's house. He stayed at his
   aunt's for 1 hour and then drove straight home in half an
   hour.
   b 1st section: 20 km per hour
   2nd section: 0 km per hour
   3rd section: 40 km per hour
   4th section: 0 km per hour
   5th section: 120 km per hour

3 a Tom set off at 08:00, travelled 45 km in 1 hour, stopped for
   20 minutes and travelled another 45 km in 40 minutes. He
   reached Sarah's house at 10:00. He then travelled back
   90 km in 1 hour.
   b Sarah set off at 08:20 and drove 30 km in 40 minutes. At
   09:00 she stopped for 10 minutes and then drove 60 km in
   40 minutes, reaching Tom's house at 09:50.
   c 09:20, 45 km from London
4 A→B: The depth of the water goes up 10 cm to 30 cm in 5
   minutes.
   B→C: The depth stays at 30 cm for 5 minutes.
   C→D: Imran gets into the bath and the depth increases to
   60 cm.
   D→E: The depth stays constant at 60 cm for 15 minutes.
   E→F: Imran takes the plug out and the depth decreases
   from 60 cm to zero in 5 minutes.

5

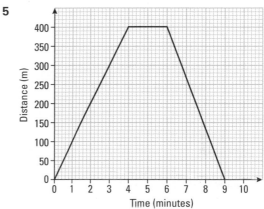

6

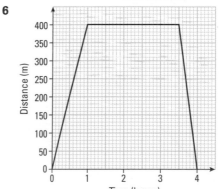

7 a

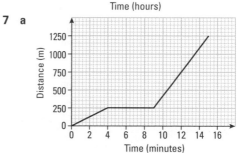

   b 167 m per minute = 10 km per hour

8

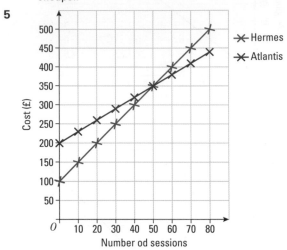

9 **a** **i** 12 m   **ii** 45 m   **b** **i** 2.8 seconds   **ii** 2.2 seconds
10 **a** **i** 1985 m   **ii** 1800 m   **iii** 1535 m
     b **i** 11:02:05   **ii** 11:02:40   **iii** 11:03:12
11 **a**

b 15.5 m per second
c 16.2 m

Review exercise

1 **a**

b Up to 20 miles

2

C	0	20	40	60	80	100
F	30	70	110	150	190	230

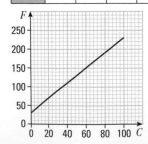

3 **a** $C = 6D + 14$
     b £74
4 **a** Tariff A: $C = 20$
       Tariff B: $C = 0.04m + 10$
     b Tariff A has a flat charge of £20, so calls don't cost any
       extra. Tariff B has a cost of £10 plus 4p per minute.
     c Josh should use Tariff B unless he uses more than
       250 minutes a month, in which case Tariff A would be
       cheaper.

5

Atlantis is cheaper if she goes to more than 50 sessions

Chapter 13 Answers

13.1 Get ready

1 £3.75   **2** £212.40   **3** 37.68

Exercise 13A

1 35 cm
2 £185
3 40 cm²
4 22 cm
5 £19.20
6 **a** Total cost = number of stamps × cost of 1 stamp
     b £5.04
7 **a** Number left = starting number − number sold
     b 29
8 £184
9 **a** Number per person = number of sweets ÷ number of
       people = 21 sweets
     b Number of people = number of slices ÷ number of
       slices received = 12
10 43 mph
11 900°
12 45°
13 16π cm² (to the nearest whole number)

13.2 Get ready

1 **a** 12 cm   **b** 16 cm   **c** 24 cm

Exercise 13B

1	a	8	b	0	c	5	d	30
	e	9	f	66	g	40	h	18
	i	9	j	26	k	9	l	54
	m	64	n	1				

2 a 1 b $3\frac{1}{2}$ c $1\frac{1}{2}$ d $\frac{1}{2}$
   e $\frac{5}{16}$ f $1\frac{1}{16}$ g 0 h 0
   i $\frac{9}{16}$ j 22 k 4 l $\frac{3}{32}$
   m 1

3	a	1	b	11	c	13.5	d	9.5
	e	9.75	f	−8.5	g	12.5	h	7
	i	1.25	j	27	k	27	l	9.25
	m	−7.875						

Exercise 13C

1	1	2	−11	3	11
4	−3	5	8	6	−1
7	−8	8	−2	9	2
10	0	11	28	12	−30
13	5	14	−36	15	2
16	−18	17	−30	18	54
19	30	20	−6	21	18
22	−2	23	−15	24	25
25	6	26	97	27	16
28	1	29	9	30	432
31	−48	32	32	33	−11
34	0.25				

13.3 Get ready

1 −2   2 15   3 2

Exercise 13D

1	a	11	b	15	c	23	d	18
2	a	21	b	63	c	22.2	d	37.8
3	a	6	b	16	c	12	d	29
4	a	12	b	33				
5	a	50	b	212	c	−22	d	32
6	a	40	b	140	c	240	d	199.206
7	a	20	b	20	c	20	d	59
8	a	90	b	135	c	195	d	285
9	a	20	b	480.2	c	26.875		

13.4 Get ready

1 9.42   2 25.1   3 31.4

Exercise 13E

1 $P = 6l$
   a 18   b 42
   c 174   d 51.6
2 $P = 4a + b$
   a 28   b 55
   c 24.6   d 27.3
3 a $P = 70n$ b i £2.80 ii £4.20 iii £8.40
4 $v = s^3$
   a 8 cm³   b 91.125 cm³
5 $A = 6s^2$
   a 24 cm²   b 121.5 cm²

6 $s = \sqrt{A}$
   a 2 cm   b 1.2 cm

Review Exercise

1 a $n = b - 3$ b $d = 2b$
2 a $b = 2a$ b $c = a - 7$
3 a $b = 3a$ b $c = a + 5$
4 a £50
   b 40 represents a fixed cost of £40
      0.05 is a cost of 5p per leaflet
5 a $A = l^2$ b 81
6 a $P = 2(l + w)$
   b i 26
      ii 20.2
7 a 5   b −23
8 a 87   b 22   c 7.5   d 8
9 a 57.6   b −133.95   c −594
10 a 300 cm² b 195 cm² c 10.585 cm² d 1.495 cm²
11 a 56   b −52   c −124
12 20
13 a 80   b 29.5
14 a −5   b 73
15 a Student's own answers
    b 3.5 hours
    c 4.5 hours
    d 4 hours 4 minutes
    e 4 hours 21 minutes
16 a £130   b $C = 90 + \frac{1}{2}m$
17 −20
18 14
19 65
20 1
21 49
22 −375
23 −27
24 20.25

Chapter 14 Answers

14.1 Get ready

1 a 90   b 180   c 45   d 30

Exercise 14A

1	a	180°	b	90°	c	360°
2	a	90°	b	135°	c	90°
3	a	180°	b	90°	c	360°
4	NE					
5	300°					

14.2 Get ready

1 a 90°   b 135°   c 180°

Exercise 14B

1 b
2 a obtuse   b acute   c right angle
   d reflex   e acute   f obtuse
   g right angle h reflex   i obtuse

Answers

14.3 Get ready

1 c
2 Students' acute angles
3 Obtuse angle

Exercise 14C

1 a = angle ABC b = angle ACB c = angle DFE
   d = angle EDF e = angle HGI f = angle HJI
   g = angle GIJ h = angle LKO i = angle MNO
   j = angle KLM k = angle QPS l = angle PQR
   m = angle PSR n = angle TVU o = angle TUV
2 **a** angle BCD, angle EAB
   b angle CDE
   c angle DEA
   d angle ABC
   e DC
   f AE
3 Students' drawings

14.4 Get ready

1 a 180°   **b** 90°   **c** 270°

Exercise 14D

1 60°   **2** 30°   **3** 155°
4 120°   **5** 45°   **6** 120°

14.5 Get ready

1 a About 60°   **b** About 130°   **c** About 40°

Exercise 14E

1 a 51°   **b** 134°
2 a 112°   **b** 61°   **c** 115°
3 a 98°   **b** 88°   **c** 124°
4 Mild steel

14.6 Get ready

Students' drawings

Exercise 14F

1 a
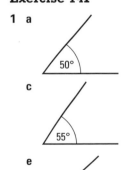

b

c

d

e
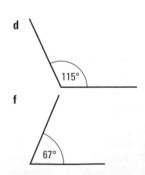

f

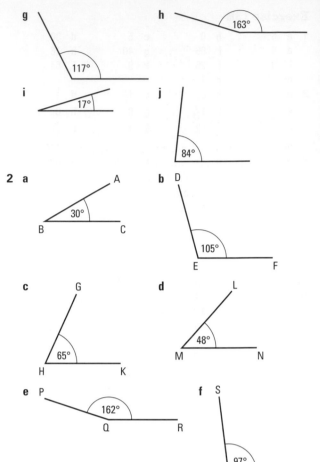

14.7 Get ready

1 Students' accurate drawings
2 60°, 60°, 5 cm
3 The sides and angles are equal.

Exercise 14G

1 a 70°   **b** 20°   **c** 43°
   d 70°   **e** 46°   **f** 71°
2 a 70°   **b** 60°   **c** 90°
3 a isosceles
   b equilateral
   c right-angled
4 55°
5 angle DEF = 25°, angle FDE = 130°

14.8 Get ready

1 a 140   **b** 113   **c** 48
2 a 313   **b** 252   **c** 113

Exercise 14H

1 a = 60° b = 85° c = 90°
   d = 100° e = 49° f = 34°
2 g = 32° h = 109° i = 30°
   j = 47° k = 133° l = 133°
   m = 119° n = 61° p = 61°

3 $a = 62°$, angles on a straight line $= 180°$
   $b = 169°$, angles around a point $= 360°$
   $c = 38°$, vertically opposite angles
   $d = 142°$, angles on a straight line $= 180°$
   $e = 142°$, vertically opposite angles
   $f = 127°$, angles on a straight line $= 180°$
   $g = 53°$, vertically opposite angles
   $h = 127°$, vertically opposite angles
   $l = 64°$, angles around a point $= 360°$, $2l = 128°$
   $m = 60°$, angles on a straight line $= 180°$, $3m = 180°$
   $n = 45°$, angles on a straight line $= 180°$, $4n = 180°$
   $o = 45°$, vertically opposite angles
   $p = 135°$, vertically opposite angles
   $q = 90°$, angles on a straight line $= 180°$
   $r = 60°$, angles on a straight line $= 180°$, $3r = 180°$

Review exercise

1 7 cm
2 Acute angle
3
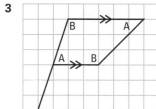
4
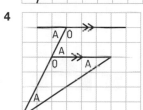
5 Students' drawings
6 **a** **i** 30°   **ii** Vertically opposite angles
   b Angles around a point should add up to 360°. These add up to 385°.
7 140°
8 **a** 4 cm   **b** 30°
9 **a** 5.5 cm   **b** 42°
10 **a** About 55°   **b** About 145°
11 Students' drawings
12 Angles on a straight line should add up to 180°. These add up to 170°.
13 **a** 69°   **b** 52   **c** 70°

Chapter 15 Answers

15.1 Get ready

1 **a** right-angled
   b obtuse
   c acute

Exercise 15A

1 **A** and **1** and **3**, **B** and **4** and **6**, **C** and **2** and **6**, **D** and **2** and **3**, **E** and **2** and **5**, **F** and **5** and **1**

2 **a** **B** and **D**   **b** **A** and **B**
   c **C** and **E**   **d** **C**, **D** and **E**
3 Students' sketch of an obtuse-angled triangle that is also isosceles.
4 This is not an isosceles triangle as all of the lengths and angles are different. An isosceles triangle has 2 equal angles and 2 sides of the same length.
5 Gerry is correct. Because all of the angles are 60°, they are therefore all less than 90°.

15.2 Get ready

1   **2**   **3**

Exercise 15B

1 **A** and **5**   **B** and **2**   **C** and **4**   **D** and **1**
   E and **2**   **F** and **3**   **G** and **4**   **H** and **6**
2 **a** **A** = Parallelogram   **B** = Trapezium   **C** = Square
      D = Rhombus   **E** = Kite   **F** = Rectangle
      G = Square
   b **A, C, D, F** and **G**
   c **E**
3 **a** Square
   b Octagon
   c Trapezium
   d Right-angled triangle
4 **a** True   **b** False   **c** True   **d** True
5 **a** FG   **b** Yes   **c** Angle EHG   **d** Yes
   e Isosceles triangle
6 Diagram **b**
7 Rhombus and kite

15.3 Get ready

1 They are the same shape.

Exercise 15C

1 **A** and **E**, **B** and **G**.
2 **a** **A** and **C**   **b** **B** and **D**   **c** **A** and **D**
3 For example
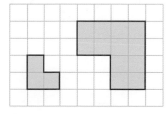

Exercise 15D

1 diameter
2 tangent
3 chord
4 radius
5 segment
6 sector

Answers

15.5 Get ready

1 a   b   c

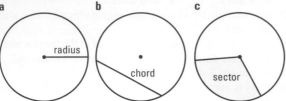

Exercise 15E

Students' accurate drawings

15.6 Get ready

1   **a**   Yes   **b**   No   **c**   No

Exercise 15F

1   **a**

     b

  c

     d

  e

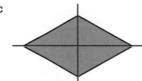

     f

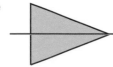

2   **a**   Yes, 1 line   **b**   No   **c**   Yes, 1 line
   d   Yes, 6 lines   **e**   Yes, 2 lines   **f**   Yes, 4 lines

3   **a**   **b**

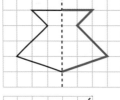

  c

4 a

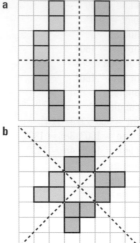

b

15.7 Get ready

1   **a**   360   **b**   90   **c**   180

Exercise 15G

1   **a**   Yes, order 2   **b**   No
   c   Yes, order 4   **d**   Yes, order 2

2   **a**   5   **b**   2   **c**   8   **d**   4

3   4

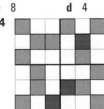

Exercise 15H

1

Shape	Name of shape	Number of lines of symmetry	Order of rotational symmetry
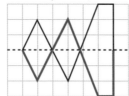	Rectangle	2	2
	Equilateral triangle	3	3
	Rhombus	2	2
	Regular hexagon	6	6
	Parallelogram	0	2

Review exercise

1 Isosceles triangle

2

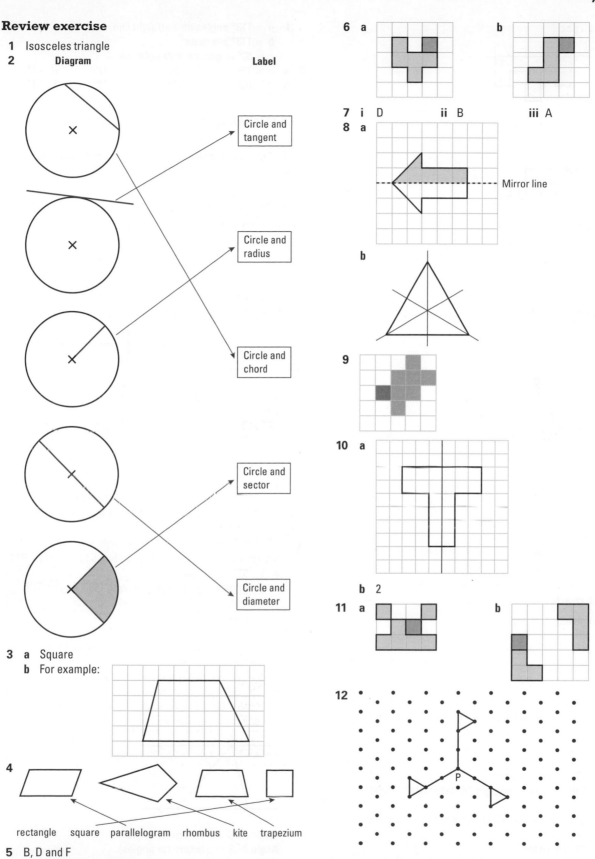

Diagram   Label

Circle and tangent

Circle and radius

Circle and chord

Circle and sector

Circle and diameter

3 a Square
  b For example:

4 rectangle   square   parallelogram   rhombus   kite   trapezium

5 B, D and F

6 a
  b

7 i D   **ii** B   **iii** A

8 a
         Mirror line

  b

9

10 a

  b 2

11 a   **b**

12

Answers

13 a

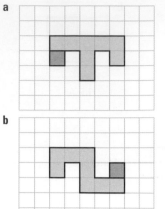

b

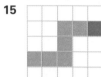

14 a A and D **b** B and C

15

Chapter 16 Answers

16.1 Get ready

1 a $a = 113°$ **b** $b = 63°$, $c = 117°$, $d = 63°$ **c** $e = 58°$

Exercise 16A

1 110° **2** 50° **3** 90°
4 144° **5** 98°

Exercise 16B

1 a A,B; C,E; D,F; G,H; I,J; K,L **b** A,G; A,H; B,G; B,H
2 Students' drawings

16.3 Get ready

1 127° (angles on a straight line)
2 59° (angles around a point)
3 38° (vertically opposite)

Exercise 16C

1 a a and d, b and e **b** b and d, c and f
2 a $a = 25°$ corresponding **b** $b = 110°$ alternate
c $a = 111°$ corresponding
   $b = 111°$ vertically opposite/alternate
d $a = 148°$ corresponding
   $b = 32°$ corresponding
   $c = 148°$ angles on straight line/alternate/vertically opposite
   $d = 32°$ alternate/angles on straight line/vertically opposite
e $a = 61°$ alternate
   $b = 119°$ angles on a straight line
   $c = 61°$ corresponding/vertically opposite
   $d = 119°$ vertically opposite

f $a = 113°$ angles on a straight line
   $b = 113°$ alternate
   $c = 67°$ angles on a straight line
3 a $a = 125°$ $b = 55°$ $c = 125°$ $d = 55°$
b $e = 108°$ $f = 72°$ $g = 72°$ $h = 108°$

16.4 Get ready

1 $x = 40°$
2 $a = 27°$ (corresponding) $b = 27°$ (vertically opposite)
3 ADC = 81° BCD and BAD = 99°

Review exercise

1 a For example:

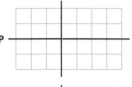

b For example:

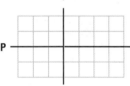

2 Students' drawings
3 110°
4 60°
5 a i 25° **ii** 130°
b 65°
6 a 30°
b 48°
7 i 127° **ii** Alternate angles
8 i 58° **ii** Alternate angles
9 i 120° **ii** Corresponding angles
10 x is 130° because angles on a straight line add up to 180°.
   y is 50° because it is a corresponding angle to angle PBC.
11 a i 63°
   ii Angles in a triangle add up to 180°. Angles ABC and ACB are the two equal angles of the isosceles triangle.
b 117°
12 a $a = 35°$ (angles on a straight line add up to 180°, corresponding angles)
   $b = 35°$ (vertically opposite angles)
b $a = 123°$ (alternate angles)
   $b = 57°$ (angles on a straight line add up to 180°)
13 You can draw parallel lines around any triangle:

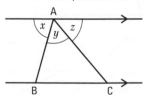

Angle ABC = x (alternate angles)
Angle ACB = z (alternate angles)
So the angle sum of the triangle is $x + y + z$.
But $x + y + z = 180°$ (angles on a straight line).
So the angle sum of the triangle is also 180°.

Chapter 17 Answers

17.1 Get ready

1 26.5 cm by 19.5 cm

Exercise 17A

1 a 2.5 cm, 25 mm   **b** 4 cm, 40 mm   **c** 5 cm, 50 mm
   d 6.5 cm, 65 mm   **e** 8.2 cm, 82 mm   **f** 4.4 cm, 44 cm
   g 7.6 cm, 76 mm   **h** 6.8 cm, 68 mm   **i** 5.7 cm, 57 mm
   j 3.3 cm, 33 mm   **k** 4.8 cm, 48 mm   **l** 6.2 cm, 62 mm

2 a

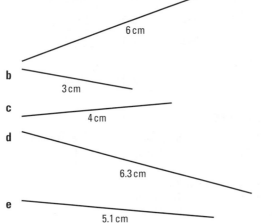

6 cm

b

3 cm

c

4 cm

d

6.3 cm

e

5.1 cm

f

6.8 cm

g

3.4 cm

h

7.9 cm

i

3.5 cm

j

4.2 cm

k

5.7 cm

l

7.6 cm

3 a

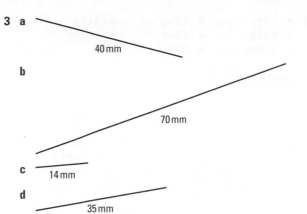

40 mm

b

70 mm

c

14 mm

d

35 mm

e

12 mm

f

31 mm

g

48 mm

h

26 mm

i

27 mm

j

59 mm

k

73 mm

l

66 mm

4 a 2.6°C   **b** 0.3°C   **c** −1.6°C
   d 3.4°C   **e** −0.7°C
5 a 1.7 litres   **b** 2.8 litres
   c 6.8 litres   **d** 6.3 litres
6 a 2.4 kg   **b** 3.6 kg
   c 9.7 kg   **d** 6.4 kg
7 a 28 mph   **b** 44 mph
   c 23 mph   **d** 67 mph

Exercise 17B

1 a 3.4 amps   **b** 8.6 amps
   c 46 amps   **d** 44 amps
2 a 2.4 m   **b** 64 m   **c** 98 m
   d 32 m   **e** 9.6 m   **f** 0.4 m
3 a 1.4°C   **b** 2.6 °C   **c** 1.2°C
   d 0.6°C   **e** 0.8°C   **f** 1.8°C

4 a 1.4 kg   b 2.6 kg   c 4.6 kg
    d 8.2 kg   e 7.8 kg   f 10.2 kg
    g 23.6 kg   h 30.8 kg

Exercise 17C

1 a 67–69 grams   b 2.3–2.4 kg
    c 17–18 grams   d 72–78 grams
2 a 2.1–2.3 amps   b 1.2–1.4 amps
    c 0.4–0.6 amps   d 1.6–1.8 amps
3 a 23–25°C, 73–77°F   b 34–36°C, 93–97°F
    c 29–31°C, 84–88 °F   d 27–29°C, 79–81°F
    e 18–19°C, 64–66°F   f 32–33°C, 89–91°F
    g 34–35°C, 93–95°F   h 14–16°C, 59–61°F
4 a 1.1–1.2 kg   b 17–19 g
    c 160–180 g   d 51–54 g
5 a 109–111 km/h, 68–70 mph
    b 164–166 km/h, 101–104 mph
    c 94–96 km/h, 57–60 mph

17.2 Get ready

1 Students' times

Exercise 17D

1 a 9.00 am, 09:00 h   b 3.30 am, 03:30 h
    c 2.00 am, 02:00 h   d 8.30 am, 08:30 h
    e 11.45 am, 11:45 h
2 a 8.15 pm, 20:15 h   b 1.15 pm, 13:15 h
    c 5.00 pm, 17:00 h   d 7.00 pm, 19:00 h
    e 5.15 pm, 17:15 h
3 a 08:00 h   b 11:15 h   c 15:40 h
    d 08:20 h   e 20:55 h   f 15:25 h
    g 02:30 h   h 17:25 h   i 22:15 h
    j 07:20 h   k 09:45 h   l 13:15 h
    m 23:25 h   n 02:50 h   o 13:50 h
    p 12:20 h
4 a 11.10 am   b 8.20 am   c 7.40 am
    d 11.35 pm   e 2.17 pm   f 9.35 am
    g 6.16 pm   h 5.25 pm   i 1.20 pm
    j 1.10 pm   k 8.30 am   l 1.35 pm
    m 3.42 am   n 10.16 pm   o 9.17 am
    p 1.37 pm
5 a 11.15 am   b 12:10 h   c 2.30 pm
    d 10.45 am   e 11:30 h   f 5.40 am
    g 1.50 pm   h 10:30 h   i 16:25 h
    j 08:40 h   k 9.45 pm   l 00:45 h
6 a 156 weeks   b 210 minutes   c 300 seconds
    d 60 months   e 96 hours   f 9 hours
    g 6 years   h 150 seconds   i $1\frac{1}{2}$ minutes
    j 208 weeks   k 480 minutes   l 60 hours
    m 182 weeks   n 730 days
7 a 156   b 20   c 12   d 168

Exercise 17E

1 a 25 mins   b 50 mins   c 2 h 10 min
    d 10 h 30 min   e 9 hours   f 4 h 10 min
    g 3h 38 min   h 18 h 32 min   i 14 h 45 min
    j 3 h 50 min   k 22 h 50 min   l 3 h 15 min

2 16 h 20 min
3 8 h 10 min
4 6 h 50 min
5

Flight number	Departure time	Arrival time	Flight time
BA52	2220	0445	**6 h 25 min**
XA160	0542	0914	**3 h 32 min**
FC492	1415	**1855**	4 h 40 min
TC223	1002	**1425**	4 h 23 min
AL517	**0222**	0759	5 h 37 min
AB614	1917	0521	**10 h 4 min**
FX910	0243	**0634**	3 h 51 min
BI451	**0945**	1217	2 h 32 min
AE105	**2056**	0225	5 h 29 min
DA452	1539	**2227**	6 h 48 min

Exercise 17F

1 a 31 min   b 2 min   c 19:40 h
    d 0735 h   e 30 min   f 40 min
    g 4   h 27   i 10 min
    j 45 min   k 07:52 h   l 11:13 h
2 a i 37 min   ii 18 min
       iii 26 min   iv 20 min
    b 41 min
    c

Manchester Victoria	09:05	11:35	Wigan Wallgate	08:30	11:55
Salford	09:08	11:38	Ince	08:35	12:00
Salford Crescent	09:10	11:40	Hindley	08:38	12:03
Swinton	09:19	11:49	Westhoughton	08:42	12:07
Moorside	09:22	11:52	Bolton	08:50	12:15
Walkden	09:26	11:56	Moses gate	08:53	12:18
Atherton	09:32	12:02	Farnworth	08:56	12:21
Hag Fold	09:35	12:05	Kearsley	08:59	12:24
Daisy Hill	09:37	12:07	Clifton	09:06	12:31
Hindley	09:42	12:12	Salford Crescent	09:10	12:35
Ince	09:45	12:15	Salford	09:12	12:37
Wigan Wallgate	09:50	12:20	Manchester Victoria	09:15	12:40

17.3 Get ready

1 Students' examples of objects

Exercise 17G

1 ml   2 mm/cm
3 kg   4 tonnes
5 litres   6 metres
7 mm/cm   8 kg
9 grams   10 mm
11 litres   12 mg/grams
13 km   14 grams
15 cm/m

Exercise 17H

1 a 400 cm   b 5 cm   c 800 cm
   d 1300 cm   e 20 cm   f 3500 cm
   g 7.4 cm   h 12.2 cm
2 a 30 mm   b 60 mm   c 220 mm
   d 400 mm   e 2000 mm   f 54 mm
   g 137 mm   h 51.5 mm
3 a 6000 m   b 5 m   c 20 000 m
   d 30 m   e 45 000 m   f 800 m
   g 1400 m   h 2450 m
4 a 2000 g   b 30 000 g   c 400 000 g
   d 250 000 g   e 2 000 000 g f 55 000 g
   g 120 g   h 4200 g
5 a 4 l b 7 l c 20 l
   d 45 l e 2.5 l f 3.7 l
   g 6.52 l h 3.13 l
6 a 3000 ml b 20 000 ml c 200 000 ml
   d 450 000 ml e 35 000 ml f 7500 ml
   g 400 ml h 1430 ml
7 a 3 km   b 8 km   c 30 km
   d 68 km   e 4.2 km   f 5.6 km
   g 5.41 km   h 2.14 km
8 a 5 t   b 6 t   c 40 t
   d 57 t   e 3.6 t   f 4.5 t
   g 7.63 t   h 4.25 t
9 a 4 kg   b 2000 kg   c 20 kg
   d 15 000 kg   e 200 kg   f 3700 kg
   g 6.4 kg   h 1.23 kg
10 500   11 90 litres
12 25   13 750 kg
14 20   15 50 days

Exercise 17I

1 6 mm, 3 cm, 60 mm, 30 cm, 4 m, 4 km
2 400 ml, 700 ml, 1 l, 3000 ml, 6 l
3 450 g, 0.5 kg, 600 g, 0.62 kg
4 0.6 cm, 370 mm, 40 cm, 55 cm, 600 mm, 1.4 m
5 0.2 cm, 0.4 cm, 9 mm, 55 mm, 6 cm, 77 mm, 46 cm
6 75 ml, 0.08 l, 260 ml, 0.3 l, 450 ml, 600 ml

17.4 Get ready

1 Students' measurements in feet and inches, and in centimetres

Exercise 17J

1 a 3 feet   b 4 gallons   c 28 inches
   d 64 ounces   e 63 inches   f 7 yards
   g 52 inches   h 48 pints   i 49 pounds
   j 18 pints   k 21 stones 6 pounds l 12 feet
   m 9 inches   n 2 gallons 6 pints o 224 ounces
2 5 foot 3 inches
3 9 stone 3 pounds
4 6 foot 2 inches

Exercise 17K

1 a 24 km   b 22 pounds   c 7 pints
   d 15 cm   e 270 cm   f 30 miles
   g 5 kg   h 100 inches   i 120 km

2 7.2 m   3 35 cm
4 4 gallons   5 3 bottles
6 640 km   7 8.75 pints
8 93.75 miles   9 4
10 4 kg (A)

17.5 Get ready

1 Explanation of how speed is measured in car

Exercise 17L

1 30 mph   2 8 mph
3 4 mph   4 2 mph
5 50 mph   6 32 km/h
7 60 mph   8 50 mph
9 80 km/h   10 400 km/h

Exercise 17M

1 80 miles   2 4 hours
3 105 miles   4 $1\frac{1}{2}$ hours
5 $2\frac{1}{2}$ hours   6 4 hours
7 12.5 km   8 $2\frac{1}{2}$ hours
9 1400 miles   10 160 miles

Exercise 17N

1 87.5 miles   2 6.5 miles
3 80 km/h   4 3 hours 12 min
5 82 km   6 60 mph
7 7 hours 48 min   8 297 miles
9 50 km/h
10 a 34 m   b 306 km/h

17.6 Get ready

1 a 7   b 52   c 14
2 a 300 cm   b 7 cm
3 a 220 mm   b 47 mm

Exercise 17O

1 12.5 cm
2 44.5 g
3 3.5 litres
4 a 9.65 cm   b 9.75 cm
5 a 1.585 m   b 1.595 m
6 The pencil could be as long as 105 mm, which is longer than the shortest possible length of the pencil case, 101.5 mm
7 The cupboard could as narrow as 81.5 cm, and the gap as wide as 81.75 cm

Review exercise

1 a metres, grams, litres
   b 400 cm
   c 1.5 kg
2 a 17.8 cm   b −2°C   c 2.8 kg
3 a i metres   ii kilograms
   b 20 mm
4 a i kilometres ii litres
   b i 50 mm   ii 4 kg

5 a 32

b

|+++++++++|+++++++++|+++++++++|+++++++++|+++++++++|
110   120   130   140   150   160

⤴ (arrow near 125)

c 4.4

d

|+++++++++|+++++++++|+++++++++|+++++++++|+++++++++|
3   3.1   3.2   3.3   3.4   3.5

⤴ (arrow near 3.15)

6 a 1.8 m   **b** 6.5 m

7 a 1.8 m   **b** 7 m

8 1.5 m

9 No, 1.5 km is 1500 m

10 a Huntingdon   **b** 3 minutes   **c** 10 05

11 a 4.6 kg

   b i 2.2 lb   **ii** 11 lb

12 a 09:50

   b i 15 minutes   **ii** 09:45

   c 1 hour 20 minutes

13 a 07:37   **b** 10 minutes   **c** 18 minutes

14

	Metric	Imperial
Distance from London to Cardiff	km	miles
Weight of a bag of potatoes	kg	pounds
Volume of fuel in a car's fuel tank	litres	gallons

15 a 57 minutes

   b

	Time
Zoe leaves the hotel	08:53
Zoe Leaves Sa Pobla on a train	09:23
Train arrives at Inca (Zoe gets off)	09:41
Zoe leaves Inca on another train	11:41
Zoe arrives at Palma	12:20

16 432 mph

17 120 litres

18 85 kg

19 80 km/h

20 a 126.5 g   **b** 127.5 g

Chapter 18 Answers

18.1 Get ready

1 3 cm, 2 cm   **2** 4 cm, 5 cm, 5 cm

Exercise 18A

1 A 16 cm   **B** 20 cm   **C** 20 cm

2 a 26 cm   **b** 15 cm   **c** 13.7 cm

3 a 12 cm   **b** 10.5 cm   **c** 11 cm

4 9 m

Exercise 18B

1 40 cm

2 a 24 cm   **b** 102 mm   **c** 32 cm

3 34 cm

4 7.2 cm

18.2 Get ready

1 a 4   **b** 5   **c** 3

Exercise 18C

1 12 cm²

2 A 9 cm²   **B** 15 cm²   **C** 22 cm²   **D** 22 cm²

3 T_1 8 cm² T_2 9 cm² T_3 10 cm² T_4 12 cm²

4 F 12 cm²   **G** 18 cm²   **H** 22 cm²

5 P 25 cm²   **Q** 13 cm²   **R** 28 cm²

Mixed Exercise 18D

1 a 18 cm   **b** 14 cm²

2 21 cm²

3 24 mm

4 13 m

18.3 Get ready

1 6 cm and 2 cm   **2** 4.5 cm and 5 cm   **3** 2 cm and 3 cm

Exercise 18E

1 a 15 cm²   **b** 14 mm²   **c** 40 m²

2 a 9 cm²   **b** 20.25 cm²   **c** 13.69 cm²

3 19.76 m²

Exercise 18F

1 a 6 cm²   **b** 48 m²   **c** 15 cm²

   d 21.6 cm²   **e** 16.74 mm²

2 30 cm²

3 4500 cm²

Exercise 18G

1 a 10 cm²   **b** 12 cm²   **c** 12 cm²

2 24 000 cm²

Exercise 18H

1 15 cm²   55 cm²   48 m²   30 m²

2 a 14 cm²   **b** 40 cm²   **c** 15 cm²   **d** 64 m²

3 4 cm

18.4 Get ready

1

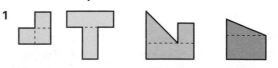

Exercise 18I

1 a 100 cm²   **b** 36 cm²   **c** 64 cm²

2 150

3 £103.80

4 a 2   **b** 220 g

5 a 88 cm²   **b** 22.5 m²   **c** 108 m²

6 a 150 cm²   **b** 36 cm²   **c** 55.5 mm²

Review exercise

1 **a** 14 cm   **b** 6 cm^2
2 **a** 60 cm   **b** 200 cm^2
3 60 m
4 7.8 m by 2.9 m
5 28 cm^2
6 **a** 11.25 cm^2
    b 18 cm
7 264 paving stones
8 3 tins needed for 63.5 m^2
9 45 cm^2
10 45 cm^2
11 56 cm^2
12 £31 425

Chapter 19 Answers

19.1 Get ready

1 Octagon, pentagon, hexagon

Exercise 19A

1 Cuboid   **2** Cone
3 Pentagonal prism   **4** Tetrahedron
5

	Shape	Object
1	sphere	football
2	sphere	tennis balls
3	cuboid	cereal cartons
4	cuboid	biscuit tin
5	cuboid	cardboard box
6	cube	choc box
7	cylinder	cake
8	cylinder	cake tin
9	cylinder	flour bin
10	cylinder	rubbish bin
11	pyramid	food cover
12	triangular prism	grater
	cone	ice cream cone
	Etc.	

Exercise 19B

1

	Shape	Faces	Edges	Vertices
A	Cube	6	12	8
B	Pentagonal prism	7	15	10
C	Triangular prism	5	9	6
D	Square-based pyramid	5	8	5
E	Cuboid	6	12	8
F	Tetrahedron	4	6	4
G	Octagonal prism	10	24	16

2 Triangle
3

4 **a** hexagon   **b** hexagonal-based pyramid

19.2 Get ready

1   **2**

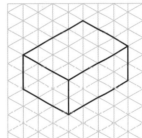

3

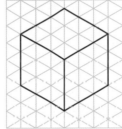

Exercise 19C

1   **2**

3

4

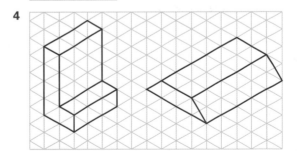

19.3 Get ready

1 12 cm^2
2 4 cm^2
3 27 cm^2
4 **a** 10 cm^2 **b** 21 cm^2 **c** 10 cm^2

Answers

Exercise 19D

1 a 24 cm³   b 6 cm³   c 4 cm³
2 a 3 cm³   b 96 cm³
3 a 224 m³   b 612 mm³
4 343 cm³
5 5 cm
6 One possibility 40 cm × 60 cm × 70 cm

Exercise 19E

1 15 m³
2 144 cm³
3 24 cm³
4 a 750 cm³   b 60 cm³   c 12   d 30 cm³
5 432 cm³

19.4 Get ready

1 a 22.94 cm²   b 8.88 m²

Exercise 19F

1 122 cm²
2 144 cm
3 3600 cm²
4 204 cm²
5 2 cm
6 3 cm

Review exercise

1 12 cm³
2 i Cylinder   ii Cone
3 20 cm³
4 i 6   ii 12   iii 8
5 i 6   ii 12   iii 8
6 420 cm³
7

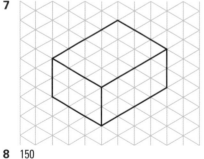

8 150

9 135 cm³
10 150 cm²
11 162 cm³

Multiplication

1 £249.50
2 Nan must give £35.80
3 £48.46

Area

1 Any two shapes made of 6 squares where some of the squares are joined at more than one edge.
2 £210
3 £248
4 Some possibilities are 74 cm, 40 cm, 30 cm and 26 cm
5 £8

2012 Olympics

1 7.17 am
2 750 euros
3

	Athlete 1	Athlete 2	Athlete 3
Swimming	843	677	1060
Shooting	1216	1252	916
Running	1272	1104	984
Show Jumping	1116	1200	1172
Fencing	1048	832	832
Total	5495	5065	4964

Athlete 1 has the highest total score.

University

1 Elaine 221 square units = £245.26
Ryan 117 square units = £129.85
Danielle 117 square units = £129.85
Rashid 331 square units = £367.34
Saria 225 square units = £249.70
2 375 g beef mince, 2 onions, 10 mushrooms, 2 cans tin tomatoes, 750 g spaghetti
3 Elaine needs a further 18.7 marks to reach 40%.
She could get this entirely from Unit 6 with a mark of $\frac{33}{60}$.
So even getting zero in Unit 5 would give her a chance of passing.

Index

Index